CISM COURSES AND LECTURES

The series presents lecture notes, monographs, edited works and proceedings in the field of Mechanics, Engineering, Computer Science and Applied Mathematics.
Purpose of the series is to make known in the international scientific and technical community results obtained in some of the activities organized by CISM, the International Centre for Mechanical Sciences.

INTERNATIONAL CENTRE FOR MECHANICAL SCIENCES

COURSES AND LECTURES - No. 446

CARDIOVASCULAR FLUID MECHANICS

EDITED BY

GIANNI PEDRIZZETTI
UNIVERSITA' DI TRIESTE

KARL PERKTOLD
GRAZ UNIVERSITY OF TECHNOLOGY

Springer-Verlag Wien GmbH

This volume contains 111 illustrations

Originally published by Springer-Verlag Wien New York in 2003
SPIN 10914571

In order to make this volume available as economically and as rapidly as possible the authors' typescripts have been reproduced in their original forms. This method unfortunately has its typographical limitations but it is hoped that they in no way distract the reader.

ISBN 978-3-211-00538-5 ISBN 978-3-7091-2542-7 (eBook)
DOI 10.1007/978-3-7091-2542-7

PREFACE

Arteriosclerosis and heart diseases represent the major cause of death in the western world. The dramatic increase in the number of symptomatic and asymptomatic patients has provoked increased attention in the development of systematic approaches to improve diagnostic and therapeutic techniques. The role of fluid dynamics derives from the fact that the onset and development of arteriosclerosis that is observed in proximity of curved and branching arteries is associated with the irregularities of the local hemodynamic pattern; at the same time most heart diseases correspond to specific variations of the flow in the heart chambers. Fluid dynamics has contributed significantly in the recent past years and most diagnoses are now made using a combination of ultrasonic imaging and Doppler measurements of flow velocity, however its role is still destined to grow with the growing capabilities of numerical and diagnostic tools.

This book follows from the lectures held by the authors at CISM during July 2002 and focussed on recent advances in the interdisciplinary field represented by the application of fluid mechanics to cardiovascular problems.

The progress in such field closely follows the growing computational capabilities that allow the direct numerical simulation of fluid flows in complex sites and conditions. Nevertheless much care and experience must be taken when evaluating the accuracy of these results that are affected either by the numerical approximations or by the model idealisation of the actual physiological conditions. The rapid advancement of diagnostic techniques allows a closer link and a completion between in vivo data and numerical models. Given the physiological differences between individuals, it is also important to build relatively simple conceptual models able to synthesise the information available, to gain some understanding of the underlying physiological and pathophysiological issues.

This book is organized in 6 parts corresponding to the different contributors. The first 4 parts (that correspond to 22 course's lectures), from the field of bio fluid mechanics, present the physical aspects and discuss recent advances in the mathematical models of flow phenomena in large blood vessels. The chapters review the essential issues in numerical simulations, conceptual models, and applications. The different aspects, new techniques, and perspectives of direct numerical simulations are carefully covered; these include advanced solution methods for complex domains, their efficiency and accuracy, the problem of grid generation for realistic geometries, and the implementation of reliable boundary conditions at the walls and at the in-outlets. Conceptual models, as well as approximate or simplified numerical models, are another important a part of the book. These topics, which include complex geometry, boundary-layer separation and vorticity dynamics, interaction between flow and elastic walls, allow qualitative syntheses of phenomena and improve interpretation of numerical results. Applications to realistic cases will be presented in connection to the different approaches presented.

The final two parts of the book (10 course lectures), which are taken from the applied medical and biological research, introduce relevant physiology and pathological aspects, and actual in vivo measurements. Non-invasive ultrasound techniques are introduced to assess mechanical properties of the blood and vessel/heart walls in humans, as well as flow characteristics and wall motion. Topics regarding the relation between intima-media thickness and wall shear stress, modifications with ageing and pathologies, and some element concerning the structure-function relation of arteries are presented. Emphasis is put on advanced diagnostic tools, related outcomes, and possible crosschecks.

The development of such an interdisciplinary field requires that the new scientists acquire competences that are not limited to a single scientific field. In this spirit we hope that this book will contribute to proceed along this difficult but stimulating direction.

Gianni Pedrizzetti, Karl Perktold

CONTENTS

Preface

Arterial and Venous Fluid Dynamics
by T.J. Pedley 1

Computational Models of Arterial Flow and Mass Transport
By K. Perktold and M. Prosi 73

Finite Difference and Finite Volume Techniques for the Solution of Navier-Stokes Equations in Cardiovascular Fluid Mechanics
by S. Tsangaris and T. Pappou 137

Fluid Flow inside Deformable Vessels and in the Left Ventricle
by G. Pedrizzetti and F. Domenichini 187

From Left Ventricular Dynamics to the Pathophysiology of the Failing Heart
by A. Barsotti and F.L. Dini 235

Element of Physiology and Mechanics of Human Arteries
by R.S. Reneman, A.P.G. Hoeks and L. Kornet 249

Arterial and Venous Fluid Dynamics

Timothy J. Pedley

DAMTP, CMS, Wilberforce Road, Cambridge CB3 0WA, UK

The majority of these lecture notes are taken from the author's chapter "Blood flow in arteries and veins", in the book "Perspectives in Fluid Mechanics" edited by G K Batchelor, H K Moffatt and M G Worster, published by Cambridge University Press, 2000. Other parts come from his own book, Pedley (1980).

1 The Cardiovascular System

Material transport from one part of the body to another, in vertebrates or other large animals, involves fluid (liquid or gas) flowing along and across the walls of systems of tubes. The most widely studied tube systems are the mammalian cardiovascular and ventilatory systems. The purposes of studying the mechanics of physiological flows can be summed up under four headings:

(1) pure physiology, or understanding how animals work;
(2) pathophysiology, or understanding how they go wrong, i.e. the origins and development of diseases;
(3) diagnosis, or working out how to infer what has gone wrong from (relatively) simple and non-traumatic measurements, for example of blood pressure;
(4) cure, which in the mechanical context usually means bioengineering, and which here includes vascular and other surgery as well as the design of prosthetic devices.

Different scientists and practitioners will have different motivations, but the basic mechanical principles are clearly the same for all. In the mechanical context (2, 3 and 4 above) it is worth remarking that disease of the arteries, notably atherosclerosis, causes around 50% of the deaths in modern western society, through heart attacks and strokes, and drastically lowers the quality of life of those elderly patients who suffer reduced blood flow to the legs through constriction or blockage (stenosis) of the femoral (thigh) arteries.

Fluid mechanical factors are strongly implicated in the development of atherosclerosis (atherogenesis) as well as its effects, because the distribution of wall shear stress (*WSS*) in arteries is linked to atherogenesis. What follows is a very brief review; for more detail see Giddens et al. (1993), Friedman (1993) and Fry (1987), for excellent treatments.

The results of several in vivo studies show that, if a normal artery, with an intact endothelium, is altered so that the mean wall shear stress (*WSS*) changes, then over a period of months the artery wall remodels itself to restore the mean *WSS* to its normal value (Zarins et al. 1987). The normal shear stress, as estimated from the assumption of Poiseuille flow, is usually in the range

1-2 N m^{-2}, the actual value depending on the species and the vessel. Such long-term adaptation is thought to be important during angiogenesis and growth, determining artery diameters in response to flow rate demands. The only signal related to flow rate that the endothelial cells could detect is the *WSS*. If it rises, then the response of the vessel is to increase its luminal diameter while retaining the same wall thickness. However, if the *WSS* falls below about 1 N m^{-2}, the diameter is reduced by means of a marked relative thickening of the intimal part of the wall.

It has been established that early atheromatous plaques are associated with intimal thickening and that they tend to develop in regions believed to have low mean wall shear stress (Caro et al. 1971; Zarins et al. 1983). This raises the interesting possibility that the initial development of an atheromatous plaque may be little more than the normal adaptive response in a region of low mean *WSS*, though it should be noted that short-term time-dependence, and in particular periodic reversal, of the *WSS* has also been observed to affect intimal thickening at certain sites (Zarins et al 1983; Ku et al. 1985). Subsequent plaque development will, of course, depend on the mechanism by which endothelial cells and the tissue behind them respond to changes in *WSS*, and will be mediated by a range of biochemical, genetic and other factors. However, a necessary precondition to understanding the process is a knowledge of the normal distribution of mean and time-dependent *WSS* in regions known to be susceptible to atherogenesis and some understanding of how that distribution will change as the plaque grows.

The original 'low wall shear stress' hypothesis, by Caro et al. (1971), was based on a visual correlation between sites in large arteries susceptible to atherogenesis and places where the wall shear stress was expected to be low. The sites in question, mostly in the aorta and its major branches, were on the inside of bends, on the outer walls of bifurcations, on the wall opposite a side branch and in regions of cross-sectional area expansion such as the 'carotid sinus', in the internal carotid artery just downstream of its bifurcation from the common carotid artery.

Simple steady flow experiments in a cast of the aorta confirmed that the wall shear is low and that flow separation may occur at the susceptible sites (Caro et al., 1971). Intuition about flow separation is often based on steady, two-dimensional flow. For example, at a realistic arterial value of the Reynolds number ($300 < Re < 2000$), the flow through a simple expansion in a two-dimensional channel (figure 1.1), experiences a deceleration which is associated with a pressure rise (Bernoulli's Theorem). This causes the flow to separate, leaving a closed recirculating eddy beneath, the velocity is low and hence so is the wall shear stress. Moreover, fluid elements remain in the eddy for a long time, so particles can escape only by diffusion and this may help to enhance the uptake of undesirable chemicals into the wall. However, the situation in arteries is much more complicated, as a result of unsteadiness and of three-dimensionality: see Chapters 3 and 4.

For steady flow in a long, straight, rigid, circular tube, every element of a Newtonian viscous fluid flows in a straight line with a constant speed, and the velocity profile is parabolic. The relationship between the driving pressure gradient and the flow rate through the tube is linear, the ratio of the two (the viscous resistance) being inversely proportional to the fourth power of the tube diameter (Poiseuille's Law). Such a smooth, laminar flow will break down into turbulence if the Reynolds number,

$$Re = \rho \bar{u} d / \mu \ , \qquad (1.\ 1)$$

exceeds a critical value of about 2000, where ρ and μ are the fluid density and viscosity, d is the tube diameter and $\bar{u}$ is the average fluid velocity within it ($\bar{u} = 4Q/\pi d^2$, where Q is the volume flow rate).

Unfortunately, Poiseuille flow is not to be found anywhere in the human body. Blood vessels (and airways) are typically short, curved, branched and elastic (figure 1.2), the flow is not steady and blood is not a Newtonian fluid, though this last is not of major importance in the large blood vessels of concern here. In short tubes, the flow does not become fully developed, and high velocity gradients occur in thin boundary layers near their entrance, which enhances the rate at which energy is dissipated and therefore increases the pressure drop for a given flow rate (Prandtl, 1952). In curved tubes, secondary motions are generated, which distort the primary velocity profile (Dean, 1927, 1928) and also enhance the energy dissipation (White, 1929). Flow in a bifurcation combines the main features of entry flow and curved tube flow (Schroter & Sudlow, 1969).

The details of unsteady viscous flow are significantly different from those of steady flow, whatever the geometry, when the dimensionless frequency parameter α is not small, where

$$\alpha^2 = \rho\omega d^2/4\mu \tag{1. 2}$$

and ω is the dominant radian frequency of the unsteadiness (Womersley, 1955). Moreover, a time-dependent pressure, such as that generated by a pumping heart, causes the dimensions of an elastic tube to vary with time, and this is responsible for the propagation of pressure waves along the tube (Young, 1809). Physiological fluid flows, in fact, are very complicated and it is a major challenge to measure or calculate the full velocity field.

In the context of large arteries, the principal topics that have been subjected to fluid dynamic study are as follows: (a) pulse-wave propagation and reflection, leading to a determination of the load against which the heart must act, the diagnosis of severe constrictions (stenoses), and the design of cardiac assist devices (McDonald, 1974; Elzinga & Westerhof, 1991); (b) the distribution in space and time of the viscous shear stress exerted by the flowing blood on the artery wall, because of its importance in atherogenesis (Fry, 1987; Giddens et al., 1993; Friedman, 1993); (c) mass transport across the arterial endothelium and within the vessel wall, the detailed mechanisms for which are also of fundamental importance in atherogenesis (Yuan et al., 1991; Weinbaum & Chien, 1993); (d) the functioning of natural and artificial heart valves (Lee & Talbot, 1979); and (e) fluid-structure interaction phenomena associated with flow in a collapsible tube under external compression. Vessel collapse is not normally of relevance in arteries, whose internal pressure considerably exceeds atmospheric and hence external pressure (except possibly some coronary arteries, squeezed by contracting heart muscle), but it does become important if the artery is actively compressed (e.g. by a blood-pressure-measuring cuff) and it can be dominant in veins above the level of the heart because of the gravitational fall of pressure with height.

It is topics (a), (b) and (e) that I shall concentrate on in these lectures, in that order: pulse wave propagation (Chapter 2), arterial wall shear stress (Chpaters 3 and 4) and flow in collapsible tubes (especially the jugular vein of the giraffe!) (Chapters 5 and 6), More mathematical detail on topics (a) and (b) can be found in my book (Pedley, 1980), as well as the many papers referred to here; however, the chapter of that book dealing with collapsible tubes has been superseded by more recent work and reference should only be made to the original papers.

Before investigating fluid *dynamics*, it is worth introducing two background aspects: the concept of blood pressure, and the elastic properties of artery walls.

1.1 Blood pressure

The pressure in the blood at any location is made up of three components: (i) atmospheric pressure, p_0, which is normally taken to be the pressure in the right atrium when its muscles are relaxed; (ii) the hydrostatic pressure, $-\rho gh$, where h is the vertical distance above the right atrium and ρ is the density of the blood; and (iii) the pressure generated by the heart, which we shall denote by p (Lighthill (1975) calls this component the 'excess pressure'). This last component is alone responsible for the motion of the blood, and is commonly called 'blood pressure' (blood pressure is always measured clinically 'at the level of the heart'); p has a mean value of 13.3 kN m^{-2} in the large arteries, but falls dramatically in the microcirculation, so that the mean pressure in the venae cavae is less than 0.6 kN m^{-2}, and in the right atrium it is even closer to zero.

The pressure p_0 relative to which blood pressures are measured should not be exactly atmospheric, but should be taken to be the pressure inside the chest, which varies with respiration and is normally subatmospheric (by as much as 2 kN m^{-2}) when the lung is expanded. This is because contraction of the ventricular muscle generates a *transmural pressure* (p_{tm}) between the inside and the outside of the ventricle, the outside being within the chest.

The hydrostatic component of pressure, $-\rho gh$, is important because it determines the transmural pressure of the blood vessels and hence, through their elastic properties, their calibre. The pressure outside most blood vessels (with the exception of those in the chest and in the skull; see below) is close to atmospheric so p_{tm} is in most cases approximately equal to $p-\rho gh$. Consider a standing man: the mean transmural pressure is about 13 kN m^{-2} at the entrance to the aorta, and about 0.5 kN m^{-2} at the exit from the venae cavae. In the large vessels of the foot(1.1 m below) these values will be increased by about 11 kN m^{-2}, so that both types of vessel will be distended, will have circular cross-section, and will be very stiff (the transmural pressures in some veins can be reduced by voluntary contraction of the skeletal muscles outside them, and can be further alleviated by the action of the valves). In vessels 30cm above the heart, however (for example in a raised arm), p_{tm} is reduced by 3 kN m^{-2}. This has little effect on arteries, which remain fully open; veins, on the other hand, experience a negative transmural pressure, and collapse therefore occurs. Venous return is maintained, either continuously, through the small side-channels, which may not completely close (figure 1.3b), or intermittently, as the build-up of upstream pressure forces the veins open. Veins in the skill do not collapse, because the skull acts as a rigid box, any fall in venous volume being accompanied by a corresponding fall in extravascular pressure.

1.2 Elastic properties of blood vessels

The walls of all arteries (and veins) have a similar structure and are made up of similar materials, although their proportions vary in different parts of the circulation. The wall is traditionally divided up into three layers, the innermost intima, the media, and the outermost adventitia. The intima consists of two parts: the endothelium, a single layer of cells, which extends as a continuous lining to all blood vessels and, surrounding it, a thin subendothelial layer containing collagen fibres. Outside this is the inner boundary of the media, formed mainly by a layer of inter-linked elastin fibres, called the internal elastic lamina. The rest of the media (usually the thickest part of the wall, and that which dominates its elastic behaviour) has a different structure in large central arteries from that in small arteries. In the former it consists of multiple concentric layers of elastic tissue (elastin), separated by thin layers of connective tissue (collagen) and occasional smooth muscle cells. In smaller arteries, the media consists almost entirely of spirally wound smooth muscle cells, arranged in layers, with small amounts of collagen and elastin between them. The outside of the media in each case consists of a thin, external elastic lamina. The adventitia is often as thick as the media, but is less important mechanically because it consists largely of loose connective tissue, containing relatively sparse elastin and collagen fibres.

The elastin, collagen and smooth muscle fibres constitute about 50% of the material of the wall; the rest is largely water, inside or outside cells, which has a negligible effect on the mechanical properties of the wall apart from being effectively incompressible. In large arteries (of dogs) elastin and collagen together constitute about 50% of the dry mass. In the intrathoracic aorta the ratio of elastin to collagen is about 1.5, while in other arteries it is about 0.5.

The elastic properties of the artery wall clearly depend both on the properties of its individual components and on how they are linked together. Elastin is an easily extensible elastic material, individual fibres having nonlinear stress-strain relations with a Young's modulus of about 300 kN m^{-2} for strains up to about 40%, but greater stiffness for larger strains (Carton et al., 1962). Collagen is much stiffer, lengths of human tendon having a Young's modulus of about 10^6 kN m^{-2}. Smooth muscle has a Young's modulus roughly similar to that of elastin, but its actual value depends on the level of physiological activity, varying from about 100 kN m^{-2} in the active state. Of these three materials, only elastin is purely elastic; both collagen and, especially, smooth muscle show marked viscoelastic properties (creep, stress relaxation and hysteresis), which are reflected in the dynamic properties of artery walls.

As the above description of arterial wall structure shows, the arrangement of the different fibres in the wall is inhomogeneous and very complicated, and to infer the elastic properties of the wall as a whole (regarded, for example, as a thin-walled cylinder) those of its constituents is virtually impossible. Instead, experiments have been performed on segments of artery, excised from newly dead animals, and with their ends occluded in some way. In these experiments the dimensions which can be accurately measured are length, l, and internal volume, V. Here the internal cross-sectional area, A, will in general be used as the important transverse dimension; this can be deduced from V and l. The forces tending to deform the artery wall in these experiments are longitudinal tension, T, and transmural pressure, p_{tm}.

Considerable detail can be derived from my book and the references given therein, and from other lectures in this course. However, from the point of view of my lectures, a good approximation to the elastic properties of arteries and veins can be given in the form of a *tube law*, an equation relating the transmural pressure p_{tm} and the local cross-sectional area, A:

$$p_{tm} = p - p_e \approx \tilde{P}(A) \; . \qquad (1.3)$$

Figure 1.3 shows typical forms for $\tilde{P}(A)$, (a) for a large artery (thoracic aorta) and (b) for a large vein (vena cava) and a latex tube of similar wall thickness to diameter ratio. In each case the curve is markedly nonlinear. Arteries become much stiffer as A exceeds its normal mean value, which would not happen in a linearly elastic material. Note that all elastic tubes can collapse to a very small cross-sectional area (even zero) when the transmural pressure becomes negative. As discussed above, this normally happens in veins above the heart, which is why figure 1.3(b) includes a region of negative p_{tm}. That figure also contains rough sketches of the tube cross-sectional shape at various values of A. More discussion of vessel collapse will be given in Chapters 5 and 6.

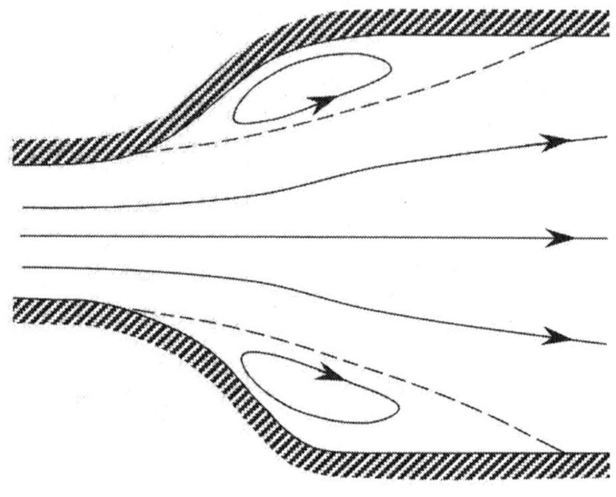

Figure 1.1 Steady flow separation at an abrupt expansion in a two-dimensional channel, illustrated by a symmetrical expansion, at which a symmetrical flow may be possible.

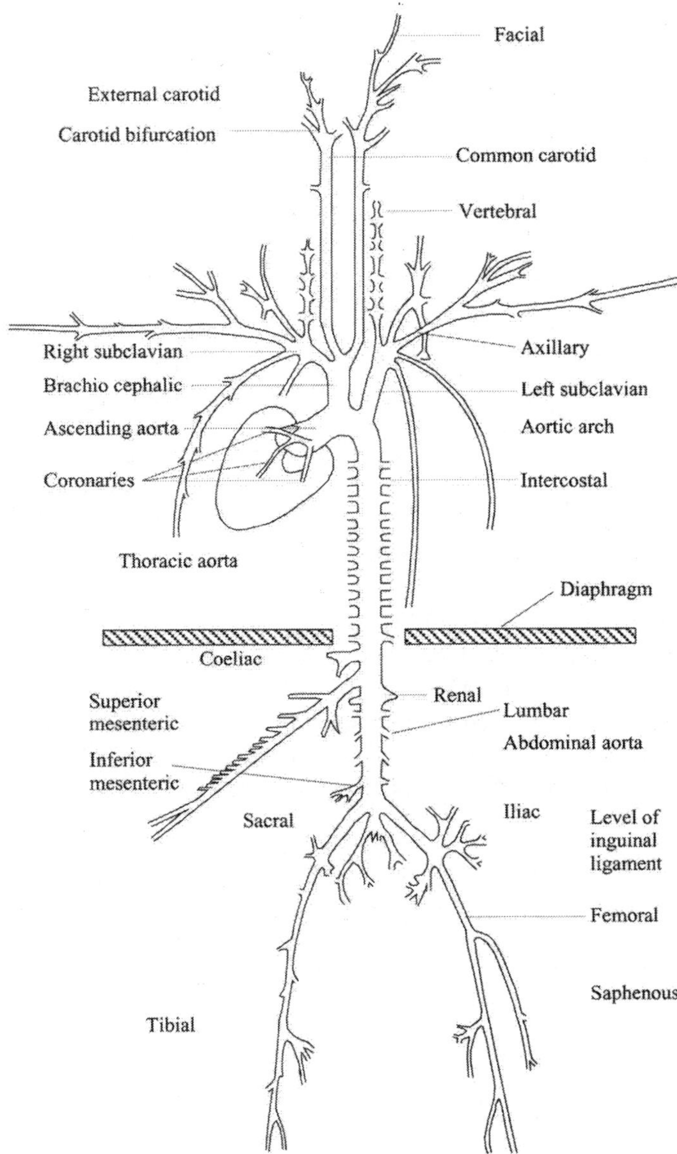

Figure 1.2 A diagrammatic representation of the major arteries in the mammalian arterial tree (after McDonald, 1974).

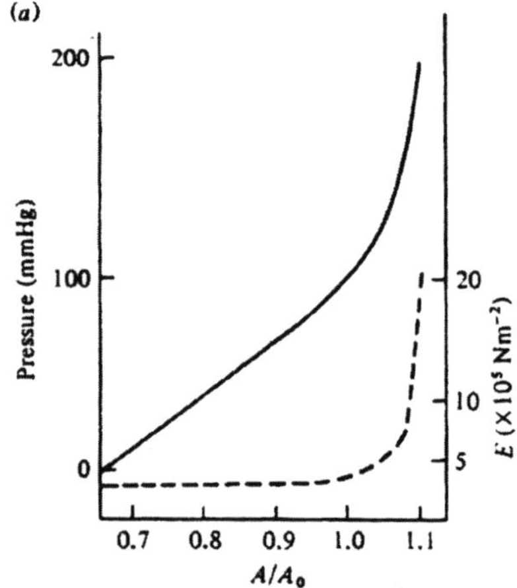

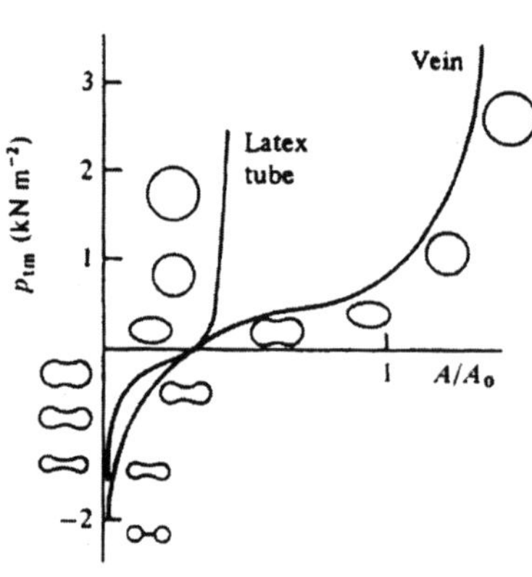

Figure 1.3(a) The pressure (continuous curve) and incremental elastic modulus (broken curve) of a segment of dog thoracic aorta, plotted against area divided by in-vivo area (100 mm Hg $\approx$ 13.3 kN m^{-2}). (After McDonald, 1974.) Figure 1.3(b) Comparison of the elastic properties of a latex tube with those of an excised segment of a canine vena cava: transmural pressure as a function of area, A, scaled with respect to the area at zero transmural pressure, A_0. Transverse cross-sections are sketched for various areas. (Data from Moreno et al., 1970.)

2 Pulse Wave Propagation

2.1 Observations

When the aortic valve opens at the beginning of the cardiac ejection phase (systole), the pressure rises at the entrance to the aorta and a volume of blood (about 80ml in man) is ejected. Because the vessel is elastic, the pressure rise locally distends it, rather than instantly setting the blood into motion in all vessels as it would if the system were rigid. The distended segment then contracts (with overshoot because of inertia), accelerating the blood a bit further downstream, and so on. In fact, a wave is generated and propagates downstream. The restoring force is provided by the elasticity of the walls, and the inertia by the mass of the blood.

Figure 2.1 shows the pressure wave form measured at a succession of equally spaced sites down a canine aorta. Propagation of the wave can be clearly seen, the front propagating at a speed of about 5ms^{-1}, as can two other phenomena: *peaking*, i.e. an increase in amplitude as the wave propagates, and *steepening* of the wave front. The wave form of flow velocity has a different shape from that of the pressure, and its amplitude decreases with distance from the heart (figure 2.2). The objective of theoretical modelling is to explain the above pressure-wave phenomena — propagation (and its speed), peaking, steepening — and the corresponding velocity-wave phenomena.

2.2 Basic theory

As with all mathematical modelling, it is sensible to start with the simplest possible model, and add complications only when it becomes clear in what respects the simple model is inadequate. We therefore begin by considering an infinitely long, straight, horizontal elastic tube, with uniform undisturbed cross-sectional area A_0 and uniform external pressure p_e, containing an inviscid incompressible fluid of constant density ρ which is initially at rest. We analyse disturbances whose wavelengths are much greater than the tube diameter, so that the time-dependent internal pressure can be taken to be a function only of longitudinal coordinate x and time t. The disturbed cross-sectional area is denoted by $A(x,t)$ and the fluid velocity (averaged across the cross-section) by $u(x,t)$. The model is thus one-dimensional.

The governing equations are those representing conservation of mass, conservation of momentum and elasticity. By considering the rate of change of volume of a thin 'slice' of the tube between two neighbouring x-locations, we can show that the equation of conservation of mass is (exactly)

$$\frac{\partial A}{\partial t} + \frac{\partial}{\partial x}(Au) = 0. \tag{2. 1}$$

The momentum (Euler) equation is

$$\frac{\partial u}{\partial t} + u\frac{\partial u}{\partial x} = -\frac{1}{\rho}\frac{\partial p}{\partial x}, \tag{2. 2}$$

and we model the vessel elasticity by means of a *tube law* relating transmural pressure difference to local cross-sectional area:

$$p - p_e = \tilde{P}(A) \tag{2. 3}$$

where the function $\tilde{P}(A)$ takes a form like figure 1.3b for most elastic tubes, including blood vessels.

We first analyse small amplitude disturbances, for which u is small (in a sense to be determined later) and

$$A = A_0 + A', \quad p - p_e = \tilde{P}(A_0) + p'$$

where $|A'| \ll A_0$, $|p'| \ll \tilde{P}(A_0)$. Substituting into the equations and neglecting all terms nonlinear in the small quantities, we can eliminate u and A' to obtain the following single equation for p':

$$\frac{\partial^2 p'}{\partial t^2} = c^2(A_0)\frac{\partial^2 p'}{\partial x^2}, \tag{2. 4}$$

where

$$c^2(A) = \frac{A}{\rho}\frac{d\tilde{P}}{dA}. \tag{2. 5}$$

Equation (2. 4) is the well-known wave equation, from which we can deduce at once that small-amplitude disturbances can propagate along the tube, in either direction, without change of shape, at speed $c_0 = c(A_0)$. The general solution of (2. 4) is

$$p' = f_1\left(t - \frac{x}{c_0}\right) + f_2\left(t + \frac{x}{c_0}\right), \tag{2. 6}$$

where f_1 and f_2 are arbitrary functions; f_2 is zero if the wave propagates only in the $+x$ direction.

If it is *supposed* that the tube wall is made from a homogeneous and isotropic Hookean solid of Young's modulus E and has small thickness-to-diameter ratio h/d, then it can be shown that

$$c_0 = \left(\frac{Eh}{\rho d}\right)^{\frac{1}{2}}, \tag{2. 7}$$

known as the *Moens–Korteweg wave speed* (after Moens 1878 and Korteweg 1878) though the above theory was first correctly performed by Thomas Young (1809). Whether calculated from equation (2. 7), or more directly from (2. 5) and measurements of $\tilde{P}(A)$, the predicted value of c_0 is about 5ms^{-1} in the ascending aorta, rising to about 8ms^{-1} in more peripheral arteries. These predictions are very close to measured values in normal subjects, either dogs or humans, at least if they are not too old. Arteries become stiffer with age.

Thus the simple theory is very successful at establishing the mechanism of wave propagation, involving only wall elasticity and blood inertia, and in predicting the pulse wave speed. However, the theory also predicts no change of shape of the wave-form as it propagates, and a velocity wave-form of the same shape as the pressure wave-form, since if $p' = P_1 f(t - x/c_0)$, where P_1 is the pressure amplitude (much less than $\tilde{P}(A_0)$), then the linearised version of (2. 2) gives

$$u = \frac{P_1}{\rho c_0} f\left(t - \frac{x}{c_0}\right). \tag{2. 8}$$

Hence the theory must be modified to account for peaking, steepening and the observed velocity waveform. It turns out that the peaking of the pressure pulse can be explained in terms of wave reflection from arterial junctions, front steepening is a nonlinear effect, and the shape of the velocity waveform can be explained if account is taken of blood viscosity. We examine these effects in order.

2.3 Wave reflection

Consider a single bifurcation from parent tube 1 to daughter tubes 2 and 3 (figure 2.3), with undisturbed cross-sectional areas $A_1,\ A_2,\ A_3$ and intrinsic wave speeds $c_1,\ c_2,\ c_3$. Let the longitudinal coordinate in each tube be x, with $x = 0$ at the bifurcation point B. An incident wave I approaches the bifurcation in tube 1 from $x = -\infty$. We suppose that the contribution of this wave to the pressure in tube 1 is

$$p'_I = P_I f\left(t - \frac{x}{c_1}\right), \tag{2. 9}$$

where P_I is an amplitude parameter, and f is a continuous, periodic function whose maximum value is 1. The corresponding contribution to the flow rate is

$$Q_I = A_1 u = Y_1 P_I f\left(t - \frac{x}{c_1}\right), \tag{2. 10}$$

where

$$Y_1 = \frac{A_1}{\rho c_1}, \tag{2. 11}$$

from (2. 8). Y_1 is called the *characteristic admittance* of tube 1.

At the bifurcation, the incident wave gives rise to a reflected wave R in the parent tube and transmitted waves $T_2,\ T_3$ in the daughters. We let the pressure contributions be

$$p'_R = P_R g\left(t + \frac{x}{c_1}\right), \quad p'_{Tj} = P_{Tj} h_j\left(t - \frac{x}{c_j}\right), \quad (j = 2, 3) \tag{2. 12}$$

where the plus sign in the function g represents the fact that the reflected wave is travelling in the $-x$ direction. The corresponding contributions to the flow rate are

$$Q_R = -Y_1 P_R g\left(t + \frac{x}{c_1}\right), \quad Q'_{Tj} = Y_j P_{Tj} h_j\left(t - \frac{x}{c_j}\right), \tag{2. 13}$$

where the minus sign in Q_R comes from the linearised equation (2. 2) (cf (2. 8)), and the Y_j are defined like Y_1 in (2. 11).

We then apply boundary conditions at the junction $x = 0$: the long wavelength approximation means that we can neglect the details of flow at the junction, which are assumed to occupy a length of $O(d)$. The boundary conditions are those of continuity of pressure (required by Newton's Law, to avoid large local accelerations), and continuity of flow rate (required by conservation of mass). The pressure and flow rate must be continuous at $x = 0$ for all time t. It follows that the functions of time, $g(t)$ and $h_j(t)$, are all equal to the incident function $f(t)$, and that

$$P_I + P_R = P_{T1} = P_{T2}, \qquad Y_1\left(P_I - P_R\right) = \sum_{j=2}^{3} Y_j P_{Tj}.$$

Solving these gives

$$\frac{P_R}{P_I} = \frac{Y_1 - \sum Y_j}{Y_1 + \sum Y_j}, \qquad \frac{P_{Tj}}{P_I} = \frac{2Y_1}{Y_1 + \sum Y_j}, \tag{2. 14}$$

and the amplitudes of the reflected and transmitted waves are determined.

The pressure perturbation in the parent tube is given by (2. 9) and (2. 12) to be

$$\frac{p'}{P_I} = f\left(t - \frac{x}{c_1}\right) + \frac{P_R}{P_I} f\left(t + \frac{x}{c_1}\right) \tag{2. 15}$$

while the flow rate, from (2. 10) and (2. 13), is

$$Q = Y_1 P_I \left[f\left(t - \frac{x}{c_1}\right) - \frac{P_R}{P_I} f\left(t + \frac{x}{c_1}\right) \right]. \tag{2. 16}$$

For a simple example, suppose that $f(t)$ is sinusoidal:

$$f(t) = \cos \omega t.$$

Then, with $\beta = P_R/P_I$, (2. 14) gives

$$\frac{p'}{P_I} = (1 - \beta) \cos \omega \left(t - \frac{x}{c_1}\right) + 2\beta \cos \omega t \cos \frac{\omega x}{c}.$$

This represents a propagating wave of uniform amplitude $(1 - \beta)P_I$ plus a *standing wave* of amplitude $|2\beta P_I \cos \frac{\omega x}{c}|$. It follows that, if β is positive, the overall pressure amplitude has a maximum value $P_I(1 + \beta)$ at the junction $x = 0$, and it falls with distance upstream, to a minimum value where $\cos \omega x/c = 0$, i.e.: at $x = -\pi c/2\omega$, one quarter-wavelength proximal to the bifurcation. The amplitude of the corresponding flow-rate wave form increases with distance upstream. If β is negative, the pressure amplitude is minimum at $x = 0$, and the flow-rate amplitude maximum. A bifurcation with $\beta > 0$ is called a *closed-end* type of reflection site, because $\beta = 1$ represents a truly closed end; $\beta < 0$ corresponds to *open-end* type of reflection, with $\beta = -1$ corresponding to a truly open end.

The fact that the pressure wave amplitude in the aorta is observed to increase with distance down the vessel can thus be interpreted as indicating that there is a closed end type of reflection at (or beyond) the iliac bifurcation (where it divides to supply the two hind limbs — see figure 1.2), as long as the length of the aorta is less than one-quarter wavelength. A wave speed of 5ms^{-1} and a human pulse frequency $\omega/2\pi$ of 1.25Hz gives a quarter wavelength of 1m, which is (a little) greater than the length of the aorta (in dogs both the wavelength and the aortic length are smaller). So is it plausible that the iliac bifurcation is a source of closed end reflection? For this to be true we would require

$$Y_2 + Y_3 < Y_1,$$

from (2. 14). Assuming that wave-speed does not vary discontinuously at a bifurcation, which is reasonable on the basis of (2. 7), then this inequality is equivalent to

$$A_2 + A_3 < A_1. \tag{2. 17}$$

Now measurements indicate that most bifurcations in the human arterial tree are well-matched ($A_2 + A_3 \approx A_1$) but that, indeed, the iliac bifurcation does satisfy (2. 17), the sum of the area ratios of daughter branches to parent normally being 0.85–0.90.

An analysis similar to the above can be performed for a system with multiple bifurcations, confirming that the peaking of the pressure pulse is primarily a consequence of closed end reflection at the end of the aorta, coupled to a gradual stiffening of the arteries (increasing c) in the peripherical direction.

2.4 Nonlinear analysis

Neglect of the nonlinear terms in equations (2. 1)–(2. 3) was based on two assumptions: (i) that the pressure amplitude be small compared with the mean and (ii) that the fluid speed be in some sense small. (i) In normal human beings (and most other mammals) the mean blood pressure, relative to atmosphere, at the level of the heart is about 100 mm Hg (13.3 kPa), and there is a cyclic variation between 80 and 120 mm Hg, so the amplitude-to-mean ratio is 0.2, which is reasonably small. (ii) If $u = f(t - x/c)$ is taken to be the solution of the linear wave problem and substituted into the nonlinear term in (2. 2), the latter is indeed seen to be negligible if u/c is small. In the ascending aorta, c is about 5ms^{-1} and the maximum value of u is (normally) about 1ms^{-1}, so u/c is also around 0.2. Thus the effect of nonlinearity is expected normally to be weak. However, there are medical conditions in which either u/c or $\Delta p/p$ (or both) become considerably larger, so it is important to consider nonlinear effects.

This can be done using equations (2. 1)–(2. 3), and defining c by (2. 5). If we add $\left(\pm\frac{c}{A}\right)$ multiplied by equation (2. 1) to equation (2. 2), we obtain the standard Riemann equations for one-dimensional nonlinear waves:

$$\left[\frac{\partial}{\partial t} + (u \pm c)\frac{\partial}{\partial x}\right]\left\{u \pm \int_{A0}^{A} \frac{c}{A} dA\right\} = 0,$$

so the quantities in the curly brackets (*Riemann invariants*) are constant along the *characteristic curves* in $x - t$ space:

$$\frac{dx}{dt} = u \pm c. \tag{2. 18}$$

If one considers a *simple wave*, propagating in the $+x$ direction into an undisturbed region, then standard theory (Lighthill, 1978) shows that the wave front will indeed continually *steepen*. Eventually, if the tube is long enough, the wave reaches a singular state in which the wave front becomes vertical and an *elastic jump* develops (analogous to a hydraulic jump in open channel flow and a shock wave in gas dynamics). The distance to the point of jump formation is increased by dissipative effects, and estimates for a normal aorta (Pedley, 1980) indicate that this distance would be greater than the length of the vessel. However, when u/c at the entrance to the aorta is much larger than normal (e.g.: in patients with chronically enlarged left ventricles consequent upon a leaky aortic valve) a jump would be predicted, and may be responsible for the 'pistol-shot pulse' reported by clinicians for this condition.

2.5 Viscous effects

Womersley (1957) performed the first thorough analysis of small-amplitude waves in an elastic tube containing a *viscous* fluid (see Pedley, 1980, chapter 2). He did not make the long-wavelength approximation but did assume axial symmetry, and his analysis has been extended in a variety of ways. Here we analyse viscous effects in a more approximate way, following Womersley (1955) and Lighthill (1975, chapter 11).

We suppose that the propagating, time-dependent, pressure-gradient wave form has been measured (figure 2.4), and we calculate the profile of axial velocity driven by that pressure gradient in a rigid tube of radius $a = d/2$. It will turn out that the viscous effects are confined to a thin boundary layer on the wall whose structure is largely unaffected by the fact that the wall is really moving in and out (remaining approximately parallel because the wavelength is long compared not only to the tube diameter but also to the axial displacement of fluid elements during one cycle).

At a given location x the pressure-gradient wave form, with fundamental radian frequency ω, can be expressed as a Fourier series:

$$\frac{\partial p}{\partial x} = -\sum_{n=0}^{N} G_n e^{in\omega t}, \tag{2. 19}$$

where the number of modes, N, required for accuracy is 4–6 for calculation of flow rate, and about 10 for wall shear rate (because the latter involves the numerically inaccurate process of differentiation). Note that the $n = 0$ term represents the mean pressure gradient, and real parts of complex quantities are assumed. The axial momentum equation for parallel flow in a rigid tube reduces at small amplitude to

$$\frac{\partial u}{\partial t} = -\frac{1}{\rho}\frac{\partial p}{\partial x} + \nu\left(\frac{\partial^2 u}{\partial r^2} + \frac{1}{r}\frac{\partial u}{\partial r}\right), \tag{2. 20}$$

where $\nu = \mu/\rho$ is the kinematic viscosity of the fluid and r is the radial coordinate. The boundary conditions on the axial velocity $u(r, t)$ are those of no slip and axial symmetry, i.e.:

$$u(a, t) = 0, \quad \frac{\partial u}{\partial r}(0, t) = 0. \tag{2. 21}$$

The problem is linear and its solution can therefore be written

$$u = u_0(r) + \sum_{1}^{N} u_n(r) e^{in\omega t},$$

yielding a separate problem for each n. Solving term by term we obtain

$$u_0(r) = \frac{G_0 a^2}{4\mu}\left(1 - \frac{r^2}{a^2}\right), \tag{2. 22}$$

i.e.: the mean flow is just steady Poiseuille flow, and

$$u_n(r) = \frac{G_n a^2}{i\mu\alpha_n^2}\left\{1 - \frac{J_0\left(i^{\frac{3}{2}}\alpha_n \frac{r}{a}\right)}{J_0\left(i^{\frac{3}{2}}\alpha_n\right)}\right\}, \tag{2. 23}$$

where J_0 is the Bessel function of the first kind of order zero and

$$\alpha_n^2 = n\omega a^2/\nu. \tag{2. 24}$$

The quantity α_n^2 represents the ratio of inertial to viscous terms in the equation for u_n derived from (2. 20). It can also be thought of as a scale for the ratio of the time for viscous diffusion of information across the radius of the tube to the period of the oscillation, or α_n as the ratio of tube radius to viscous diffusion distance in one period. The quantity α_1, corresponding to the fundamental oscillation, is the same as α defined in equation (1. 3), and is often referred to as the Womersley parameter. In the normal human aorta, α is quite large (15–20), and remains significantly greater than 1 in all large arteries (and note that if α is large, α_n for $n > 1$ is even larger). In the microcirculation (arterioles and capillaries) α is very small.

Appropriate asymptotic expansion of the Bessel functions shows that, when α_n is small, equation (2. 23) reduces to

$$u_n(r) \approx \frac{G_n a^2}{4\mu}\left(1 - \frac{r^2}{a^2}\right),$$

i.e.: the flow is *quasi-steady* at low frequency (cf (2. 1)). When α_n is large, on the other hand, we have

$$u_n(r) \approx \frac{G_n a^2}{i\mu\alpha_n^2}\left\{1 - \exp\left[-\sqrt{\frac{\omega}{2\nu}}(1+i)(a-r)\right]\right\},$$

which is the velocity profile in the Stokes boundary layer on a *plane* wall in an oscillating flow, to be expected if the boundary layer thickness is much less than the tube radius.

When the velocity profile (2. 22) is integrated across the cross section to give the volume flow rate $Q(t)$, and differentiated at the wall to give the wall shear rate $\tau(t)$, the following results are obtained:

$$Q(t) = \int_0^a 2\pi r u dr = \pi a^2 \left\{\frac{G_0 a^2}{8\mu} + \frac{a^2}{i\mu}\sum_1^\infty \frac{G_n}{\alpha_n^2}[1 - F(\alpha_n)]e^{in\omega t}\right\} \tag{2. 25}$$

and

$$\tau(t) = -\frac{\partial u}{\partial r}\,|_{r=a} = \frac{G_0 a}{2} + \frac{a}{2}\sum_1^\infty G_n F(\alpha_n)\, e^{in\omega t} \tag{2. 26}$$

where

$$F(\alpha) = \frac{2J_1(i^{\frac{3}{2}}\alpha)}{i^{\frac{3}{2}}\alpha J_0(i^{\frac{3}{2}}\alpha)} \approx 1 - \frac{i\alpha^2}{8} \text{ as } \alpha \to 0 \tag{2. 27a}$$

$$\approx \frac{2}{i^{\frac{1}{2}}\alpha}(1 + \frac{1}{2\alpha}) \text{ as } \alpha \to \infty. \tag{2. 27b}$$

It turns out that the first two terms of the small α expansion, (2. 27*a*), are quite accurate when $\alpha < 4$, while the first two terms of the large α expansion, (2. 27*b*), are accurate for $\alpha > 4$. Figure 2.4 shows that, if viscous effects are ignored, the flow-rate wave form, calculated using (1. 3) with the measured pressure-gradient wave form, agrees much less well with the measured flow-rate wave form than that calculated using (2. 5). Thus viscous effects as well as wave reflections are required to explain the shape of the flow-rate (or average velocity) wave form.

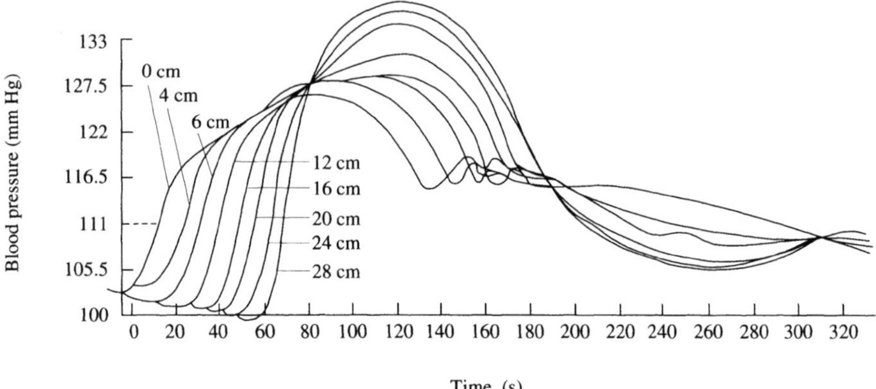

Figure 2.1 Instantaneous blood pressure records made at a series of sites along the aorta in a dog; 0 cm is at the start of the descending aorta (after Olson, R. M. *J. Appl. Physiol.* **24**: 563–569, 1968).

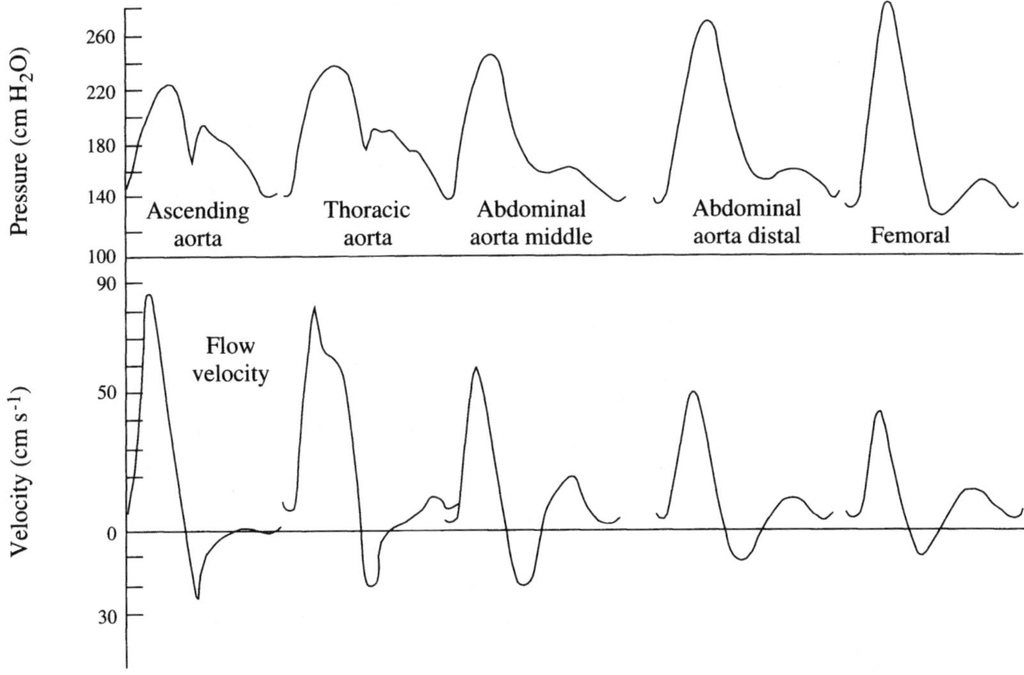

Figure 2.2 Matched records of (a) pressure and (b) average velocity at different sites in the arteries of a dog (from McDonald, 1974).

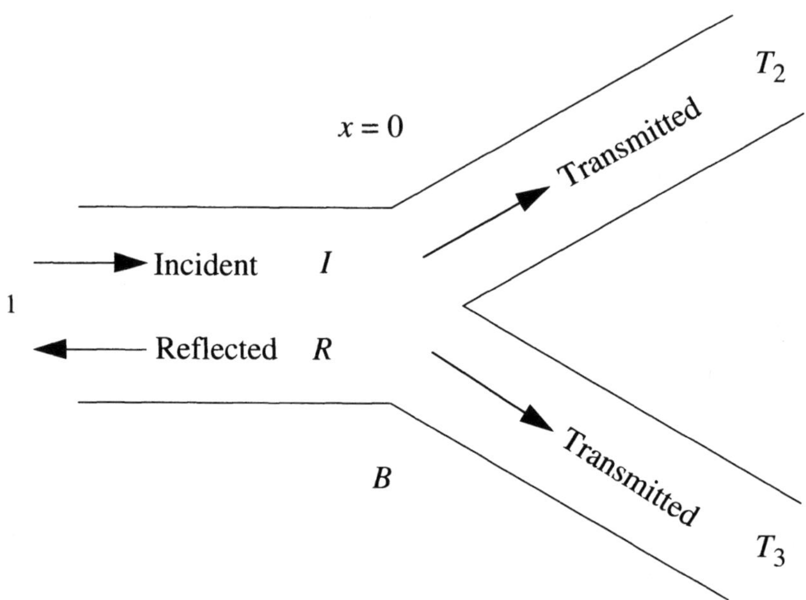

Figure 2.3 Sketch of an arterial bifurcation. The incident wave is partially reflected in the parent tube 1 and partially transmitted in the daughter tubes 2 and 3.

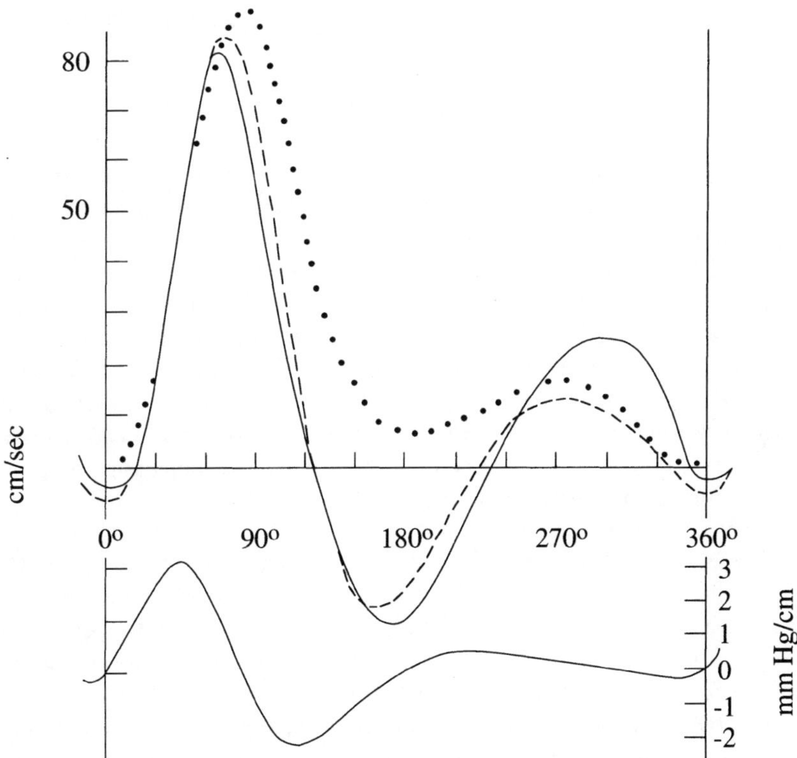

Figure 2.4 Wave forms of average velocity (upper traces) in a dog's femoral artery. Continuous curve: measured with an electromagnetic flow meter; broken curve: calculated by the viscous theory from the measured pressure gradient (lower trace); dotted curve: calculated from inviscid theory. (After McDonald, 1974.)

3 Flow Patterns in Complex Internal Geometry: Flow in Curved Tubes

The geometry of arteries is three-dimensional, and the flow resembles its two-dimensional counterpart (cf figure 1.1) only on planes of symmetry, if they exist. Off the plane of symmetry, however, the flow is quite different in three dimensions. Secondary motions arise in a uniform curved tube, since the faster-moving fluid near the centre is swept towards the outside of the bend, to be replaced at the inside by the slower-moving fluid near the walls. The consequence is a relatively high wall shear near the outside wall, but a much reduced wall shear on the inside part of the wall. Steady flow in a symmetrical bifurcation resembles that in two curved tubes stuck together; the wall shear stress at the outside of the bends (the inner wall of the bifurcation) is enhanced further because new boundary layers with steep velocity gradients develop downstream from the flow divider as it splits the oncoming velocity profile (figure 3.1). Separation may be seen if the area ratio (of combined daughter tubes to parent tube) exceeds 1, so that there is an overall expansion, or if the curvature of the outer wall is relatively sharp, or if the distribution of flow rate in the daughter tubes is markedly unequal (Zeller et al., 1970; Walburn and Stein, 1980). When the geometry of the bifurcation is not symmetrical, flow separation is more likely. The extreme case is that of a perpendicular side branch of a parent artery (figure 3.2). Regions of separated flow are seen just inside the daughter tube and on the wall of the parent tube opposite the branch. The trajectories of individual fluid elements can be very complex (Karino et al., 1979), as can the shear stress distribution (Lutz et al., 1977). Another asymmetric geometry that has been widely studied is the carotid bifurcation, where the carotid sinus is particularly prone to separation and to atheroma (Ku, 1988).

In all the above cases, the location of flow separation and the WSS distribution may be sensitively dependent on small geometrical details, but within separated eddies the fluid velocity and WSS remain low. In other circumstances that fact cannot be relied on. One example is under the longitudinal 'horseshoe' (or 'hairpin') vortices that stretch away downstream of an abrupt protuberance on an otherwise smooth boundary (Fukushima & Azuma, 1982). Other examples arise in non-uniform conduits, particularly when the flow is time-dependent and even (or especially) when it is two-dimensional. The time-dependence arises either from wall motion (Pedley & Stephanoff, 1985; Ralph & Pedley, 1988, 1990) or from an unsteady driving pressure gradient (Sobey, 1985; Tutty & Pedley, 1993). In these cases waves (so called 'vorticity waves') are generated in the channel downstream of an indentation and separated eddies occur under the wave crests and above the troughs, at locations quite distant from the indentation itself. Moreover, the WSS is largest, not smallest, under the recirculating flow in the eddies. Neither of these flow features is consistent with steady-flow intuition, providing a further demonstration of how hard it will be to assess WSS and predict sites of atherogenesis in individual subjects. An analysis of such two-dimensional, time-dependent flows will be given in Chapter 4. In the remainder of this lecture we discuss steady flow in curved tubes.

3.1 Steady Flow in Curved Tubes

As stated in chapter 1, the complicating feature of curved tube flow is that the motion cannot be everywhere parallel to the curved centreline because secondary (transverse) motions are inevitably generated. These secondary motions themselves influence the distribution of axial velocity and result in a complicated interaction of the two.

We begin by writing down the full equations of motion for viscous flow in a curved tube, without yet imposing the restriction that the flow should be steady or fully developed. We also retain the possibility that the curvature of the centre line is non-uniform, although we do restrict it to lie in a single plane. These equations may be found to be a useful reference. Coordinates (ar, θ, as) are chosen, where a is the radius of the tube cross-section, assumed uniform (so the tube boundary is $r = 1$); as is the distance measured along the centre line and (ar, θ) are polar coordinates in the cross-section (figure 3.3). We take the velocity field to be $\hat{W}_0(u, v, w)$, where $\hat{W}_0$ is a suitable velocity-scale. Let the radius of curvature of the centre line at a particular cross-section be R, and define the dimensionless ratio $\delta(s) = a/R$, which depends on s if R is not uniform. Finally, we define the parameter

$$h = 1 + \delta(s) r \cos\theta \ ,$$

which is $1/R$ times the distance from the centre of curvature of the centre line (at a particular s) to the projection on the plane containing the centre line of a general point (r, θ, s). In this orthogonal coordinate system, the continuity equation $div\mathbf{u} = 0$ is

$$u_r + \frac{u(1 + 2\delta r \cos\theta)}{rh} + \frac{v_\theta}{r} - \frac{v\delta \sin\theta}{h} + \frac{w_s}{h} = 0 \ , \tag{3. 1}$$

and the three momentum equations are:

$$\begin{aligned} & u_t + uu_r + \frac{v}{r}u_\theta + \frac{w}{h}u_s - \frac{v^2}{r} - \frac{\delta\cos\theta}{h}w^2 \\ &= -p_r + \frac{Re^{-1}}{rh}\left\{\frac{\partial}{\partial s}\left[\frac{r}{h}\left(u_s - hw_r - \delta\cos\theta w\right)\right]\right. \\ &\quad \left. -\frac{\partial}{\partial \theta}\left[h\left(v_r + \frac{v}{r} - \frac{u_\theta}{r}\right)\right]\right\} \ , \end{aligned} \tag{3. 2}$$

$$\begin{aligned} & v_t + uv_r + \frac{v}{r}v_\theta + \frac{w}{h}v_s + \frac{uv}{r} + \frac{\delta\sin\theta}{h}w^2 \\ &= -\frac{1}{r}p_\theta + \frac{Re^{-1}}{h}\left\{\frac{\partial}{\partial r}\left[h\left(v_r + \frac{v}{r} - \frac{u_\theta}{r}\right)\right]\right. \\ &\quad \left. -\frac{\partial}{\partial s}\left[\frac{1}{rh}\left(hw_\theta - \delta r\sin\theta w - rv_s\right)\right]\right\} \ , \end{aligned} \tag{3. 3}$$

$$\begin{aligned} & w_t + uw_r + \frac{v}{r}w_\theta + \frac{w}{h}w_s + \frac{\delta\cos\theta}{h}uw - \frac{\delta\sin\theta}{h}vw \\ &= -\frac{1}{h}p_s + \frac{Re^{-1}}{r}\left\{\frac{\partial}{\partial \theta}\left[\frac{1}{rh}\left(hw_\theta - \delta r\sin\theta w - rv_s\right)\right]\right. \\ &\quad \left. -\frac{\partial}{\partial r}\left[\frac{r}{h}\left(u_s - hw_r - \delta\cos\theta w\right)\right]\right\} \ . \end{aligned} \tag{3. 4}$$

The time has been non-dimensionalised with respect to $a/\hat{W}_0$ and the pressure with respect to $\rho\hat{W}_0^2$; the Reynolds number $Re = \hat{W}_0 a/\nu$.

We now restrict attention to steady ($\partial\mathbf{u}/\partial t = 0$), fully developed ($\partial\mathbf{u}/\partial s = 0$), flow in a uniformly ($\delta$ = constant) and weakly ($\delta \ll 1$) curved tube.

When δ is uniform, and hence the velocity field independent of s, the only way that the transverse velocities are affected by the axial velocity is through the 'centrifugal force' terms that drive it; these are the last terms on the left-hand sides of (3. 2) and (3. 3), proportional to δw^2. The axial velocity is itself influenced by the transverse motions through the convective inertia terms in (3. 4). In this case (3. 4) shows that the pressure gradient $-p_s$ is independent of s. Thus the pressure can be written in the form of $-Gs + p'(r,\theta)$; from (3. 2) and (3. 3) G is seen to be independent of r and θ is therefore a constant. It also proves to be convenient to take $\hat{W}_0 = \nu/a$ so, formally, $Re = 1$. With $\delta \ll 1$ we neglect all terms that are $O(\delta)$ compared with the leading term in any equation. First, however, we rescale w so that the centrifugal force terms in (3. 2) and (3. 3) are of the same order of magnitude as the viscous and inertial terms (this is necessary because the centrifugal force terms *drive* the secondary motion). To this end we replace w by $(2\delta)^{-1/2}w'$ (the factor of $2^{-1/2}$ is introduced for consistency with other authors). Equations (3. 1) – (3. 4) then reduce to

$$u_r + \frac{u}{r} + \frac{v_\theta}{r} = 0 \ , \tag{3. 5}$$

$$uu_r + \frac{v}{r}u_\theta - \frac{v^2}{r} - \frac{1}{2}w'^2\cos\theta = -p'_r - \frac{1}{r}\frac{\partial}{\partial\theta}\left(v_r + \frac{v}{r} - \frac{u_\theta}{r}\right) \ , \tag{3. 6}$$

$$uv_r + \frac{v}{r}v_\theta + \frac{uv}{r} + \frac{1}{2}w'^2\sin\theta = -\frac{1}{r}p'_\theta + \frac{\partial}{\partial r}\left(v_r + \frac{v}{r} - \frac{u_\theta}{r}\right) \ , \tag{3. 7}$$

$$uw'_r + \frac{v}{r}w'_\theta = D - p''_s + \left(w'_{rr} + \frac{1}{r}w'_r + \frac{1}{r^2}w'_{\theta\theta}\right) \ . \tag{3. 8}$$

The boundary conditions are that $u = v = w' = 0$ on $r = 1$, and that there should be no singularity at $r = 0$.

The parameter D is given by

$$D = (2\delta)^{1/2}G = (2\delta)^{1/2}\frac{\hat{G}a^2}{\mu}\cdot\frac{a}{\nu} \ , \tag{3. 9}$$

where $-\hat{G}$ is the dimensional pressure gradient. $\hat{G}a^2/\mu$ is four times the peak velocity in Poiseuille flow in a straight tube, driven by the same pressure gradient $-\hat{G}$, so $D = (2\delta)^{1/2}\cdot 4Re'$, where Re' is the Reynolds number of the straight-tube flow. D is called the Dean number, because Dean (1928) was the first to realise its significance in curved-tube flow.

The flow is entirely governed by the single dimensionless parameter D. We shall examine the flow both for small D (an analytical solution) and for a wide range of D, becoming quite large (numerical solutions). For all D, the continuity equation (3. 5), admits of the existence of a secondary stream function, Ψ, such that

$$u = (1/r)\Psi_\theta, \quad v = -\Psi_r \ , \tag{3. 10}$$

and then the other equations reduce to the pair of coupled equations:

$$\nabla_1^2 w' + D = (1/r)\left(\Psi_\theta w'_r - \Psi_r w'_\theta\right) \ , \tag{3. 11}$$

$$\nabla_1^4 \Psi + \frac{1}{r}\left(\Psi_r \frac{\partial}{\partial\theta} - \Psi_\theta \frac{\partial}{\partial r}\right)\nabla_1^2 \Psi = -w'\left(\sin\theta w'_r + \frac{\cos\theta}{r} w'_\theta\right) \ , \tag{3. 12}$$

where

$$\nabla_1^2 \equiv \partial^2/\partial r^2 + (1/r)\partial/\partial r + (1/r^2)\partial^2/\partial\theta^2 \ .$$

The boundary conditions at $r = 1$ are $\Psi = \Psi_r = w' = 0$.

3.2 Small D

Dean (1928) gave the first few terms of a series solution of (3. 11) and (3. 12) in powers of D; the leading term for $w'(= O(D))$ is just Poiseuille flow in a straight tube, and the leading term for Ψ is $O(D^2)$ from (3. 12). The procedure is equivalent to the successive approximation of inertia terms in lubrication theory, only the lubrication theory is particularly simple because the cross-sectional area of the tube is uniform. If we set

$$w' = D\sum_{n=0}^{\infty} D^{2n} w_n(r,\theta), \quad \Psi = \sum_{n=1}^{\infty} D^{2n}\Psi_n(r,\theta) \ ,$$

then

$$\left.\begin{aligned} w_0 &= \tfrac{1}{4}\left(1 - r^2\right) \ , \\ \Psi_1 &= \left[2\sin\theta/(9\times 8^3)\right] r\left(1 - \tfrac{1}{4}r^2\right)\left(1 - r^2\right)^2 \ , \\ w_1 &= \left[\cos\theta/(45\times 8^5)\right] r\left(1 - r^2\right)\left(19 - 21r^2 + 9r^4 - r^6\right) \ , \end{aligned}\right\} \tag{3. 13}$$

etc. Dean's main application was to calculate the dimensionless flow rate, Q, corresponding to a particular value of D. The term w_1 does not contribute to it, and, in fact, Dean calculated the contributions from all the terms up to $O(D^8)$ (i.e. w_4) to obtain

$$\begin{aligned} Q &= \int_0^{2\pi}\int_0^1 w' r dr d\theta \\ &= (\pi D/8)\left[1 - 0.0306(D/96)^4 + 0.0120(D/96)^8 + O(D^{12})\right] \ , \end{aligned}$$

where $\frac{1}{8}\pi D$ is the Poiseuille flow value in the present scaled variables. The size of the coefficients in this series suggests that the small-D expansion is valid for values of D up to about 100. In the canine aorta , however, where $\delta \approx 0.2$ and the mean Reynolds number is approximately 800, the mean value of D is greater that 2000 (and the peak value is greater still), so this expansion can be useful only for much smaller blood vessels. Note that the leading term in the secondary flow, represented by Ψ_1 in (3. 13), takes the form of a pair of vortices, as shown schematically in figure 3.1. They are symmetrical about $\theta = \pm\frac{1}{2}\pi$, and the flow is towards the outside of the curve ($\theta = 0$) in the core of the tube and towards the inside near the walls.

The quantity of greatest interest in the context of blood vessels is the shear stress on the vessel walls. This has two components which in non-dimensional variables are $-w'_r|_{r=1}$ in the axial ($s-$) direction and $-v_r|_{r=1}$ in the tangential ($\theta-$) direction. To the order of this calculation, these quantities are:

$$-w'_r|_{r=1} = \frac{1}{2}D + [D^3 \cos\theta/(30 \times 8^4)] \ , \quad -v_r|_{r=1} = [D^2 \sin\theta/(6 \times 8^2)] \ . \qquad (3.\ 14)$$

These results show (i) that curvature increases axial wall shear on the outside wall ($|\theta| < \frac{1}{2}\pi$) and decreases it on the inside, and (ii) that curvature generates a positive secondary shear in the θ-direction (consistent with the qualitative diagram in figure 3.3).

3.3 *Intermediate and Large D*

The small-D expansion is inadequate to describe arteries of interest such as the aorta. Equations (3. 11) and (3. 12) have been solved numerically for values of D up to about 600 by McConalogue & Srivastava (1968), subsequently up to $D = 5000$ by Collins & Dennis (1975) and to $D = 6000$ by Daskopoulos & Lenhoff (1989). The flow pattern is depicted by plotting curves of constant Ψ (secondary streamlines) and of constant w' (axial velocity contours). Results for $D = 96$, 606 and 5000 are shown in figure 3.4 Figure 3.4(a) shows that for $D = 96$ the secondary motions are approximately symmetrical about the line $\theta = \frac{1}{2}\pi$ (as predicted by Dean's expansion) and that the point of peak axial velocity ($w'_{\text{max}} = 23.4$) is shifted from the tube axis to about $r = 0.20$ (compared with 0.24, which is the value predicted above by the first term in Dean's expansion). Figure 3.4(b) shows that at the intermediate value of D, a boundary layer appears to be developing (on the outside wall of the bend) in which the axial shear is high, whereas in the core the secondary flow appears to be approximately uniform while the contours of axial velocity are approximately straight lines across the tube. This view of the axial velocity field is confirmed at D=5000 by figure 3.4(c), but the secondary streamlines at this value of D are quite complicated (figure 3.4(d)), suggesting that an asymptotic theory for large D may not be straightforward. Collins & Dennis (1975) showed how well their computations agreed with experiment; both between the computed position and value of the maximum axial velocity and the same quantities as measured by Adler (1934), and between theoretical and experimental values of the ratio $f(D)$ of the flow-rate in a straight tube to that in a curved tube subject to the same pressure gradient. Collins & Dennis were also concerned to confirm the asymptotic structure of the solution for large D, as proposed by Ito (1969). This structure is discussed by Pedley (1980); suffice it here to say that the numerical solutions confirmed the predicted structure in that (i)

$$f(D) \sim D^{-1/3}\left[a + bD^{-1/3} + O(D^{-2/3})\right] \ , \qquad (3.\ 15)$$

where a and b are constants (equal to 8.12 and -16.7 respectively); and that (ii) $D^{-2/3}w'$ is approximately independent of D across the plane of symmetry of the tube (the maximum value of $D^{-2/3}w'$ appears to tend to a constant value around 1.6 as $D \to \infty$), except near the outside wall, where (iii) there is a boundary layer of thickness proportional to $D^{-1/3}$, so that the wall shear, $-w_r$ at $r = 1, \theta = 0$, is proportional to $D (\approx 0.85D)$.

Other studies on flow in a curved tube of uniform circular cross-section include: Pedley (1980) for steady flow with weakly non-uniform curvature, Zabielski & Mestel (1998) for steady and unsteady flow with helical geometry, Singh (1974) and Agrawal et al. (1978) for entry flow, Lyne (1971) and Smith (1975) for oscillatory flow, Lynch et al. (1996) and Waters & Pedley

(1999) for flow in tubes of time-dependent curvature (applicable to coronary arteries). Another general reference is Berger et al. (1983). There are many papers in which computations are described for flow in more complex geometries, as in real arterial bifurcations (see Pedley (1995) for some references). However, given the possibility of non-uniqueness and the sensitivity to small geometric and temporal changes, it is not clear how valuable they are!

A particularly interesting feature of curved tube flow was found by Daskopoulos and Lenhoff (1989): non-uniqueness. When D is sufficiently small, the steady-flow equations have just one solution and there is a single secondary-flow vortex in each half of the tube, as in figure 3.4(a). A similar two-vortex flow has been computed for quite large values of D, up to at least 6000, as shown in figure 3.4(b). However, there is a critical value of D, say D_1, above which more than one steady solution to the equations exists; Dennis and Ng (1982) found D_1 to be about 956, a value confirmed by Daskopoulos and Lenhoff (1989) and Yanase et al. (1989). Daskopoulos and Lenhoff computed two more solutions for D in excess of D_1, and two more again for D greater than another critical value D_2 (approximately 2500). The contour and streamline plots for all four solutions are given in figure 3.5, of which A and B are the two solutions that exist for $D > D_1$ (at D=3000) and C and D are those that exist only for $D > D_2$ (at D=3100). All these solutions are four-vortex in character (two vortices in each half of the tube) and do not look very different from each other. However, Daskopoulos and Lenhoff investigated their stability, and found that the only stable four-vortex flow for $D > D_1$ is that depicted in figure 3.5A; the others are all unstable. This result suggests that, for $D > D_1$, the flow that will actually be observed in an experiment will be either the two-vortex solution (figure 3.4) or the four-vortex solution such as that of figure 3.5A. Presumably, if the Dean number is gradually increased from a low value, the two-vortex flow will persist, but a suitably violent perturbation could result in the four-vortex flow being set up. [We should note that Yanase et al. (1989) developed a slightly more general stability analysis and concluded that the four-vortex flow is unstable after all, and therefore not observable. In contrast, in the experiments of Cheng and Mok (1986), the four-vortex flow was observed, and hence presumably stable, for a range of values of Dean number when the curvature ratio δ was not very small.]

It may be thought that whether a unique, stable steady flow exists for a certain geometry and Reynolds (or Dean) number is not particularly important when the real flows of interest are time-dependent anyway. Starting from a given initial state, with a given time-dependent pressure gradient, surely the flow will be well determined as a solution of Navier-Stokes equations? In principle that is correct, but the existence of more than one steady solution, stable or unstable, means that the outcome can depend very sensitively on the initial or boundary conditions. A small change in geometry, or a small change in the pressure-gradient waveform, may lead to a substantially different pattern of flow and wall shear. The pattern could vary from one cycle to the next, in a matter described technically as chaotic. As far as I know, there have not yet been any studies that examine flow in branched tubes from this point of view, and perhaps there should be.

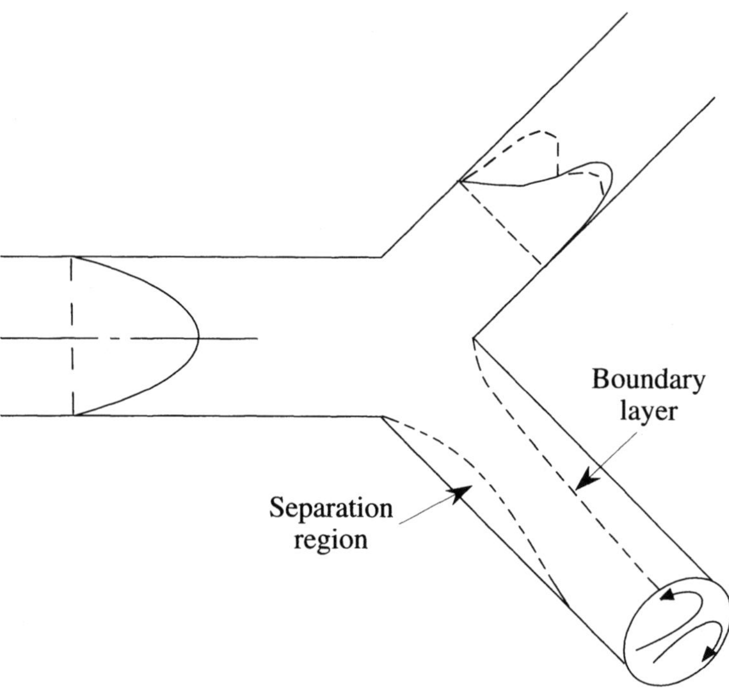

Figure 3.1 Sketch of flow downstream of a symmetrical bifurcation with Poiseuille flow in the parent tube. Secondary motion, a new boundary layer and a possible separation region are indicated in the lower branch; distorted velocity profiles in the plane of the junction (solid curve) and in the perpendicular plane (broken curve) are sketched in the upper branch. (From Pedley et al., 1977.)

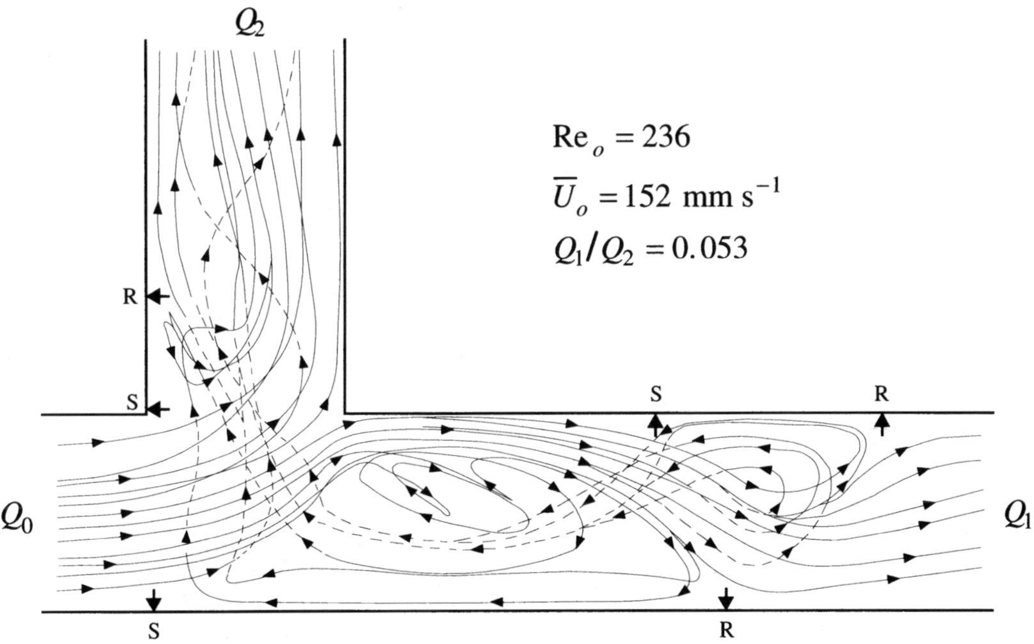

Figure 3.2 Flow patterns in an asymmetric, three-dimensional branch. Projections of three-dimensional particle paths in a right-angled branch when most of the flow goes down the side branch. The subscript 'o' denotes parameter values in the parent tube. (From Karino et al., 1979.)

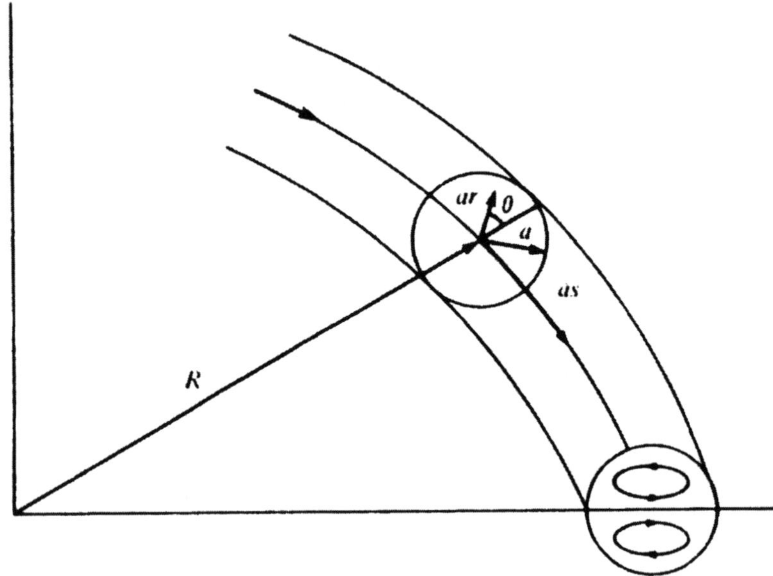

Figure 3.3 Coordinate system to be used in analysing flow in curved tubes (a is the radius of the tube, R is the radius of the curvature of the axis). A qualitative sketch of the secondary motions to be expected in steady flow is included.

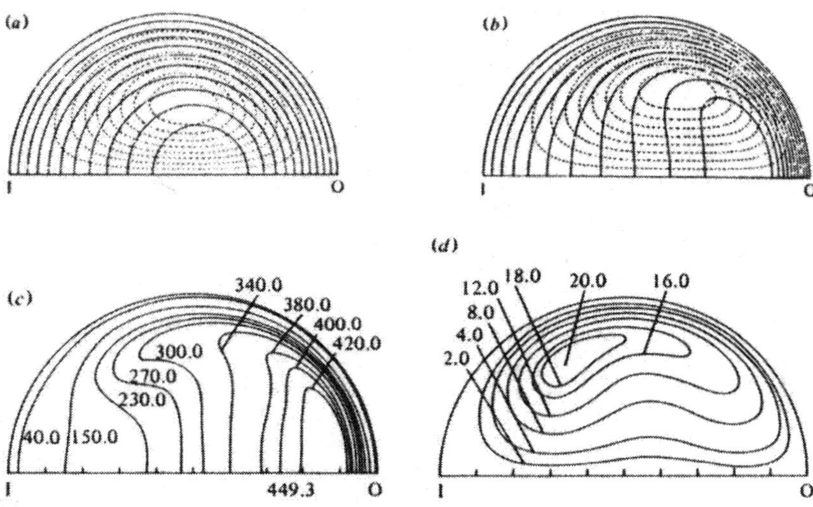

Figure 3.4 Contour plots of axial velocity (continous curves) and secondary streamlines (broken curves), (a) $D = 96$, (b) $D = 606$, (c) Contour plots of axial velocity, $D = 5000$, (d) Secondary streamlines $D = 5000$. I is the inside of the bend, O the outside. (After McConalogue & Srivastava (1968) and Collins & Dennis (1975).)

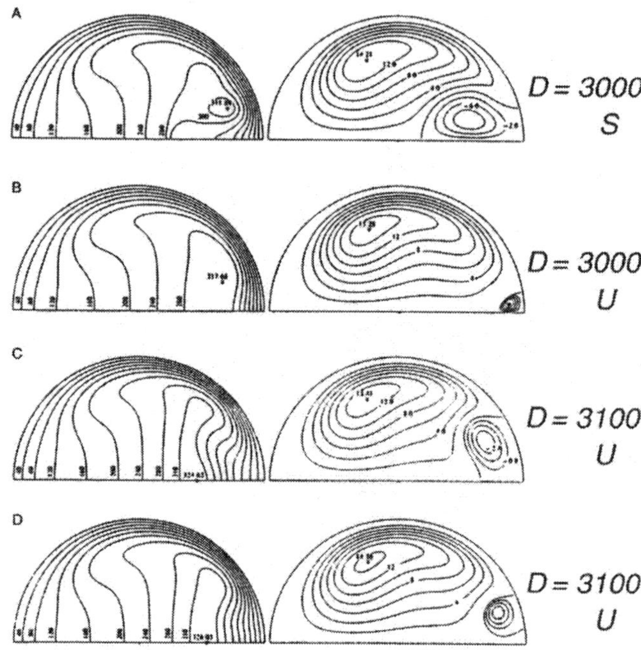

Figure 3.5 Computed axial velocity contours (left) and secondary flow streamlines (right) for steady flow in a curved tube of small curvature at comparable values of the Dean number D, showing the non-uniqueness of the flow. The flow marked S are stable, those marked U are unstable.

4 Steady and Unsteady Flow in Indented, Two-Dimensional Channels: Interactive Boundary Layer Theory

This lecture introduces the methods of high-Reynolds-number asymptotics as applied to two-dimensional, steady or unsteady, internal flow. The methods were pioneered by K. Stewartson and F. T. Smith in the 1960s and 1970s in a series of important papers; the following analysis is based in particular on Smith (1976 a, b).

4.1 Governing equations and scaling

Consider high-Reynolds-number flow in a two-dimensional channel with a long, slender indentation in one wall (figure 4.1). The channel width is a, the upstream flow is steady Poiseuille flow with average velocity $\overline{U}$, the incompressible fluid has density ρ and kinematic viscosity ν, and the indentation has length λa ($\lambda \gg 1$), height $O(\epsilon a)$ ($\epsilon \ll 1$) and may be time-dependent with timescale $\overline{T}$. We introduce Cartesian coordinates $a(\lambda x, y)$, corresponding velocity components $\overline{U}(u, v/\lambda)$, pressure $\rho\overline{U}^2 p$, and time $\overline{T}t$. The shape of the wall with the indentation is given by

$$y = \epsilon F(x,t) \tag{4. 1}$$

where $F(x,t) = 0$ for $x \leq 0$ and $x \geq 1$, while the upper wall is at $y = 1$ for all x. The governing continuity and Navier-Stokes equations, in dimensionless form, are

$$u_x + v_y = 0 \tag{4. 2}$$

$$St\, u_t + \lambda^{-1}\left(uu_x + vu_y\right) = -\lambda^{-1}p_x + Re^{-1}\left(\lambda^{-2}u_{xx} + u_{yy}\right) \tag{4. 3}$$

$$St\, v_t + \lambda^{-1}\left(uv_x + vv_y\right) = -\lambda p_y + Re^{-1}\left(\lambda^{-2}v_{xx} + v_{yy}\right), \tag{4. 4}$$

where suffixes represent partial differentiation and the Reynolds and Strouhal numbers are respectively defined by

$$Re = \overline{U}a/\nu \gg 1, \tag{4. 5}$$

$$St = a/\overline{U}\,\overline{T} \ll 1. \tag{4. 6}$$

The boundary conditions are those of no slip and no penetration,

$$u = v = 0 \quad \text{on} \quad y = 1 \tag{4. 7}$$

$$u = 0,\ v = \epsilon\left(uF_x + StF_t\right) \quad \text{on} \quad y = \epsilon F, \tag{4. 8}$$

together with the condition of Poiseuille flow upstream,

$$u \sim u_0(y) \equiv 6y(1-y) \tag{4. 9}$$

as $x \to -\infty$. The uF_x term is retained in (4. 8) because we shall need the boundary conditions for an inviscid fluid.

The objective of this section is to find the orders of magnitude of ϵ, λ and St, in terms of Re, for which self-consistent and non-trivial solutions can be found. We take the Reynolds number to be sufficiently large that $Re \gg \lambda \gg 1$, so that the viscous terms in (4. 3) and (4. 4) are negligible except (probably) in boundary layers on the walls. We also assume that the time-dependence is sufficiently weak that $\lambda St \ll 1$. Given that $\epsilon \ll 1$, we seek an expansion for the velocity components and pressure in powers of ϵ which we expect to be valid in the core of the channel, outside any boundary layers there might be. We therefore try

$$\begin{aligned} u &= u_0(y) + \epsilon\, u_1(x,y,t) + \epsilon^2 u_2(x,y,t) + \cdots \\ v &= \epsilon\, v_1(x,y,t) + \epsilon^2 v_2(x,y,t) + \cdots \\ p &= \frac{-12\lambda x}{Re} + \epsilon\, p_1(x,y,t) + \epsilon^2 p_2(x,y,t) + \cdots, \end{aligned}$$

where the first term in p is small, but is required to drive the basic Poiseuille flow. Since the viscous terms are absent from the equations in the core, we must also drop the no-slip condition on the boundaries, retaining only the kinematic boundary conditions on v.

The $O(\epsilon)$ terms in (4. 2) – (4. 4) are

$$u_{1x} + v_{1y} = 0 \tag{4. 10a}$$

$$u_0 u_{1x} + u_{0y} v_1 = -p_{1x} \tag{4. 10b}$$

$$u_0 v_{1x} = -\lambda^2 p_{1y}. \tag{4. 10c}$$

Since $\lambda \gg 1$, the last of these shows that p_1 is independent of y and then evaluating (4. 10*b*) at $y = 1$ shows that $p_{1x} = 0$. The solution of (4. 10*a*, 4. 10*b*) that satisfies the conditions $v_1 = 0$ on $y = 0$ and $y = 1$ is then

$$v_1 = -A_x(x,t)u_0(y), \;\; u_1 = A(x,t)u_0'(y), \tag{4. 11}$$

where $A(x,t)$ is an $O(1)$ unknown function. This represents a lateral displacement of all oncoming streamlines by a distance $-A$ in the y-direction.

The $0(\epsilon^2)$ terms in (4. 4) then give

$$p_{2y} = \lambda^{-2}\epsilon^{-1} A_{xx} u_0^2(y)$$

so

$$p_2 = \tilde{P}(x,t) + \lambda^{-2}\epsilon^{-1} A_{xx} \int_0^y u_0^2(y')dy', \tag{4. 12}$$

where $\tilde{P}(x,t)$ is another $O(1)$ unknown function. The product $\lambda^{-2}\epsilon^{-1}$ is taken to be $O(1)$ or smaller, as $\epsilon \to 0$. The pressure at $y = 1$ is given by (4. 12) to be

$$p_2 = P(x,t) = \tilde{P}(x,t) + \sigma A_{xx}, \tag{4. 13}$$

where $\sigma = 6/(5\lambda^2\epsilon)$. The parameter σ represents the importance or otherwise of the cross-stream pressure gradient p_{2y}.

The solution (4. 11) for u_1 does not satisfy the no-slip condition at the walls. There must therefore be boundary layers on the walls, which we take to have thickness-scale $\delta \ll 1$. For the boundary layer near $y = 1$ we introduce new variables as follows:

$$1 - y = \delta z, \quad u = \delta U, \quad v = -\delta^2 V, \quad p = \epsilon^2 P(x,t).$$

The y-momentum equation (4. 4) of course gives $p_y = 0$. The other equations, (4. 2) and (4. 3), become

$$U_x + V_z = 0 \tag{4. 14a}$$

$$\lambda St \delta U_t + \delta^2 (UU_x + VU_z) = -\epsilon^2 P_x + \frac{\lambda}{Re\delta} U_{zz}. \tag{4. 14b}$$

For viscous and inertial terms to balance in the boundary layer (as they must) δ must be given by

$$\delta = (\lambda / Re)^{1/3} \tag{4. 15}$$

and for the pressure gradient also to be important we need

$$\epsilon = \delta. \tag{4. 16}$$

In both (4. 15) and (4. 16) a multiplicative constant has without loss of generality been taken equal to 1. Finally, for an unsteady indentation in which local accelerations are important in the boundary layer (they are not important in the core, according to (4. 11)), it would be necessary that

$$\lambda St \delta^{-1} = \beta = O(1), \text{ i.e. } St = \beta (\lambda^2 Re)^{-1/3}. \tag{4. 17}$$

So far λ and Re have remained independent dimensionless parameters. However, if the cross-stream pressure gradient is important, so $\sigma = O(1)$, it is necessary that

$$\lambda = \left(\frac{6}{5\sigma} \right)^{3/7} Re^{1/7}, \tag{4. 18}$$

with the consequence that

$$\delta = \epsilon = \left(\frac{6}{5\sigma} \right)^{1/7} Re^{-2/7}, \; St = \beta \left(\frac{6}{5\sigma} \right)^{-2/7} Re^{-3/7}. \tag{4. 19}$$

If $\sigma \ll 1$, then $\lambda \gg Re^{1/7}$; $\lambda = O(Re^{1/7})$ represents the shortest indentation for which the present scaling is consistent.

When all the above scalings are incorporated, the boundary-layer equations (4. 14*a, b*) become:

$$U_x + V_z = 0 \tag{4. 20a}$$

$$\beta U_t + UU_x + VU_z = -P_x + U_{zz}, \tag{4. 20b}$$

to be solved subject to the boundary conditions

$$U = V = 0 \text{ at } z = 0, \tag{4. 21a}$$

$$\text{and} \quad U \sim 6(z - A) \text{ as } z \to \infty, \tag{4. 21b}$$

where (4. 21b) comes from matching to the core flow.

The boundary layer on the indented wall $y = \epsilon F$ can be similarly analysed, if a Prandtl transformation is first made. The new variables here are given by

$$y - \epsilon F = \epsilon \tilde{z}, \quad v = \epsilon \tilde{U}, \quad \nu = \epsilon^2 \tilde{V} + \epsilon^2 \tilde{U} F_x + \lambda St \epsilon F_t, \quad p = \epsilon^2 \tilde{P}(x,t),$$

where $\tilde{P}$ is already known in terms of P and A (4. 13). The resulting boundary-layer problem is the same as that on the other wall, (4. 20a, b) and (4. 21a, b), except that the variables are replaced by those with a tilde over them, and the outer boundary condition (4. 21b) is replaced by

$$\tilde{U} \sim 6(\tilde{z} + F + A) \text{ as } \tilde{z} \to \infty. \tag{4.21c}$$

Two comments can be made here. Firstly, if the boundary-layer problems were 'classical' ones, with P and $\tilde{P}$ prescribed in advance, then the relevant outer boundary condition in each case would be

$$U_z \to 6 \quad \text{as} \quad z \to \infty, \quad \text{or} \quad \tilde{U}_{\tilde{z}} \to 6 \quad \text{as} \quad \tilde{z} \to \infty. \tag{4. 22}$$

Solution of the problem would then determine the displacement functions $(-A)$ and $(F + A)$. Since A appears in both of these, however, and in (4. 13), the problems are coupled and 'interactive', and P, $\tilde{P}$ and A will normally be determined only after the boundary-layer problems are solved.

Secondly, A can be determined explicitly without solving the boundary-layer problem in some circumstances, namely when the cross-stream pressure gradient is negligible, $\sigma \ll 1$. In that case $P = \tilde{P}$ from (4. 13) so the two boundary-layer problems, with (4. 22) in place of (4. 21b) or (4. 21c), are really identical. Thus, if the boundary-layer problem has a unique solution, it follows that (4. 21b) and (4. 21c) are also identical, i.e.

$$A = -\frac{1}{2}F. \tag{4. 23}$$

The displacement of the core-flow streamlines is the average of the displacements of the two walls. Smith (1976a) derived this result for steady flow, extending it to unsteady flow in (1976b). He assumed that the boundary-layer problem would have a unique solution, which was confirmed in his linearised analysis for a very small indentation (see below). However, Borgas & Pedley (1990) demonstrated, in a similarity solution for steady geometries in which $F \propto x^{1/3}$, that the solution of the boundary-layer problem is not unique if F is negative and decreases sufficiently rapidly, i.e. the channel expands with a sufficiently adverse pressure gradient. This was confirmed by a fully numerical solution in a similar geometry by Bujurke et al. (1996) and is related to the well-known non-uniqueness of separated flow in a symmetrically expanding channel (or 'diffuser' — see Reneau et al. (1967)).

4.2 *Linearised solution for small, time-independent indentations*

We now proceed to an explicit solution of the time-independent boundary-layer problems for very small indentations with $F \ll 1$. The longitudinal velocity is taken to be a small perturbation to the oncoming shear flow, so that

$$U = 6z + U', \quad \tilde{U} = 6\tilde{z} + \tilde{U}',$$

where $U' \ll 6z$, $\tilde{U}' \ll 6\tilde{z}$ everywhere in the boundary layers. All unknowns (U', V, P, A etc) are taken to be of the same, small, order of magnitude as F. Then the problem near $y = 1$ becomes

$$U'_x + V_z = 0 \tag{4. 24a}$$
$$6zU'_x + 6V = -P_x + U'_{zz} \tag{4. 24b}$$
$$U' = V = 0 \quad \text{on} \quad z = 0, \quad U' \sim -6A \quad \text{as} \quad z \to \infty. \tag{4. 24c}$$

The problem can be solved using Fourier transforms in x, for which we define

$$\widehat{U}(k, z) = \int_{-\infty}^{\infty} e^{-ikx} U'(x, z) dx, \tag{4. 25}$$

and similar transforms $\widehat{V}$, $\widehat{P}$, $\widehat{A}$, $\widehat{F}$ for V, P, A, F. The transforms of (4. 24*a,b*) then give

$$ik\widehat{U} + \widehat{V}_z = 0 \tag{4. 26a}$$
$$6ikz\widehat{U} + 6\widehat{V} = -ik\widehat{P} + \widehat{U}_{zz}. \tag{4. 26b}$$

Eliminating $\widehat{V}$ and $\widehat{P}$ results in a single equation for $\widehat{U}_z$ which is essentially Airy's equation:

$$\widehat{U}_{zzz} - 6ikz\widehat{U}_z = 0;$$

the solution that satisfies $\widehat{U}_z \to 0$ as $z \to \infty$ is

$$\widehat{U}_z = \widehat{B}(k) Ai(\zeta) \tag{4. 27}$$

where $\zeta = (6ik)^{1/3} z$ and $\widehat{B}$ is an unknown function. [Note that to ensure that the Airy function $Ai(\zeta) \to 0$ as $z \to \infty$ for all k on the inversion contour between $-\infty$ and $+\infty$ it is necessary to choose the correct branch of $(6ik)^{1/3}$. This is achieved by means of a branch cut in the k-plane along the positive imaginary axis, so that $-3\pi/2 < \arg k < \pi/2$.]

Applying (4. 26*b*) at $z = 0$ gives

$$ik\widehat{P} = \widehat{U}_{zz}|_{z=0} = (6ik)^{1/3} \widehat{B}(k) Ai'(0).$$

Moreover

$$\widehat{U}(\infty) = \frac{\widehat{B}(k)}{(6ik)^{1/3}} \int_0^{\infty} Ai(\zeta) d\zeta = \frac{\widehat{B}(k)}{3(6ik)^{1/3}}$$

and (4. 24*c*) tells us that this is equal to $-6\widehat{A}$. Eliminating $\widehat{B}$, we obtain one relationship between $\widehat{P}$ and $\widehat{A}$:

$$\widehat{P} = \frac{-108 Ai'(0)}{(6ik)^{1/3}} \widehat{A}. \tag{4. 28}$$

A similar analysis for the boundary layer on the other wall gives

$$\widehat{\tilde{P}} = \frac{+108 Ai'(0)}{(6ik)^{1/3}} \left(\widehat{F} + \widehat{A} \right), \tag{4. 29}$$

while (4. 13) gives

$$\widehat{P} - \widehat{\tilde{P}} = -\sigma k^2 \widehat{A}. \tag{4. 30}$$

Thus all unknowns can be determined in terms of $\widehat{F}(k)$. In particular

$$\widehat{A} = -\frac{\widehat{F}}{2 + \frac{\sigma k^2}{\gamma}(ik)^{1/3}}, \tag{4. 31}$$

where $\gamma = -18.6^{2/3} Ai'(0) \approx 15.383$. Clearly $\widehat{A} = -\frac{1}{2}\widehat{F}$ when $\sigma = 0$, as expected from (4. 23). One further quantity of interest is the wall shear rate; on the wall opposite the indentation, for example, this is equal to

$$-u_y|_{y=1} = 6 + U'_z|_{z=0}.$$

Writing $\tau = U'_z|_{z=0}$, we see from (4. 27) that

$$\widehat{\tau} = \widehat{B}(k)Ai(0) = \frac{\gamma'(ik)^{1/3}\widehat{F}}{2 + \frac{\sigma k^2}{\gamma}(ik)^{1/3}},$$

where $\gamma' = 18.6^{1/3} Ai(0) \approx 11.612$.

The inverses of the Fourier transforms, with and without cross-stream pressure gradient, can be expressed as convolutions and evaluated by standard methods (analytical or numerical) for particular functions $F(x)$. The results for $\tilde{P}(x)$ and $\tilde{\tau}(x)$ (pressure and shear rate perturbation on the wall with the indentation) are plotted for several values of σ in figure 4.2, for the case in which

$$F(x) = hx(1-x),$$

where $h \ll 1$. Unsurprisingly, the pressure rises and the shear rate falls ahead of the maximum indentation, and vice versa downstream. The reduction in shear rate downstream would tend to cause flow reversal if h were not small. Note also the increasing extent of upstream influence as σ increases. All these features are confirmed by numerical solution of the nonlinear boundary-layer problem. Similar results were presented by Smith (1976a).

Linearised solutions can be obtained in a similar way for time-dependent indentations. Suppose that the indentation height, and hence all perturbations to the oncoming flow, are proportional to e^{it} and define the Fourier transform in x by

$$e^{it}\widehat{U}(k,z) = \int_{-\infty}^{\infty} e^{-ikx} U'(x,z,t)\,dx$$

etc, instead of (4. 25). Then the solution for $\widehat{U}_z$ is again given by (4. 27) except that ζ is redefined as

$$\zeta = (6ik)^{1/3} z + \zeta_0,$$

where

$$\zeta_0 = i\beta/(6ik)^{2/3}. \tag{4. 32}$$

The final result for $\widehat{A}$ can again be written in the form of (4. 31), but now with

$$\gamma = \frac{-6^{5/3} Ai'(\zeta_0)}{\int_{\zeta_0}^{\infty} Ai(\zeta)\,d\zeta}. \tag{4. 33}$$

This is much more difficult to invert than before, because of the infinite number of poles associated with the zeros in the k-plane of the denominator of (4. 31) (see Bogdanova & Ryzhov, 1983). We discuss the solution no further here, except to say that, when $\beta > 21.7\sigma^{-2/7}$ (in the present notation), one of the poles has a negative imaginary part, which means that non-decaying oscillations will be found downstream of the moving indentation. The downstream oscillations take the form of propagating waves, the physics of which are more easily understood in the context of indentations that oscillate with relatively large amplitude, for then the waves appear in the inviscid core flow.

4.3 Vorticity waves

In this section we suppose that the indentation height scale, ϵ, is much greater than the boundary-layer thickness, δ, while remaining small compared with one. We then perform a weakly nonlinear analysis of the inviscid core flow, keeping $\sigma = O(1)$, so the cross-stream pressure gradient is important in (4. 12), and taking $St = \beta\epsilon/\lambda$ (eq. 4. 17), where $\beta = O(1)$, so that the time-dependence is important at $O(\epsilon^2)$ in the x-momentum equation (4. 3). That equation then gives

$$u_0 u_{2x} + u_0' v_2 + \beta A_t u_0' + AA_x\left(u_0'^2 - u_0 u_0''\right) = -\tilde{P}_x - \frac{5\sigma}{6} A_{xxx} \int_0^y u_0^2(y')dy', \qquad (4.\ 34)$$

after use is made of (4. 11) and (4. 12). The kinematic boundary conditions on v, from (4. 7) and (4. 8), give

$$\begin{aligned} & v_2(1) = 0 \\ \text{and}\quad & v_2(0) = u_0'(0)\left(FA_x + AF_x + FF_x\right) + \beta F_t. \end{aligned}$$

We can therefore evaluate (4. 34) at $y = 1$ and $y = 0$. Putting $y = 1$ gives

$$-\tilde{P}_x - \sigma A_{xxx} = -6\beta A_t + 36AA_x, \qquad (4.\ 35)$$

where we recall that $u_0'(1) = -u_0'(0) = -6$. Evaluating (4. 34) at $y = 0$ gives another equation involving $\tilde{P}$ and A, and if $\tilde{P}$ is eliminated between that equation and (4. 35), a single partial differential equation is obtained for $A(x, t)$:

$$A_{xxx} - 2\beta' A_t = \beta' F_t + \frac{36}{\sigma}\left(FA_x + AF_x + FF_x\right), \qquad (4.\ 36)$$

where $\beta' = 6\beta/\sigma$.

It will be noticed that equation (4. 36) is a linear equation for A despite the fact that the derivation is weakly nonlinear. The AA_x term in (4. 35) cancels the corresponding term at $y = 0$, because its coefficient, $[u_0'(1)]^2$, has the same value at $y = 0$. If the magnitudes of the unperturbed shear rates at the two walls were different, as would be the case in an annular channel for example, the nonlinear term would not disappear. In the regions where $F = 0$ ($x < 0$ or $x > 1$) the equation would then be the Korteweg-de Vries (KdV) equation, according to which disturbances could propagate upstream as well as downstream (Borgas, 1986).

As it is, however, equation (4. 36) becomes the linearised KdV equation when $F = 0$, with zero on the right hand side. This supports only downstream propagating waves: if A is taken proportional to exp $[\mathrm{i}(\omega t - kx)]$ then

$$\omega = k^3/2\beta'. \tag{4. 37}$$

This result also tells us that the group velocity, $d\omega/dk$, is three times the phase velocity, ω/k. Thus, if the indentation were started moving from rest, the speed of the wave front would be three times the phase speed of any disturbances. The prediction of such waves in inviscid core flow was first made by Secomb (1979).

Equation (4. 36) was solved numerically by Pedley & Stephanoff (1985) for a particular function $F\left[= \frac{1}{2}\left(1 - \tanh\left(\alpha\left(x - x_1\right)\right)\right)\right]$, and with a particular value of α, chosen because it was a close approximation to the downstream end of the moving indentation in laboratory experiments performed by those authors, as described below.

Results of the computations for $\sigma = (2.17)(2\beta')^{-2/3}$ (also chosen to match one of the experimental runs), and with x rescaled to $x' = x(2\beta')^{-1/3}$, are shown in figure 4.3. The development of a wave-train can be clearly seen. If a displacement becomes observable when $|A| = 0.02$ (say) then the first wave-crest will be noticed at $t \approx 0.2$ ($t = 0$ is the start of the cycle, when the wall is flat, and $t = 1$ at the end). As time goes on, this crest will grow and propagate downstream, and a new trough will become observable further downstream. Later still, another crest will be seen, etc: the wave front clearly propagates faster than the crests/troughs.

Equation (4. 36) can be derived from the vorticity equation rather than the momentum equation, and that derivation makes it clear that the essential feature of the oncoming flow $u_0(y)$ is a non-zero vorticity gradient u_0''. The waves can therefore be called vorticity waves. The mechanism of their generation is essentially the same as that of Rossby waves in the atmosphere, for which the gradient of background vorticity is provided by the variation with latitude of the vertical component of the earth's angular velocity.

In the experiments of Pedley & Stephanoff (1985), the flow conduit (figure 4.4) was a long, rectangular channel (height, a, = 1cm; width = 10cm). Far from the entrance a section of one wall was replaced by a square piston, of side nearly 10cm, wrapped round by a stiff elastic membrane which permitted it to be pushed in and out by a motor in a prescribed sinusoidal oscillation. Steady (Poiseuille) flow was introduced upstream and it was confirmed that the flow across the middle of the channel was close to two-dimensional. The flow was visualised using small pearl-essence flakes (Mearlmaid AA) and the centre plane at and downstream of the piston was illuminated and photographed; one set of photographs, taken at different times during the cycle, is shown in figure 4.5. A wave train is indeed generated in the core, and the wave front clearly propagates more quickly than the wave crests, in a manner very similar to that shown in figure 4.3. Confirmation that the theory, albeit small-amplitude and non-viscous, is relevant to the experiments is provided by figure 4.6, where the positions and times of the first few crests/troughs are plotted, the curves arising from the solution of (4. 36) and the points from the experiments, for two particular sets of parameters. Agreement is quite good for the first crest and first trough, deteriorating further downstream where viscous effects are presumably more significant. Agreement became less good for lower frequencies (smaller St).

A feature of the experiments that is not predicted by the simple theory is the recirculating flow in the separated flow zones beneath the wave crests and above the troughs. These do not appear to be regions of very sluggish flow, as in steady two-dimensional separation, but instead consist of a tightly rotating vortex near the downstream end of the separated flow region, sometimes accompanied by another, co-rotating vortex just upstream. These, and all other, features of the observed flow were faithfully reproduced in solutions of the two-dimensional Navier-Stokes equations by Ralph & Pedley (1988). Unlike the experiments, however, the computations were able to provide values for the wall shear stress (WSS) distributions, and revealed that the walls next to the tightly rotating vortices under the waves experienced the largest values of shear stress in the whole flow, greater even than the maximum value above the moving indentation. This finding is totally inconsistent with intuition based on steady two-dimensional separated flow and reinforces the warning that intuition cannot be relied on for assessing the distribution of WSS in flow-conveying conduits such as arteries.

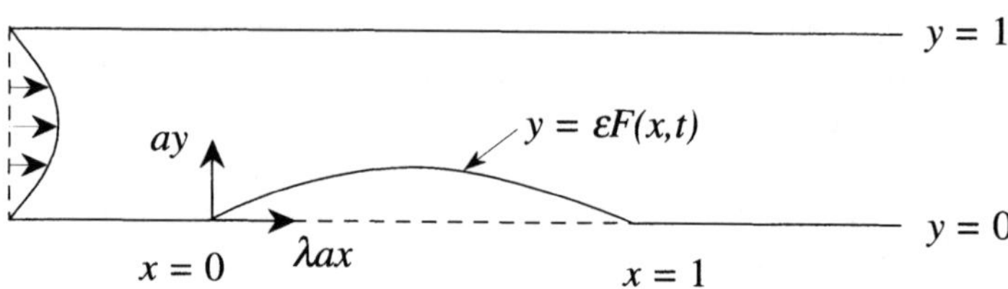

Figure 4.1 Sketch of a two-dimensional channel with time-dependent indentation $y = \epsilon F(x, t)$ in one wall. Steady Poiseuille flow enters far upstream.

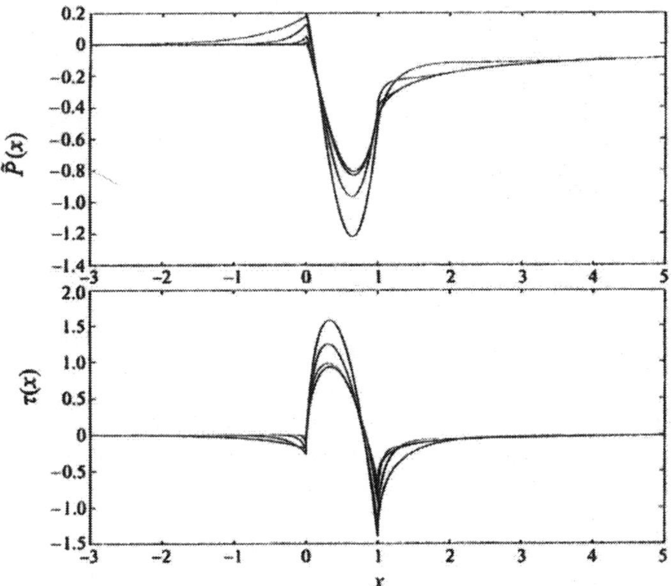

Figure 4.2 Graphs of perturbations to (a) pressure and (b) shear-rate at the indented wall for various values of σ, according to linear theory (curves computed by J. C. Guneratne). The ordinate scale is arbitrarily set to correspond to $h = 0.5$.

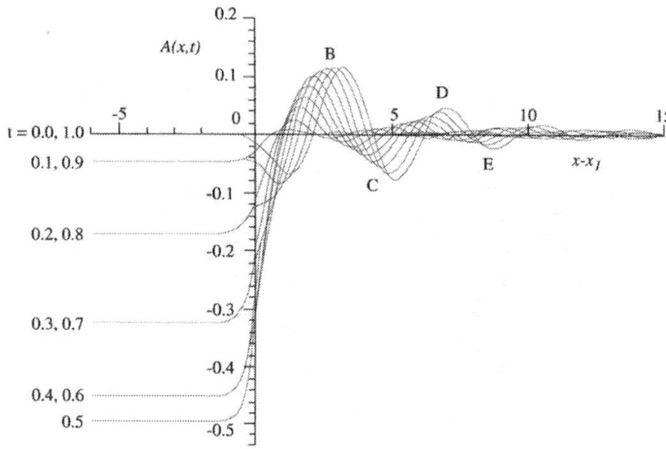

Figure 4.3 Graphs of A against $(x - x_1)(2\beta')^{-1/3}$ at different times t during a cycle, computed from (4. 36), with $F(x,t)$ given by a tanh profile and with $\sigma = 2.17(2\beta')^{-2/3}$. From Pedley & Stephanoff (1985).

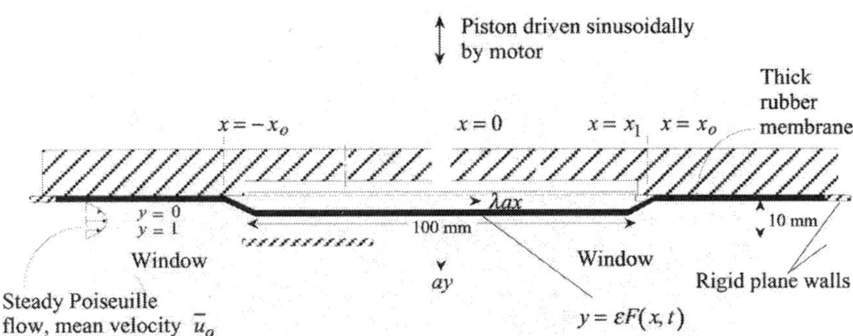

Figure 4.4 Experimental apparatus for observing flow in an approximately two-dimensional channel with a section of one wall which can be moved in and out; there is steady, Poiseuille flow upstream. Dimensionless variables are also marked (see text). From Pedley & Stephanoff (1985).

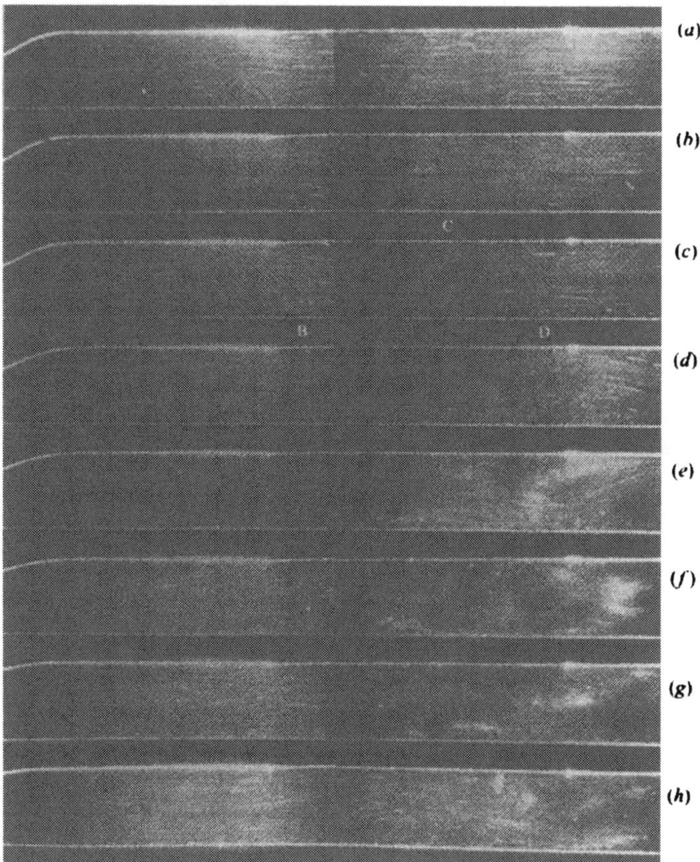

Figure 4.5 Photographs of the midplane, taken from above, showing the development of the flow field downstream of the oscillating indentation. The mean flow is from left to right, and the governing parameters are $Re = 610$, $St = 0.038$, $\epsilon = 0.38$. The dimensionless times at which the photographs were taken are (*a*) 0.41; (*b*) 0.48; (*c*) 0.55; (*d*) 0.62; (*e*) 0.69;(*f*) 0.76; (*g*) 0.84; (*h*) 0.91. The second eddy B is first visible in (*a*), the third C in (*b*) and the fourth D in (*c*). Eddy doubling has occurred in (*d*) (eddy B) and (*e*) (eddy C). From Pedley & Stephanoff (1985.)

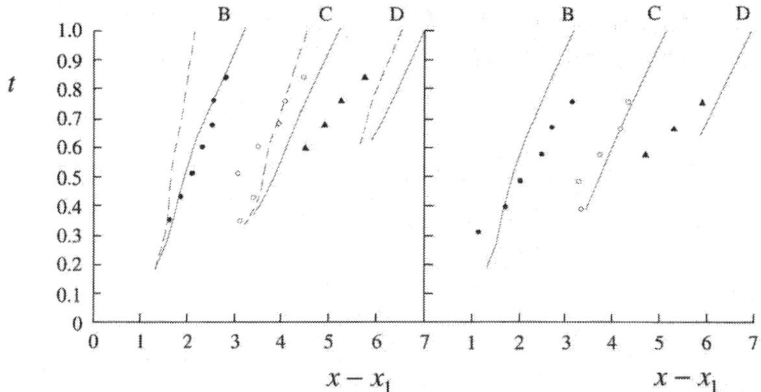

Figure 4.6 Predicted and measured positions of the wave crests and troughs, plotted against time, for two runs, both with $\epsilon = 0.38$ and (*a*) $St = 0.057$; (*b*) $St = 0.037$. The solid curves are theoretical, beginning at the time when the magnitude of the relevant maximum or minimum of A equals 0.02. The points are experimental: ● , wave B; ○, C; △, D. The broken curves in (*a*) represent the predicted positions when $F(x,t) \propto t^2 f(x)$. From Pedley & Stephanoff (1985).

5 Flow in Collapsible Tubes: Physiological and Laboratory Data. The Giraffe Jugular Vein

5.1 Collapse

Any elastic tube will collapse if it is squeezed hard enough. If a long segment of uniform elastic tube is subjected to different levels of transmural (internal minus external) pressure, p_{tm}, the cross-sectional shape and area, A, will vary as sketched in Figure 1.3b. When p_{tm} is large and positive, the cross-section will be circular and rather stiff because the perimeter must be stretched in order to increase A. As p_{tm} is lowered, a critical value is passed at which the circular cross-section buckles, becoming at first elliptical and then more significantly deformed. During this phase a thin-walled tube is very compliant (large area change for small pressure change) because only wall bending is required for a change of shape and hence area. At very low values of A the tube is almost totally collapsed and becomes stiff again. During the compliant phase, even the small pressure changes associated with flow through the tube (viscous or inertial) can be enough to cause collapse.

The collapse of compressed elastic tubes conveying a flow occurs naturally in several physiological applications. Examples include: (i) Blood flow in veins, either above the level of the heart where the internal pressure may be subatmospheric because of the effect of gravity, or being squeezed by contracting skeletal muscle as in the 'muscle pump' used to return blood to the heart from the feet of an upright mammal. (ii) Blood flow in arteries, such as intra-myocardial coronary arteries during the contraction of the left ventricle, or when actively squeezed by an external agency such as a blood-pressure cuff. (iii) Air flow in the large intrathoracic airways of the lung during a forced expiration, cough or sneeze, because an increase in alveolar air pressure, intended to increase the expiratory flow rate, is also exerted on the outside of the airways. In this case, increasing alveolar pressure above a certain level does not increase the expiratory flow rate, a process known as *flow limitation.* (iv) Urine flow in the urethra during micturition, where flow limitation is again commonplace. These and other examples are discussed in greater detail by Shapiro (1977 a,b). Note that in all the cases mentioned the Reynolds number of the flow (Re) is in the hundreds or higher. Moreover the flow is essentially steady, except in (ii): the beating of the heart is not transmitted back along the veins in the limbs, for example, because of the presence of valves in those veins (Caro, et al., 1978); there are also valves in the jugular veins of giraffes.

5.2 Experiments

Many workers have performed laboratory experiments on nominally steady flow through collapsible tubes. In the standard experiment a segment of collapsible (e.g. rubber) tube is mounted at its ends on rigid tubes and contained in a chamber whose pressure, p_e, can be independently controlled; the behaviour of the system depends on two independent pressure *differences*, e.g. $p_u - p_d$ and $p_e - p_d$, where p_u, p_d are the fluid pressures far up- and downstream (see figure 5.1). Some early experimental studies sought to characterise the collapsible tube by plotting the pressure difference along it ($\Delta p = p_1 - p_2$) against the flow-rate, Q. There was some confusion in the literature because it was not always clear which of the controlled pressure differences was being varied, as flow rate was varied, and which was held constant. Three different examples, in each of which the shape of the $\Delta p - Q$ curve is quite different, are shown in figure 5.2 (a,b,c), taken from Brecher (1952), Bertram (1986) and Conrad (1969), respectively. The following explanations of the three different curves are taken from Kamm & Pedley (1989).

For the case depicted in figure 5.2(a), $p_1 - p_2$ is increased while $p_1 - p_e$ is held constant. This can be accomplished either by reducing p_2 with p_1 and p_e fixed, or by simultaneously increasing p_1 and p_e while p_2 is held constant. With either manoeuvre, Q at first increases but above a critical value it levels off and exhibits flow limitation: however much the driving pressure is increased the flow rate remains constant or may even fall (so-called 'negative effort dependence' (Mead et al., 1967)), as a result of increasingly severe tube collapse. This version of the experiment is directly relevant to forced expiration from the lung (Elad & Kamm, 1989; Lambert, 1989) to venous return (Guyton, 1962) and to micturition (Griffiths, 1971).

Different results are obtained if $p_1 - p_2$ or Q is increased while $p_2 - p_e$ is held constant at some negative value (Fry, 1958; Brower & Noordergraaf, 1973; Bonis & Ribreau, 1978). In this case the tube is collapsed at low flow rates, but starts to open up from the upstream end as Q increases above a critical value, so that the resistance falls and $p_1 - p_2$ ceases to rise; so called 'pressure-drop limitation' (figure 5.2(b)). This experiment is not directly applicable to any particular physiological condition, but it turns out that $p_2 - p_e$ is a natural control parameter for at least one of the types of theoretical model that have been proposed (Shapiro, 1977a). Figure 5.2(b) represents some experimental results of pressure-flow relations for several values of $p_e - p_2$.

In a third type of experiment, $p_1 - p_2$ is held constant while $p_2 - p_e$ is decreased from a large positive value. The tube first behaves as though it were rigid and the flow rate is nearly constant. Then, as $p_2 - p_e$ becomes sufficiently negative to produce partial collapse, the resistance rises and Q begins to fall. This is analogous to what happens in the pulmonary capillaries toward the apex of the lung (Permutt et al., 1963). One notable variation on these experiments is that pioneered by Conrad (1969) who held p_e and the pressure downstream of a flow resistance constant; upstream of the flow resistance the pressure, p_2, varies with Q as does the degree of tube collapse. Thus at high flow rates the tube is distended and its resistance is low, but as the flow rate is reduced below a critical value the tube starts to collapse and both its resistance and $p_1 - p_2$ increase as Q is decreased. Only when the tube is severely collapsed along most of its length does $p_1 - p_2$ start to decrease again as Q approaches zero (figure 5.2).

In almost all such collapsible-tube experiments with $Re \gtrsim 200$, ranges of parameters were found in which steady flow could not be achieved but instead large-amplitude, flow-induced oscillations were observed. Bertram and his colleagues (1990, 1991) have made probably the most systematic series of experiments on self-excited oscillations in collapsible tubes, recording as functions of time the pressures (p_1, p_2) and flow rates (Q_1, Q_2) at the upstream and downstream ends of the collapsible segment, and the cross-sectional area A_n at the narrowest point. Examples of some of the measurements of $p_2(t)$ for various parameter values are shown in figure 5.3; a great variety of oscillatory behaviour is exhibited. Bertram et al. (1990) have tried to map out the associated control space diagrams, to identify regions in which different types of oscillation arise. All that can be said in summary is that a finite length of compressed collapsible tube conveying a flow represents a dynamical system of remarkable richness and complexity. It would clearly be of great interest to be able to model the system theoretically and hence understand it physically. That interest is independent of any physiological relevance, though it should be noted that flow-induced oscillations do arise in some of the physiological applications: wheezing during forced expiration; the Korotkov sounds listened for during blood pressure measurement with a cuff; and 'cervical venous hum' (Danaky & Ronan, 1974) are but three examples. However, the self-excited oscillations are not normally important in veins, and in this lecture we shall largely restrict attention in this section to the analysis of steady flow in collapsible tubes, using one-dimensional models. We will review the modelling of instabilities and self-excited oscillations in Chapter 6, while models of deterministic unsteady flows, arising from time-dependent external conditions such as driving pressure, external pressure or posture, can be found in Kamm & Shapiro (1979), Jan et al. (1983) and Brook (1997). For an initial analysis of fully three-dimensional solid and fluid mechanics in this context, see Heil (1997).

5.3 The (giraffe) jugular vein

The cardiovascular system of the giraffe is of interest because of the large range of intravascular pressures caused by the gravitational pressure gradient in an upright animal. The mean central aortic pressure in a 4m animal is 250mm Hg (33kPa) (Goetz & Keen, 1957) so the mean arterial pressure in the head is 75mm Hg, and in the feet it is 400mm Hg (53 kPa). All pressures quoted here are taken relative to atmosphere. One might ask why the pressure generated by the heart needs to be so high, since a higher pressure p_a at the root of the aorta has at least two disadvantages. There are both a greater tendency to oedema in the feet ('swollen ankles'), which has to be countered by anatomical and physiological adaptations such as tight skin and fascia in the legs (an 'anti-gravity suit'; Hargens et al., 1987), and greater energy demands on the left ventricle of the heart: if the mean volume flow rate (cardiac output) is Q, the work done by the ventricle per unit time is $W = p_a Q$. Hence a greater muscle mass is required: heart mass is 2.3% of body mass in adult giraffes, compared to about 0.5% in other mammals (Mitchell & Skinner, 1993; see also Goetz et al., 1960), with a correspondingly greater oxygen requirement.

It might be supposed (and, indeed, has often been proposed — e.g.: Badeer & Rietz, 1979), that a siphon mechanism could operate in the head and neck of an upright giraffe. Then p_a would not have to be much larger than in man, but the effect of gravity would mean that the internal pressure would be sub-atmospheric in the upper regions, and that is physiologically undesirable. Moreover, blood vessels are elastic, not rigid, so sub-atmospheric internal pressure, with approximately atmospheric pressure externally, would result in collapse of the vessels.

Indeed, the giraffe jugular vein is normally partially collapsed, as indicated both by direct observation (Goetz et al., 1960) and, especially, by inference from measurements of intravascular pressure which is positive and increasing with height above the heart (Hargens et al., 1987), not negative and decreasing as the gravitational gradient and the siphon concept would lead one to expect (Pedley, 1987). The measurements of Hargens et al. (1987) show the internal pressure to be about 7mm Hg at 0.3m above the heart, rising to 16mm Hg at 1.2m above the heart. If the vein were an uncollapsed cylinder, of diameter 2.5cm, the measured pressure distribution would require an absurdly high flow rate of 27 litres/sec, by Poiseuille's Law. The necessary increase in resistance can be achieved only if the vessel is quite severely collapsed. Laboratory studies using highly collapsible tubes (Hicks & Badeer, 1989) show internal pressure to be uniformly zero, i.e.: atmospheric. That this is approximately true for the human jugular vein has also been well-known for many years (Guyton, 1962).

Pedley et al. (1996) gave a thorough discussion of why the siphon concept could not be valid for a laboratory or living system modelled as an inverted U-tube with a collapsible downflow arm. Their steady-flow model was very simplistic, in that blood inertia was totally neglected, but we outline it briefly here, with reference to figure 5.4. Suppose that the inverted U-tube has uniform viscous resistance R per unit length, so that the variation of pressure p with distance x along it is given by

$$\frac{dp}{dx} = -RQ - \rho g \frac{dz}{dx}, \tag{5. 1}$$

where Q is the flow rate, g is the gravitational acceleration, and z is the height above the inlet pump, where $x = z = 0$ and pressure $p = P_1$. If the U-tube is rigid everywhere (figure 5.4a) and the outlet, where pressure is atmospheric (zero), is at the same level as the inlet, so that $z = 0$ while $x = L + h$, then (5. 1) gives

$$P_1 - 0 = (L + h)RQ.$$

This means that some flow can be generated whenever $P_1 > 0$. However, at point (2), where $z = h$ and $x = L$, (5. 1) also gives

$$P_1 - P_2 = LRQ + \rho g h, \tag{5. 2}$$

so P_2 is subatmospheric if $RQ < \rho g$. If the downflow arm is replaced by a highly collapsible segment (figure 5.4c) or is absent altogether (figure 5.4b) then P_2 must be zero, and (5. 2) shows that a positive flow rate is achievable only if $P_1 > \rho g h$. In other words, the heart has to 'pump the blood uphill' if the jugular vein can collapse.

Pedley et al. (1996) also discussed the possible reasons for the difference between the simple laboratory experiment (atmospheric pressure everywhere in the collapsed downflow arm) and the giraffe jugular vein (non-uniform pressure, slightly above atmospheric, rising with height). The main difference is probably non-uniform elastic properties in the vein. However, they also considered the effect of inertia on steady flow, following the one-dimensional model of Shapiro (1977a), and predicted a possible form of flow limitation that had apparently not been considered before. We now turn to the one-dimensional model, retaining fluid inertia as well as permitting non-uniform tube properties.

5.4 *One-dimensional models*

The traditional basic equations for one-dimensional flow in smoothly-varying elastic tubes are those already introduced in Chapter 2, equations (2. 1)–(2. 3), except that (a) the external pressure $p_e(x)$ and the tube-law function $\tilde{P}(A, x)$ may independently vary with x, and (b) the momentum equation (2. 2) is modified by the explicit inclusion of gravity and the addition of a term representing viscous resistance (Shapiro, 1977a):

$$\frac{\partial u}{\partial t} + u\frac{\partial u}{\partial x} = -\frac{1}{\rho}\frac{\partial p}{\partial x} - g\frac{dz}{dx} - \frac{R(A,u)uA}{\rho}. \tag{5. 3}$$

The convective inertia term does not include the contribution from integrating the non-flat velocity profile across the tube, but this can in general be incorporated into the resistance term RuA; R is assumed to be positive, and increases rapidly as A decreases. The equations (2. 1), (2. 3) and (5. 3) are exactly analogous to those for water flow in a shallow channel with a free surface.

Steady flow Let us first consider steady flow in a horizontal tube with uniform elastic properties and external pressure. Equation (2. 1) gives

$$uA = Q, \tag{5. 4}$$

where Q is the constant volume flow rate. If we now combine the remaining equations into a single equation for $\partial A/\partial x$, we obtain

$$\frac{1}{A}\frac{\partial A}{\partial x} = \frac{-RQ}{\rho\,(c^2 - u^2)}, \tag{5. 5}$$

where c^2 is again the speed of propagation of small-amplitude long waves, as given by (2. 5).

Now consider how A will vary with x, starting from an upstream location at which the tube is distended (i.e. the area A_1 is such that $\tilde{P}(A_1)$ is greater than the value, close to zero, at which the cross-section buckles — see figure 1.3b — so that $d\tilde{P}/dA = \tilde{P}'(A_1)$ is relatively large) and the fluid velocity is less than the wave speed; the flow is said to be *subcritical*. Since RQ is positive, it follows that $\partial A/\partial x$ is negative. The decrease of A with x will then have two consequences: u will increase because uA is constant (eq.5. 4); and c will decrease, because A and $\tilde{P}'(A)$ will both decrease (eq.(2. 5) and figure 1.3b). Hence $c^2 - u^2$ will decrease and $(1/A)\partial A/\partial x$ will become more negative.

Eventually, if the tube is long enough, one of two things will happen. If inertia is weak, so that $u_1 (= Q/A_1)$ is very much less than $c(A_1)$, the area may become small enough for c to increase again while u is still less than c. Equation (5. 5) can then be integrated forward in x and the collapse process would be smooth; this process is termed *viscous collapse*. On the other hand, if u_1 is large enough compared with $c(A_1)$, a point will be reached at which $c^2 = u^2$ and $\partial A/\partial x$ is predicted to be infinitely negative. Clearly this is impossible. What it means is that steady flow at the flow rate Q with upstream area A_1 cannot be achieved: either the flow will remain unsteady or the upstream conditions will change so that u does not exceed c anywhere. In the latter case, the flow is said to be *choked*, by analogy with compressible gas flow in a nozzle.

Are there circumstances in which supercritical flow ($u > c$) can occur in a tube which has subcritical flow upstream? Clearly not according to equation (5. 5), but it is possible in tubes which are not horizontal and not intrinsically uniform. To allow the potential phenomena to be investigated in general, it is helpful to follow Shapiro (1977a) and recast the tube law (2.3) to reflect the facts that the undisturbed cross-sectional area may vary with x [$A_0(x)$, say] and that the elastic properties may also vary. I.e:

$$p - p_e(x) = K_p(x)\tilde{P}(\alpha) \tag{5. 6}$$

where

$$\alpha = A/A_0(x) \tag{5. 7}$$

is the scaled area, and K_p is the 'stiffness' of the tube. This is not quite the most general form of tube law, because the shape of $\tilde{P}(\alpha)$ could also depend on x, but it is sufficiently general to describe all relevant phenomena. We also introduce the *speed index*,

$$S = u/c, \tag{5. 8}$$

where [cf (2. 5)]

$$c^2 = c_0^2(x)\alpha\tilde{P}'(\alpha) \tag{5. 9}$$

and

$$c_0^2(x) = K_p(x)/\rho. \tag{5. 10}$$

The speed index is analogous to the Froude number in free-surface channel flow. Combining these equations with (5. 3) and (5. 4), we obtain, in place of (5. 5):

$$\left(1 - S^2\right)\frac{\alpha_x}{\alpha} = \frac{S^2}{A_0}A_{0x} - \frac{1}{\rho c^2}\left[p_{ex} + g\rho z_x + K_{px}\tilde{P}(\alpha) + RQ\right] \tag{5. 11}$$

where the suffix x represents differentiation. We can also derive the following equation for the variation of the speed index:

$$\begin{aligned}\left(1 - S^2\right)\frac{\left(S^2\right)_x}{S^2} &= \frac{A_{0x}}{A_0}\left[-2 + (2 - M)S^2\right] \\ &+ \frac{K_{px}}{\rho c^2}\left[M\tilde{P}(\alpha) - \left(1 - S^2\right)\alpha\tilde{P}'(\alpha)\right] \\ &+ \frac{M}{\rho c^2}\left[p_{ex} + g\rho z_x + RQ\right],\end{aligned} \tag{5. 12}$$

where

$$M(\alpha) = 3 + \frac{\alpha\tilde{P}''(\alpha)}{\tilde{P}'(\alpha)} = \frac{1}{\alpha^2\tilde{P}'}\left(\alpha^3\tilde{P}'\right)'. \tag{5. 13}$$

It can be seen that, although the right hand sides of (5. 11) and (5. 12) are different in general, that of (5. 12) is precisely $-M$ times that of (5. 11) when $S = 1$. It follows that it is possible for α_x or $(S^2)_x$ to be non-zero when $S = 1$, as long as the right-hand sides are zero at $S = 1$. Since RQ is always positive, a necessary condition is that at least one of $(p_e + g\rho z)_x$, K_{px} or $-A_{0x}$ should be negative; i.e. the external pressure, the height or the stiffness should decrease with x, or the undisturbed cross-sectional area should increase.

In the case of a given flow rate Q in a vertical collapsible tube, or the jugular vein of an upright giraffe (figure 5.4c), it is clear that dz/dx is negative ($= -1$), whether or not the other quantities vary. Certainly in the laboratory experiment with a uniform collapsible tube subjected to a uniform (atmospheric) external pressure, the flow in the highly collapsed tube is supercritical $(S > 1)$. In this case equation (5. 11) gives

$$(1 - S^2)\frac{\alpha_x}{\alpha} = \frac{gp - RQ}{\rho c^2}, \tag{5. 14}$$

and a steady state is possible in which the cross-sectional area is given by the gravity-viscous balance $R(\alpha)Q = g\rho$. Note too that the steady state is stable when $S > 1$, in the sense that an increase of α above its equilibrium value leads to negative α_x because R increases as α decreases; the corresponding subcritical $(S < 1)$ steady state would be unstable, implying that α_x would not become zero in subcritical flow.

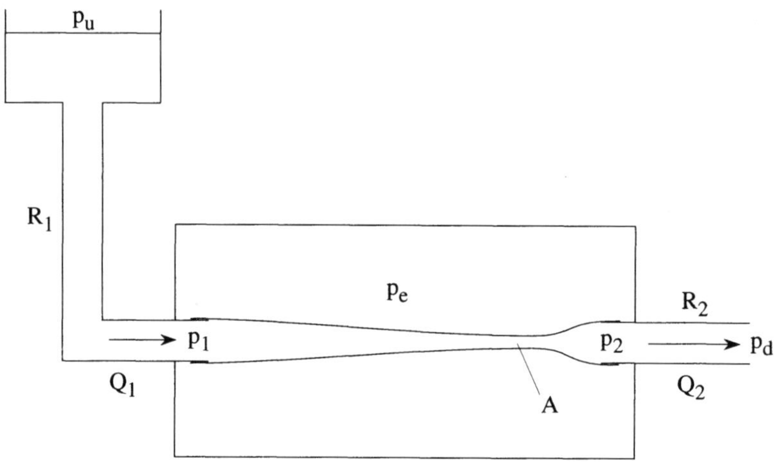

Figure 5.1 Sketch of a standard laboratory experiment. p_1, Q_1 are pressure and flow rate upstream of the collapsible segment; p_2, Q_2 are pressure and flow rate downstream; p_u is total pressure far upstream; p_e is pressure in the chamber surrounding the collapsible segment. R_1 and R_2 represent the rigid pipes up- and downstream, whose resistance can be prescribed.

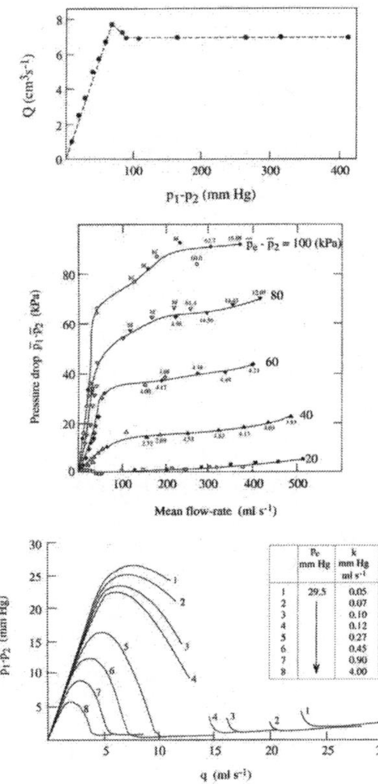

Figure 5.2 Pressure drop $p_1 - p_2$ along the collapsible segment of the apparatus sketched in figure 5.1, plotted against flow-rate q for three different conditions: (a) $p_u - p_e$ held constant (from Brecher, 1952); (b) $p_e - p_2$ held constant (from Bertram, 1986); (c) $p_e - p_d$ held constant (from Conrad, 1969).

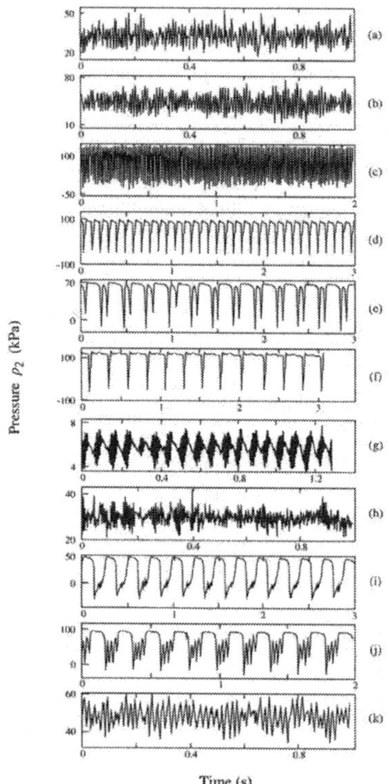

Figure 5.3 Pressure p_2 at the downstream end of the collapsible segment, plotted against time t during self-excited oscillations for various values of the governing parameters (from Bertram et al., 1991).

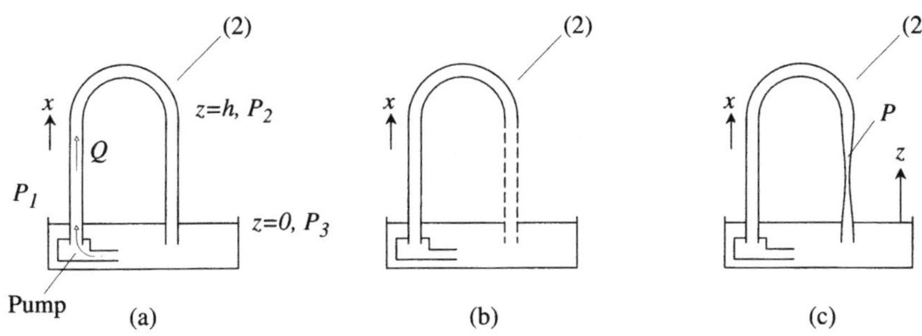

Figure 5.4 Sketch of possible experiments with an inverted U-tube, supplied with flow rate Q by a pump at $z = 0$ generating pressure P_1. The point (2) is at height $z = h$ and the pressure there is P_2; the pumped fluid is collected in a reservoir at level $z = 0$, where pressure is atmospheric, P_3. (a) Complete rigid tube; P_2 is subatmospheric. (b) Tube is cut off at (2); P_2 is atmospheric. (c) Tube replaced by collapsible drain tubing below (2); P_2 is atmospheric. (See text for explanation.) From Pedley et al. (1996).

6 Instabilities. Flow in a Two-Dimensional Collapsible Channel

We return now to the experiment depicted in figure 5.1, in which the collapsible tube has finite length, L. Even if choking is not predicted, the model leading to equation (5. 5) must break down near the downstream end, because dA/dx would have to become positive again near $x = L$. Following Cancelli & Pedley (1985), we add two new features to the model, both of which should be important in the region downstream of the narrowest point. One is longitudinal tension T in the tube wall, the simplest model for which causes equation (5. 6) to be replaced by

$$p - p_e = K_p \tilde{P}(\alpha) - T\frac{d^2A}{dx^2}. \tag{6. 1}$$

In the highly collapsed region, the tube wall resembles two flattish membranes under tension, with longitudinal curvature roughly proportional to d^2A/dx^2. It was felt that the addition of extra x-derivatives would enable more boundary conditions to be applied, such as $A(L) = A(0) = A_0$. The other new feature is the recognition that flow through a constriction will separate, a process leading to enhanced energy loss and therefore substantially incomplete pressure recovery in the region downstream of the narrowest point. The energy loss near the narrowest point had already been identified as important in lumped-parameter models (Pedley, 1980). Crude momentum arguments can be used to suggest that a reasonable, yet still simple, model of the energy loss in steady flow can be achieved by replacing equation (5. 3), downstream of the narrowest point, by

$$\chi u\frac{du}{dx} = -\frac{1}{\rho}\frac{dp}{dx}, \tag{6. 2}$$

where χ is a non-negative quantity, less than 1; in their (unsteady) calculations Cancelli & Pedley took $\chi = 0.2$. It proves not to be crucial what model for energy loss is assumed, as long as there is some; equation (6. 2) is more amenable to analysis than (5. 3), say.

The steady flow model described by equations (5. 4), (6. 1) and (6. 2) with

$$\begin{aligned} \tilde{P}(\alpha) &= (1 - \alpha^{-3/2}) \quad \text{for} \quad \alpha < 1 \\ &= k(\alpha - 1) \quad \text{for} \quad \alpha > 1 \end{aligned} \tag{6. 3}$$

can be analysed using phase-plane methods. Jensen & Pedley (1989), for example, took $\chi = 1$ for $0 < x < x_s$ (the unknown point of flow separation, taken to be identical with the narrowest point at sufficiently high Reynolds number) and $\chi = \text{constant} \leq 1$ for $x_s < x < L$. The principal results can be summarised as follows:

(i) When $\chi = 1$ everywhere, i.e. there is no energy loss in the collapsible tube downstream of the narrowest point, then there exists a critical value of the flow rate Q, dependent on the longitudinal tension T, above which the steady problem has no solution. In other words, the presence of longitudinal tension alone does not abolish choking. This should not have been a surprise; in the same way, surface tension does not abolish critical behaviour in shallow-water channel flow.

(ii) However, whenever there is any downstream energy loss, i.e. $\chi < 1$ for $x_s < x < L$, then a steady solution exists for all positive values of flow rate Q and tension T. Since some such energy loss is inevitable, it follows that the breakdown of steady flow is not caused by choking, i.e. the non-existence of a steady flow at the chosen parameter values, but must arise through instability of the steady solution.

Jensen (1990) gave a detailed linear, and weakly nonlinear, analysis of the instability of the steady flow. He used the same one-dimensional model but with the time derivatives, $\partial A/\partial t$ and $\partial u/\partial t$, restored. The elasticity equations (6. 1) and (6. 3) remained unchanged. When appropriately non-dimensionalised, Jensen's model has two principal governing dimensionless parameters, in addition to χ (which was fixed at a value of 0.2 in all numerical computations): $\widehat{Q}$, which is proportional to the flow rate Q, and $\widehat{P}$, proportional to the transmural pressure, $p_e - p_2$, at the downstream end of the collapsible segment when the flow is steady. Other parameters describe the resistance and inertance of the upstream and downstream rigid segments; these were kept fixed throughout. Figure 6.1 shows the computed stability boundaries in the $\widehat{P} - \widehat{Q}$ plane, for the first two instability modes found. The general shape, showing stable steady flow for sufficiently small $\widehat{P}$ at all $\widehat{Q}$ (the tube remaining effectively open), and for sufficiently small $\widehat{Q}$ at all $\widehat{P}$ (the tube being collapsed when $\widehat{P}$ is large enough), is in qualitative agreement with the control diagrams plotted by Bertram et al. (1990). So too are the presence of mode-crossing points (i) and (ii) and the existence of several regions in parameter space in which different behaviour of the system is to be expected. Jensen's weakly nonlinear analysis showed that both modes become unstable through supercritical Hopf bifurcations everywhere except for the small segments of the stability boundaries marked as dotted in figure 6.1, where they are subcritical Hopf bifurcations.

In a subsequent paper Jensen (1992) showed some results of a numerical integration of the fully nonlinear one-dimensional equations, at a few selected points in parameter space, near the upper left mode-crossing point in figure 6.1. Some of the computed time series, of $p_2(t) = p(L,t)$ for example, look quite similar to the measurements of Bertram, et al. (1990) shown in figure 5.3. It is clear that this one-dimensional model contains much that is relevant to the self-excited oscillations of real collapsible tubes in the laboratory. It would be possible to extend Jensen's (1992) fully nonlinear computations to cover the whole of parameter space, and map out the behaviour in as much (or more) detail as has been done experimentally.

However, this has not been done, and should not, because of the severe oversimplifications in the one-dimensional model, both in the crude solid mechanics of equations (6. 1) and (6. 3) and in the crude ad hoc model of flow separation and the associated energy loss (equation 6. 2), which is especially weak in unsteady flow. What is required is a solution of the unsteady, three-dimensional Navier–Stokes equations, coupled to the equations for the unsteady, three-dimensional, large-deformation theory of highly compliant shells, but numerical codes for the solution of such problems are not yet available in any branch of computational mechanics, and would require resources in excess of any available to us. Instead, we have sought a sound scientific solution for a simpler, two-dimensional configuration which is nevertheless in principle realisable experimentally, Pedley & Luo (1998). The configuration is sketched in figure 6.2. A two-dimensional channel consists of two parallel, rigid planes, distance h_0 apart, from one of which a segment of length Lh_0 has been removed and replaced by a thin membrane, with no bending stiffness or inertia but under longitudinal tension T. Steady, plane Poiseuille flow with flow rate q enters far upstream. The external pressure takes a constant value, p_e, referred to the pressure at the far end of the downstream rigid segment.

In all the following discussion, lengths are made dimensionless with respect to h_0, and the position of the membrane is given by

$$y = h(x,t) \quad , 0 \le x \le L \ , \tag{6. 4}$$

where $h(0,t) = h(L,t) = 1$.

The first approach to this problem (Pedley, 1992) was based on lubrication theory, assuming negligible fluid inertia, steady flow and small wall slope: a one-dimensional model for low Reynolds number flow, but rationally derivable from the full equations of motion. The main innovation of that paper was its inclusion of the fact that the longitudinal tension in the membrane falls with downstream distance as a consequence of the viscous shear stress exerted by the fluid. However, the results were not qualitatively very different between the constant and variable tension cases, except when T fell close to zero. The main conclusion was that the steady problem has a solution, for all values of q and p_e, as long as T remains positive everywhere. For given positive values of longitudinal tension T_D and transmural pressure $p_e - p_D$ at the downstream end of the membrane (see figure 6.2), the membrane is collapsed everywhere ($h < 1$ for all $0 < x < L$) for sufficiently small flow rate q, but exhibits a bulge outwards at its upstream end when q exceeds a critical value q_b. In these respects, the conclusions are the same as from the high Reynolds number one-dimensional model discussed above.

Even for low Reynolds number flow, the lubrication theory analysis was not uniformly valid because the wall slope became large at the downstream end in cases for which T_D was small. The next stage was therefore a numerical solution of the Stokes equations, coupled to the membrane equations. This was performed iteratively by Lowe & Pedley (1995), who used the finite element method to solve for the flow with the membrane position assumed given, calculated the pressure and shear stress exerted on the membrane, and then updated the membrane position by requiring that the membrane equilibrium equations be satisfied, and so on. This procedure let to predictions of membrane shape, for given values of q, T_D and $p_e - p_D$, which agreed remarkably well with the lubrication theory results even when the wall slope was quite large, but failed to give a solution for sufficiently small (but positive) values of membrane tension. We attribute this failure at small T_D to a poor iteration scheme for very compliant boundaries.

The next computations to be described will be for steady and unsteady flow at non-zero Reynolds number (Luo & Pedley, 1995). The full governing equations and boundary conditions, in dimensionless form, are as follows, where velocities are made non-dimensional with $U_0 = q/h_0$, time with h_0/U_0, stresses with ρU_0^2 (ρ is fluid density), and wall tension with $\rho U_0^2 h_0$; the Reynolds number is $Re = \rho U_0 h_0/\mu$ (μ is fluid viscosity) and the summation convention is used over suffixes $i, j = 1, 2$.

$$\text{Navier-Stokes} \quad u_{i,t} + u_j u_{i,j} = -p_{,i} + Re^{-1} u_{i,jj} \tag{6.5a}$$

$$\text{Conservation of mass} \quad u_{i,i} = 0 \tag{6.5b}$$

Boundary conditions (refer to figure 6.2):

$$\text{on } AB(x = -L_u) u_1 = 6y(1-y), \quad u_2 = 0 \tag{6.6a}$$

$$\text{on } EF(x = L + L_d) - p + Re^{-1} \quad u_{1,1} = 0, \; u_2 = 0 \tag{6.6b}$$

$$\text{on } BC, DE, AF (y = 1 \text{ for } x < 0 \text{ or } x > L; y = 0 \text{ for all } x) \quad u_1 = u_2 = 0 \tag{6.6c}$$

$$\text{on } CD(y = h(x,t), 0 \leq x \leq L) \quad u_1 = u_2 = 0 \text{ (steady)} \quad (6.7a)$$

$$u_i = \text{velocity of membrane (unsteady)} \quad (6.7b)$$

$$p_e - \sigma_n = T h_{xx}(1 + h_x^2)^{-3/2} \quad (6.7c)$$

$$-\sigma_t = \partial T/\partial s \ . \quad (6.7d)$$

In the membrane equations (6.7c, d), σ_n and σ_t are the normal and tangential components of the stress exerted by the fluid on the membrane and s is the distance measured along the membrane. There should, in addition, be an equation relating the tension of each element of the membrane to its extension, but in this work we have assumed that T is independent of time, t, which is equivalent to assuming that the tension is sufficiently large for length variations to cause negligible changes in T. That suggests that the tension is also sufficiently large for the longitudinal variation to be negligible, so from henceforth we ignore condition (6.7d) and take T in (6.7c) to be a constant. The computations have confirmed that the overall length changes are no more than $\pm\, 4\%$, even during the most vigorous oscillations found (Luo & Pedley, 1996).

Although the non-dimensionalisation described above is the most convenient for numerical solution, it is not convenient for the presentation of results because U_0 appears in the scalings for p_e and T. In presenting the results, therefore, we shall take

$$T = T_0/\beta Re^2, \ p_e = p_{e0}/\gamma Re^2 \ , \quad (6.8)$$

where T_0 and p_{e0} are reference values, and increasing β or γ alone is equivalent to decreasing T or p_e at fixed Reynolds number Re.

Steady flow at finite Re was computed independently by Luo & Pedley (1995) and by Rast (1994), who both used the finite element method for the fluid flow, but used quite different techniques for coupling it to the membrane displacement. Luo & Pedley (1995) used the commercial flow solver FIDAP and iterated for the wall position in the manner described above in the context of Stokes flow. Rast (1994), on the other hand, used a finite element mesh which was coupled automatically to the membrane displacement by the method of spines (see Ruschak, 1980), and the membrane equation (6.7c) was discretised and solved simultaneously with the flow equations using Newton's method. The authors of both papers reported extensive accuracy tests, such as the effect of mesh refinement and adjustment of the location of the downstream boundary (i.e. the value of L_d), not only on membrane shape but also on the wall vorticity distribution, always one of the most sensitive tests of a CFD code. The best tests of all were agreement (a) between the results of the two computations and (b) with those of Lowe & Pedley (1994) at low Re (see Lowe, Luo & Rast, 1996).

Both approaches to the steady problem, like Lowe & Pedley (1995), failed to find a convergent solution for sufficiently small, but positive, values of T (or sufficiently large β: equation (6.8). Luo & Pedley (1995) discussed whether the breakdown was associated with the corner singularity at the upstream end of the membrane (point C on figure 6.2) when the membrane began to bulge out there. However, Rast (1994) and more recent computations of our own (Luo & Pedley, 1996) have found converged solutions with upstream bulging; breakdown occurs at a much lower tension (for given Re) than bulging. We have concluded that such problems are extremely ill-conditioned when the boundaries are highly compliant.

6.1 Steady Flow Results

Just the main features of the results will be presented here; more details can be found in the original papers. One general finding is that qualitatively similar behaviour is obtained when Re is increased at fixed tension (β) as when tension is decreased (β increased) at fixed Re. In what follows we fix Re at the value 300 and vary β. Other dimensionless parameters (chosen for comparison with previous papers) are taken to be: $L = 5, L_u = 5, L_d = 30, T_0 = 1.61 \times 10^7$, $p_{e0} = 9.3 \times 10^4$ and $\gamma = 1$.

The membrane displacement for various values of β is plotted in figure 6.3. At small β (large T) the membrane is stretched tight and is not deformed. As β is increased, the deformation increases, the minimum channel width h_{min} occurring close to the mid-point of the membrane. As the constriction becomes more severe, it tends to move downstream and a point of inflection appears in the upstream half. When β increases above about 30 two, possibly independent, phenomena are seen: the upstream part of the membrane begins to bulge out and the constriction, while continuing to move downstream, ceases to become more severe. In fact, h_{min} increases somewhat as β increases. The membrane slope becomes very large.

Both the above phenomena are also seen in the corresponding high-Reynolds-number one-dimensional model, which is exactly that of Jensen & Pedley (1989) described above but with h for A and $\tilde{P}(A) \equiv 0$. Indeed, the shape of the graph of h_{min} against β predicted by that model is very similar to that given by the full computation, as shown in figure 6.4; the value of β at which bulging is first predicted is particularly close. The same is true at all Reynolds numbers from 50 to 500 (Luo & Pedley, 1995) though, as Re is decreased, h_{min} also falls, and occurs at larger β (smaller T). The one-dimensional model appears to be better than it deserves to be, at least in steady flow.

6.2 Unsteady Results

Extension of the above studies to time-dependent flow and membrane displacement required extensive development of the computational scheme.

The fully-coupled finite element method of Rast (1994) was extended to deal with time dependence. The mesh was taken to be time-dependent, but based on fixed spines; Newton's method was used to obtain convergence at each time step. Details are given in Luo & Pedley (1996). The main difficulty concerned the kinematic boundary condition (6.7b), because it is necessary to track boundary points as they move, and that is not possible in the absence of a description of membrane elasticity. The boundary condition was eventually based on the assumption that elements of the membrane always move in a normal direction; this is not strictly true, but is reasonable. To check the importance of this boundary condition, we compared the results with those obtained with the even simpler assumption that boundary points move only in the y-direction. This is clearly less satisfactory (e.g. near the downstream end of the membrane) but fortunately there was not much difference in the results.

The unsteady code was used to investigate the stability of the steady solutions already computed (Luo & Pedley, 1996). The procedure was to start with a steady solution at a particular value of β, then increase the value of β by a small amount and start the computation; the initial condition was therefore a small displacement from the steady solution at the new value of β. For values of β less than a critical value $\beta_c (\approx 27.5$ for $Re = 300)$ the perturbation dies away, revealing the steady solution to be stable. For $\beta > \beta_c$, the perturbation grew and finite-amplitude oscillations ensued, showing that there had been a Hopf bifurcation. Examples of the behaviour are given in figure 6.5, which shows the wall displacement h as a function of time at a fixed value of $x(x = 3.5$, close to the site of greatest constriction in the steady solution) and for three values of β. For $\beta = 30.0$, figure 6.5(a) shows an approximately sinusoidal oscillation, as is to be expected for a slightly supercritical value of β, with period 11.7 time units. However, for $\beta = 32.5$, figure 6.5(b) shows a few cycles of adjustment, followed by a (nearly) periodic oscillation, of period 21-25, in which large maxima and minima alternate with small ones. It seems clear that the system has gone through a period-doubling bifurcation. Finally, figure 6.5(c) shows the wall motion for $\beta = 35.0$; the wave-form is again more complex, indicating that at least one further bifurcation has occurred. We conclude that even this simple, two-dimensional, constant-tension model is an interesting dynamical system which may well incorporate some of the complexities of real collapsible tube flow.

6.3 Streamlines and Energy Dissipation

We revert now to the case of no wall inertia, in an attempt to understand the mechanism of the instability and oscillations. In figure 6.6 we show the streamlines of the flow at various times during the oscillation cycle in just one case, that of $Re = 300, \beta = 32.5$ (figure 6.5b). The important point to note is that the flow separation downstream of the narrowest point does not occur always at or near that point, as it would if the flow were quasi-steady. Moreover, waves are seen to be generated and to propagate downstream in the rigid channel downstream of the oscillatory membrane. These are clearly the same as the vorticity waves observed and analysed by Pedley & Stephanoff (1985); not only do they look the same, but the wavelength $\lambda \approx 3.6$ of the nearly sinusoidal oscillations of period ≈ 11.5 at $\beta = 30$ is comparable in magnitude to those measured by Pedley & Stephanoff. Their run 5 with $Re = 487$ and the inverse of the dimensionless oscillation period, $St, = 0.77$ had wavelength ≈ 2.6; that was the shortest wavelength observed by those authors, corresponding to the highest value of St - theory suggests that $\lambda \propto St^{-1/3}$. The streamline plots make it look as if the coupling between the vorticity waves and the flow separation process is somehow important for the latter, and hence for the separated flow energy loss that, according to the one-dimensional model, is a crucial feature in the system.

However, if we compute the rate of energy dissipation per unit volume, $\Phi = \mu u_{i,j}(u_{i,j}+u_{j,i})$, we are led in a different direction. At almost all times, the highest rates of energy dissipation occur in viscous boundary layers, on the membrane and on the opposite wall *upstream* of the point of greatest constriction, not downstream as postulated by Cancelli & Pedley (1985) and used in the subsequent one-dimensional models. There are occasional pockets of high dissipation, at the edges of the primary separated eddy and associated with the vorticity waves, but most of the dissipation is upstream.

A good physical explanation for the above findings still eludes us. Part of the difficulty is that the full time-dependent computations require very large computer resources, so we cannot yet examine parameter space in any detail.

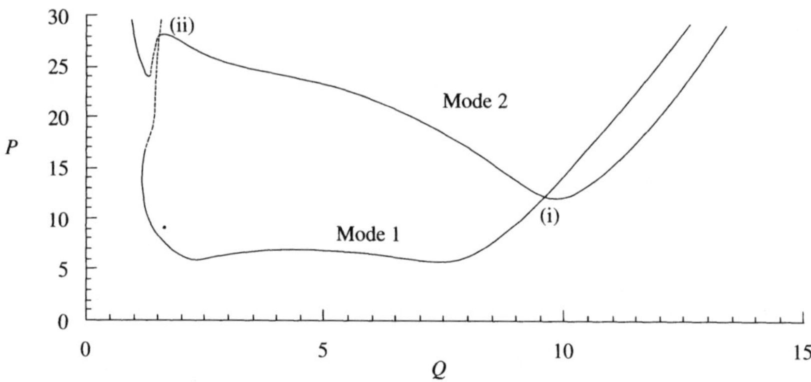

Figure 6.1 Stability boundaries for the first two modes of instability, plotted on the dimensionless $P-Q$ plane $(P \propto p_e - p_2; Q \propto q)$, as predicted by the one-dimensional model of Jensen (1990). The Hopf bifurcations are subcritical where the curves are dotted, supercritical elsewhere; (i) and (ii) are mode crossing points.

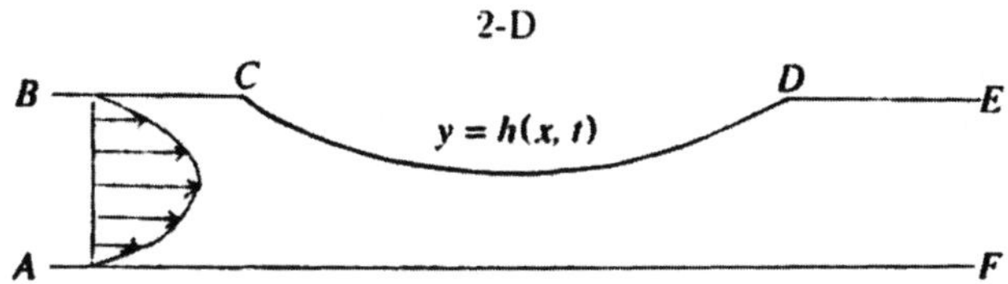

Figure 6.2 Sketch of the two-dimensional model problem.

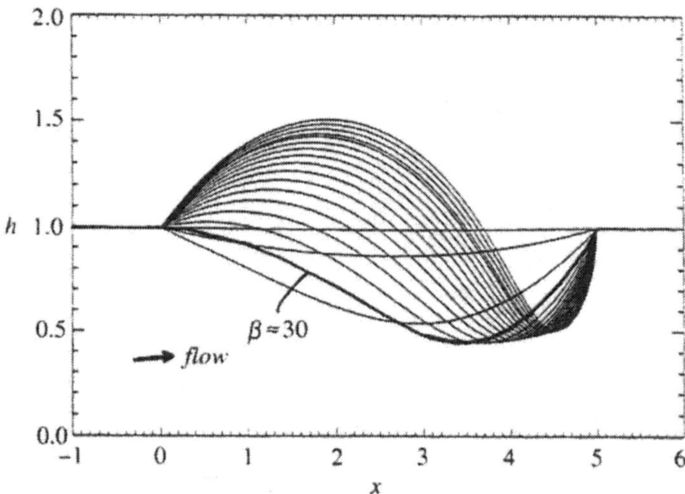

Figure 6.3 Predictions of steady membrane shape at $Re = 300$ and various values of tension parameter $\beta(\alpha 1/T)$, from Luo & Pedley, 1996.

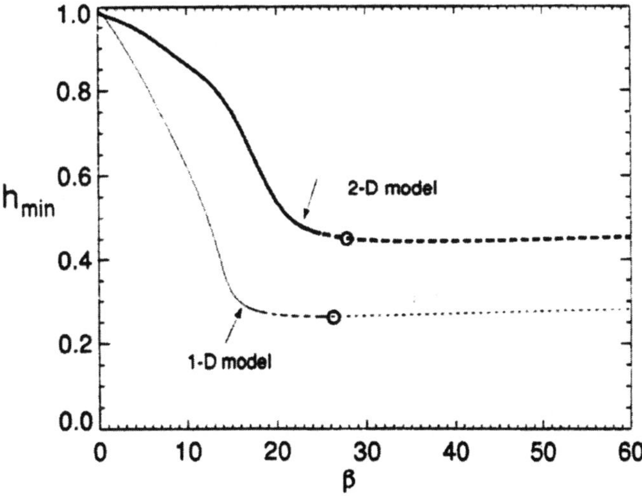

Figure 6.4 Predictions of minimum channel width during steady flow, plotted against β for fixed $Re(=300)$. Bold solid and broken curves, from the two-dimensional computations; fine solid and broken curves, from the one-dimensional model. The broken curves represent steady states that are subsequently found to be unstable. Circles make the value of β at which upstream bulging first appears.

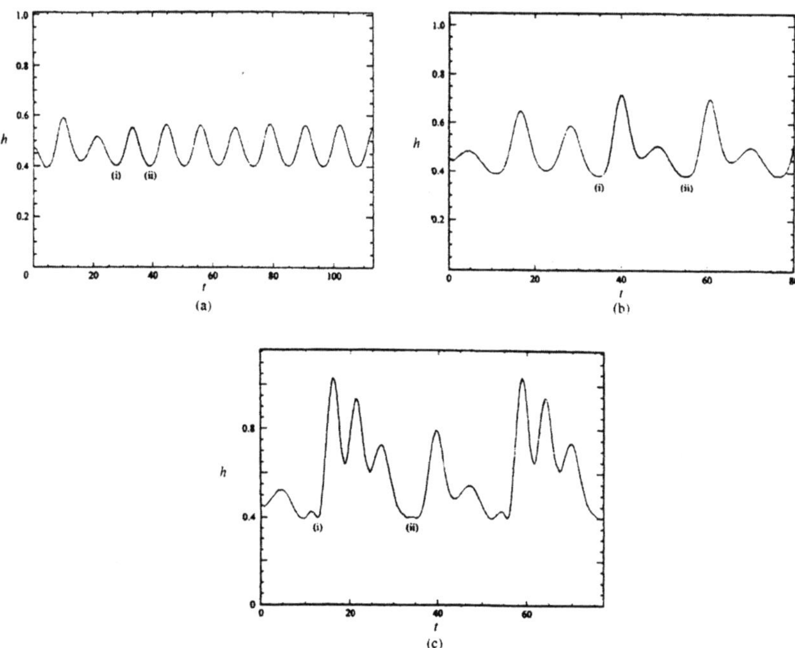

Figure 6.5 Membrane displacement h at fixed $x = 3.5$ as a function of time during self-excited oscillations. $Re = 300$; (a) $\beta = 30.0$, (b) $\beta = 32.5$, (c) $\beta = 35.0$.

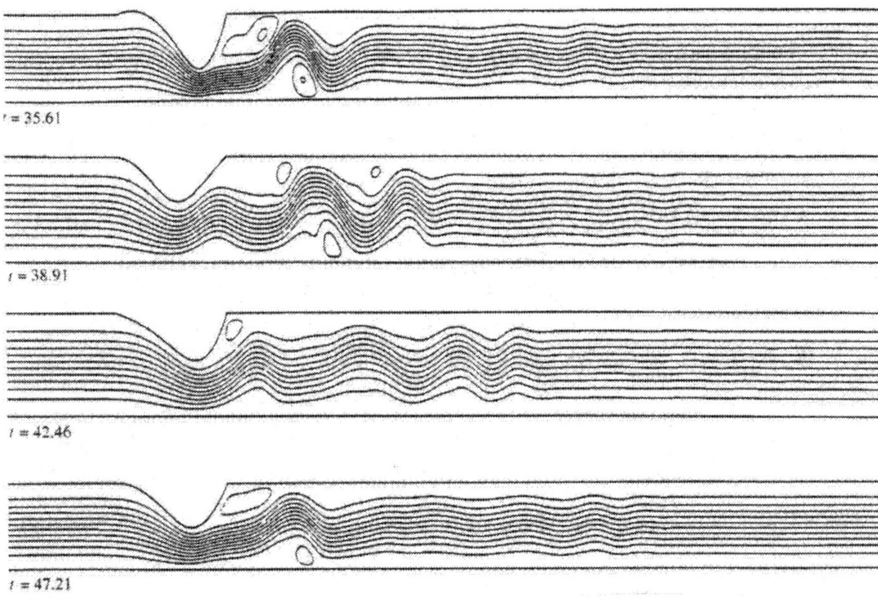

Figure 6.6 Streamline plots at various times during self-excited oscillations for $Re = 300$, $\beta = 32.5$.

Acknowledgements

The author is deeply indebted to all his present and former students, post-docs and colleagues whose work forms the basis of most of these lectures. He is also grateful to the EPSRC for financial support over many years.

References

ADLER, M. 1934 Strömung in gekrümmten Rohren. *Z. Angew. Math. Mech.* **14**, 257–75.

AGRAWAL, Y., TALBOT, L. & GONG, K. 1978 Laser anemometer study of flow development in curved circular pipes. *J. Fluid Mech.* **85**, 497–518.

BADEER, H. S. & RIETZ, R. R. 1979 Vascular hemodynamics: deep-rooted misconceptions and misnomers. *Cardiology* **64**, 197–207.

BERGER, S. A., TALBOT, L. & YAO, L. S. 1983 Flow in curved pipes. *Ann. Rev. Fluid Mech.* **15**, 461–512.

BERTRAM, C. D. 1986 Unstable equilibrium behaviour in collapsible tubes. *J. Biomech.* **19**, 61–69.

BERTRAM, C. D., RAYMOND, C. J. & PEDLEY, T. J. 1990 Mapping of instabilities during flow through collapsed tubes of differing length. *J. Fluids & Structures* **4**, 125–154.

BERTRAM, C. D., RAYMOND, C. J. & PEDLEY, T. J. 1991 Application of nonlinear dynamics concepts to the analysis of self-excited oscillations of a collapsible tube conveying a flow. *J. Fluids & Structures* **5**, 391–426.

BOGDANOVA, E. V. & RYZHOV, O. S. 1983 Free and induced oscillations in Poiseuille flow. *Q. J. Mech. Appl. Math.* **36**, 271–287.

BONIS, M. & RIBREAU, C. 1978 Etude de quelques propriétés de l'écoulement dans une conduite collabable. *La Houille Blanche* **3/4**, 165–173.

BORGAS, M. S. 1986 Waves, singularities and non-uniqueness in channel and pipe flows. PhD Thesis, Cambridge University.

BORGAS, M. S. & PEDLEY, T. J. 1990 Non-uniqueness and bifurcation in annular and planar channel flow. *J. Fluid Mech.* **214**, 229–250.

BRECHER, G. A. 1952 Mechanism of venous flow under different degrees of aspiration. *Am. J. Physiol.* **169**, 423–433.

BROOK, B. S. 1997 The effect of gravity on the haemodynamics of the giraffe jugular vein. PhD Thesis, University of Leeds.

BROWER, R. W. & NOORDERGRAAF, A. 1973 Pressure flow characteristics of collapsible tubes: a reconciliation of seemingly contradictory results. *Ann. Biomed. Eng.* **1**, 333–335.

BUJURKE, N. M., PEDLEY, T. J. & TUTTY, O. R. 1996 Comparison of series expansion and finite-difference computations of internal flow separation. *Phil. Trans R. Soc. Lond. A* **354**, 1751–1773.

CANCELLI, C. & PEDLEY, T. J. 1985 A separated flow model for collapsible tube oscillations. *J. Fluid Mech.* **157**, 375–404.

CARO, C. G., FITZ-GERALD, J. M. & SCHROTER, R. C. 1971 Atheroma and arterial wall shear: observation, correlation and proposal of a shear dependent mass transfer mechanism for atherogenesis. *Proc. R. Soc. Lond. B* **177**, 109–159.

CARO, C. G., PEDLEY, T. J., SCHROTER, R. C. & SEED, W. A. 1978 *The Mechanics of the Circulation*, Oxford University Press.

CARTON, T. W., DAINAUSKAS, J. & CLARK, J. W. 1962 *Elastic properties of single elastic fibres. J. Appl. Physiol.* **17**, 547–51.

CHENG, K. C. & MOK, S. Y. 1986 In *Fluid Control and Measurement*, vol. 2 (ed. M. Harada), pp. 765–773. New York: Pergamon.

COLLINS, W. M. & DENNIS, S. C. R. 1975 The steady motion of a viscous fluid in a curved tube. *Q. Jl. Mech. appl. Math.* **28**, 133–156.

CONRAD, W. A. 1969 Pressure-flow relationships in collapsible tubes. *IEEE. Trans. Bio-Med. Eng.* **BME-16**, 284–295.

DANAKY, D. T. & RONAN, J. A. 1974 Cervical venous hums in patients on chronic hemodialysis. *New Engl. J. Med.* **291**, 237–239.

DASKOPOULOS, P. & LENHOFF, A. M. 1989 Flow in curved ducts: bifurcation structure for stationary ducts. *J. Fluid Mech.* **203**, 125–148.

DEAN, W. R. 1927 Note on the motion of fluid in a curved pipe. *Phil. Mag. Ser. 7* **4**, 208–223.

DEAN, W. R. 1928 The streamline motion of fluid in a curved pipe. *Phil. Mag. Ser. 7* **5**, 673–695.

DENNIS, S. C. R. & NG, M. 1982 Dual solutions for steady laminar flow through a curved tube. *Q. Jl. Mech. appl. Math.* **35**, 305–324.

ELAD, D. & KAMM, R. D. 1989 Parametric evaluation of forced expiration using a numerical model. *ASME J. Biomech. Eng.* **111**, 192–199.

ELZINGA, G. & WESTERHOF, N. 1991 Matching between ventricle and arterial load: an evolutionary process. *Circulation Res.* **68**, 1495–1500.

FRIEDMAN, M. H. 1993 Atherosclerosis research using vascular flow models: from 2-D branches to compliant replicas. *ASME J. Biomech. Eng.* **115**, 595–601.

FRY, D. L. 1958 Theoretical considerations of the bronchial pressure-flow-volume relationships with particular reference to the maximum expiratory flow volume curve. *Phys. Med. Biol.* **3**, 174–194.

FRY, D. L. 1987 Mass transport, atherogenesis and risk. *Arteriosclerosis* **7**, 88–100.

FUKUSHIMA, T. & AZUMA, T. 1982 The horseshoe vortex: a secondary flow generated in arteries with stenosis, bifurcations and branchings. *Biorheology* **19**, 143–154.

GIDDENS, D. P., ZARINS, C. K. & GLAGOV, S. 1993 The role of fluid dynamics in the localisation and detection of atherosclerosis. *ASME J. Biomech. Eng.* **115**, 588–594.

GOETZ, R. H. & KEEN, E. N. 1957 Some aspects of the cardiovascular system in the giraffe. *Angiology* **8**, 542–564.

GOETZ, R. H., WARREN, J. V., GAVER, O. H., PATTERSON, J. L., DOYLE, J. T., KEEN, E. N. & MCGREGOR, M. 1960 Circulation of the giraffe. *Circulation Res.* **8**, 1049–1058.

GRIFFITHS, D. J. 1971 Hydrodynamics of male micturition I. Theory of steady flow through elastic-walled tubes. *Med. Biol. Eng.* **9**, 581–588.

GUYTON. A. C. 1962 'Venous Return': Handbook of Physiology, Section 2, Circulation, Vol. II ed. by W. F. Hamilton and P. Dow, American Physiological Society, Washington D. C.

HARGENS, A. R., MILLARD, R. W., PETTERSSON, K. & JOHANSEN, K. 1987 Gravitational haemodynamics and oedema prevention in the giraffe. *Nature* **329**, 59–60.

HEIL, M. 1997 Stokes flow in collapsible tubes: computation and experiment. *J. Fluid Mech.* **353**, 285–312.

HICKS, J. W. & BADEER, H. S. 1989 Siphon mechanism in collapsible tubes: application to circulation of the giraffe head. *Am. J. Physiol.* **256**, R567–R571.

ITO, H. 1969 Laminar flow in curved pipes. *Z. Angew. Math. Mech.* **49**, 653–63.

JAN, D. L., KAMM, R. D. & SHAPIRO, A. H. 1983 Filling of partially collapsed compliant tubes. *ASME J. Biomech. Eng.* **105**, 12–19.

JENSEN, O. E. 1990 Instabilities of flow in a collapsed tube. *J. Fluid Mech.* **220**, 623–659.

JENSEN, O. E. 1992 Chaotic oscillations in a simple collapsible-tube model. *ASME J. Biomech. Eng.* **114**, 55–59.

JENSEN, O. E. & PEDLEY, T. J. 1989 The existence of steady flow in a collapsed tube. *J. Fluid Mech.* **206**, 339–374.

KAMM, R. D. & PEDLEY, T. J. 1989 Flow in collapsible tubes: a brief review. *ASME J. Biomech. Eng.* **111**, 177–179.

KAMM, R. D. & SHAPIRO, A. H. 1979 Unsteady flow in a collapsible tube subjected to external pressure or body forces. *J. Fluid Mech.* **95**, 1–78.

KARINO, T., KWONG, H. H. M. & GOLDSMITH, H. L. 1979 Particle flow behaviour in models of branching vessels. I. Vortices in 90° T-junctions. *Biorheology* **16**, 231–248.

KORTEWEG, D. J. 1878 Ùber die Fortpflanzungesgeschwindigkeit des Schalles in elastischen Rohren'. *Ann. Phys. Chem.* Ser 3, **5**, 525–542.

KU, D. N. 1988 A review of carotid duplex scanning. *Echocardiography* **5**, 53–69.

KU, D. N., GIDDENS, D. P., ZARINS, C. K. & GLAGOV, S. 1985 Pulsatile flow and atherosclerosis in the human carotid bifurcation: positive correlation between plaque location and low oscillating shear stress. *Arteriosclerosis* **5**, 293–302.

LAMBERT, R. K. 1989 A new computational model for expiratory flow from nonhomogeneous human lungs. *ASME J. Biomech. Eng.* **111**, 200–205.

LEE, C. S. F. & TALBOT, L. 1979 A fluid-mechanical study of the closure of heart valves. *J. Fluid Mech.* **91**, 41–63.

LIGHTHILL, J. 1975 *Mathematical Biofluiddynamics,*Soc. Ind. Appl. Math, Philadelphia.

LIGHTHILL, J. 1978 *Waves in Fluids,*Cambridge University Press.

LOWE, T. W. & PEDLEY, T. J. 1995 Computation of Stokes flow in a channel with a collapsible segment. *J. Fluids & Structures* **9**, 885–905.

LUO, X. -Y. & PEDLEY, T. J. 1996 A numerical simulation of steady flow in a 2-D collapsible channel. *J. Fluids & Structures* **9**, 149–174.

LUO, X. -Y. & PEDLEY, T. J. 1996 A numerical simulation of unsteady flow in a 2-D collapsible channel. *J. Fluid Mech.* **314**, 191–225. (corrigendum **324**, 408–409.)

LUTZ, R. J., CANNON, J. N., BISCHOFF, K. B. & DEDRICK, R. L. 1977 Wall shear stress distribution in a model canine artery during steady flow. *Circulation Res.* **41**, 391–399.

LYNE, W. H. 1971 Unsteady viscous flow in a curved pipe. *J. Fluid Mech.* **45**, 13–31.

MCCONALOGUE, D. J. & SRIVASTAVA, R. S. 1968 Motion of fluid in a curved tube. *Proc. R. Soc. Lond.* **A307**, 37–53.

MCDONALD, D. A. 1974 *Blood Flow in Arteries*(2nd edn.)London: Edward Arnold.

MITCHELL, G. & SKINNER, J. D. 1993 How giraffe adapt to their extraordinary shape. *Trans. R. Soc. S. Africa* **48**, 207–218.

MOENS, A. I. 1878 *Die Pulskurve.*Brill, Leiden.

MORENO, A. H., KATZ, A. I., GOLD, L. D. & REDDY, R. V. 1970 Mechanics of distension of dog veins and other very thin-walled tubular structures. *Circulation Res.* **27**, 1069–1080.

OLSON, R. M. 1968 Aortic blood pressure and velocity as a function of time and position. *J. Appl. Physiol.* **24**, 563–569.

PEDLEY, T. J. 1980 *The Fluid Mechanics of Large Blood Vessels.* Cambridge University Press.

PEDLEY, T. J. 1987 How giraffes prevent oedema. *Nature* **329**, 13–14.

PEDLEY, T. J. 1992 Longitudinal tension variation in collapsible channels: a new mechanism for the breakdown of steady flow. *ASME J. Biomech. Eng.* **114**, 60–67.

PEDLEY, T. J. 1995 High Reynolds number flow in tubes of complex geometry with application to wall shear stress in arteries. In: *Biological Fluid Dynamics*, ed. by C. P. Ellington & T. J. Pedley, Soc. Exp. Biol. Symposium no. 49, pp. 219–241.

PEDLEY, T. J., BROOK, B. S. & SEYMOUR, R. S. 1996 Blood pressure and flow rate in the giraffe jugular vein. *Phil. Trans. R. Soc. Lond. B* **351**, 855–866.

PEDLEY, T. J. & LUO, X.-Y. 1998 Models of flow and oscillations in collapsible tubes. *Theor. & Comput. Fluid Dyn.* **10**, 277–294.

PEDLEY, T. J., SCHROTER, R. C. & SUDLOW, M. F. 1977 Gas flow and mixing in the airways. In: *Bioengineering aspects of the lung*, ed. by J. B. West. Marcel Dekker, New York, pp. 163–265.

PEDLEY, T. J. & STEPHANOFF, K. D. 1985 Flow along a channel with a time-dependent indentation in one wall: the generation of vorticity waves. *J. Fluid Mech.* **160**, 337–367.

PERMUTT, S., BROMBERGER-BARNEA, B. & BANE, H. N. 1963 Hemodynamics of collapsible vessels with tone. The vascular waterfall. *J. Appl. Physiol.* **18**, 924–932.

PRANDTL, L. 1952 The Essentials of Fluid Mechanics 3rd edn., translated. Glasgow: Blackie & Son.

RALPH, M. E. & PEDLEY, T. J. 1988 Flow in a channel with a moving indentation. *J. Fluid Mech.* **190**, 87–112.

RALPH, M. E. & PEDLEY, T. J. 1990 Flow in a channel with a time-dependent indentation in one wall. *ASME J. Fluids Eng.* **112**, 468–475.

RAST, M. P. 1994 Simultaneous solution of the Navier-Stokes and elastic membrane equations by a finite-element method. *Int. J. Numer. Methods. Fluids* **19**, 1115–1135.

RENEAU, L. R., JOHNSTON, J. P. & KLINE, S. J. 1967 Performance and design of straight, two-dimensional diffusers. *ASME J. Basic Eng.* **89**, 141–150.

SCHROTER, R. C. & SUDLOW, M. F. 1969 Flow patterns in models of the human bronchial airways. *Respir. Physiol.* **7**, 341–355.

SECOMB, T. W. 1979 Flows in tubes and channels with indented and moving walls. PhD Thesis, Cambridge University.

SHAPIRO, A. H. 1977a Steady flow in collapsible tubes. *ASME J. Biomech. Eng.* **99**, 126–147.

SHAPIRO, A. H. 1977b Physiologic and medical aspects of flow in collapsible tubes. *Proc. 6th Can. Congr. Appl. Mech.*, 883–906.

SINGH, M. P. 1974 Entry flow in a curved pipe. *J. Fluid Mech.* **65**, 517–39.

SMITH, F. T. 1976a Flow through constricted or dilated pipes and channels: Part I. *Q. J. Mech. Appl. Math.* **29**, 343–364.

SMITH, F. T. 1976b Flow through constricted or dilated pipes and channels: Part II. *Q. J. Mech. Appl. Math.* **29**, 365–376.

SMITH, F. T. 1975 Pulsatile flow in curved pipes. *J. Fluid Mech.* **71**, 15–42.

SOBEY, I. J. 1985 Observation of waves during oscillatory channel flow. *J. Fluid Mech.* **151**, 395–426.

TAYLOR, C. A., HUGHES, T. J. R. & ZARINS, C. K. 1996 Computational investigations in vascular disease. *Computers in Physics* **10**, 224–232.

TUTTY, O. R. & PEDLEY, T. J. 1993 Oscillatory flow in a stepped channel. *J. Fluid Mech.* **247**, 179–204.

WALBURN, F. J. & STEIN, P. D. 1980 Flow in a symmetrically branched tube simulating the aortic bifurcation: the effects of unevenly distributed flow. *Ann. Biomed. Eng.* **8**, 159–173.

WEINBAUM, S. & CHIEN, S. 1993 Lipid transport aspects of atherogenesis. *ASME J. Biomech. Eng.* **115**, 602–610.

WHITE, C. M. 1929 Streamline flow through curved pipes. *Proc. R. Soc. Lond. A* **123**, 645–663.

WOMERSLEY, J. R. 1955 Method for calculation of velocity, rate of flow and viscous drag in arteries when the pressure gradient is known. *J. Physiol. Lond.* **127**, 553–563.

WOMERSLEY, J. R. 1957 The mathematical analysis of the arterial circulation in a state of oscillatory motion. *Wright Air Development Center, Tech. Rep.* WADC-TR 56-614.

YANASE, S., GOTO, N. & YAMAMOTO, K. 1999 Dual solutions of the flow through a curved tube. *Fluid Dyn. Res.* **5**, 191–201.

YOUNG, T. 1809 On the functions of the heart and arteries. *Phil. Trans. R. Soc. Lond.* **99**, 1–31.

YUAN, F., CHIEN, S. & WEINBAUM, S. 1991 A new view of convective-diffusive transport processes in the arterial intima. *ASME J. Biomech. Eng.* **113**, 314–329.

ZARINS, C. K., GIDDENS, D. P., BHARADVAJ, B. K., SOTTIURAI, V. S., MABON, R. F. & GLAGOV, S. 1983 Carotid bifurcation atherosclerosis: quantitative correlation of plaque localization with flow velocity profiles and wall shear stress. *Circulation Res.* **53**, 502–514.

ZELLER, H., TALUKDER, N. & LORENZ, J. 1970 Model studies of pulsating flow in arterial branches and wave propagation in blood vessels. In *Fluid Dynamics of Blood Circulation and Respiratory Flow,* AGARD Conference Proceedings no **65**.

Computational Models of Arterial Flow and Mass Transport

Karl Perktold[1] and Martin Prosi[1]

[1] Institute of Mathematics, Graz University of Technology, Graz, Austria

Abstract. Numerical simulation of arterial hemodynamics and mass transport have become an important tool in recent years due to the significant advances in numerical mathematics, scientific computation and due to the increased power of computers. Over the past years increasingly more elaborate models have been developed in order to gain a better insight into the physiological processes in the vascular system and the initiation and development of arterial diseases. Hemodynamic factors apparently play an important role in the development of these diseases, and therefore, local arterial flow dynamics, such as flow separation, flow recirculation, low and oscillatory wall shear stress, and the influence on mass transport in the arterial lumen and in the artery wall are subjects of intensive research. Corresponding studies include rheological effects in blood flow resulting from the interactions between blood phases up to the transport processes of macromolecules in the arteries and in the artery wall layers. The problems discussed are mathematically described by systems of coupled non-linear partial differential equations mostly in large parameter range. The numerical approach uses the finite element method, which is often the most suitable approximation technique due to its high flexibility.

1 Mathematical Modeling of Local Blood Flow in Arteries

The complex vascular system can be subdivided into several groups with descending diameter, the large arteries, medium-sized arteries, and the arterioles and capillaries. The physical characteristics of blood change in dependence of the type of the vessel. In large and medium-sized arteries the blood rheology can be described using macroscopic models. In arterioles and capillaries the blood particles must to be taken into account.

1.1 Blood Rheology

Blood is a concentrated suspension of blood cells in plasma and has non-Newtonian properties. The main determinants of rheologic behavior are plasma viscosity and the concentration and properties of the suspended particles. The particles are the red blood cells (RBC), the white blood cells (WBC) and the platelets, where the WBC and the platelet contribution is insignificant compared to the RBC concentration of about 40 to 45 percent of the blood volume, representing the so called hematocrite. The RBCs affect blood rheology mainly by two properties, the aggregability and deformability. At rest a RBC has a biconcave shape with a diameter of about

---

The authors acknowledge the support by the Austrian Science Foundation (FWF) through the Projects P 11982-TEC and P 14321-TEC.

7μm. At high flow rate the RBCs exist as individual particles and take a thinner shape without changing the surface size nor the volume. Therefore, at these high flow rates blood viscosity is influenced only by the inner viscosity of the single RBC. At lower flow rates ($\gamma < 10\ s^{-1}$) RBCs aggregate to form rouleaux resulting in an increase of viscosity. If further decrease of the shear rate ($\gamma < 1 s^{-1}$) is applied the RBCs rouleaux form three-dimensional structures inducing additional viscosity increase. Higher shear stress causes disaggregation, and the RBCs align with flow, and thus, a decrease of viscosity occurs. While the breakdown of rouleaux happens almost instantaneously the time for reaggregation is mainly determined by shear rate and hematocrite (Schmid-Schoenbein et al., 1980). Normal human blood in a shear field of 10 s^{-1} shows a characteristic aggregation time of about 10 s. Due to the high shear rates during the pulse cycle and due to the fact that the pulse period is about ten times shorter than the aggregation time, RBCs can hardly form rouleaux in relatively undisturbed large and medium-sized artery flow (Sharp et al., 1996).

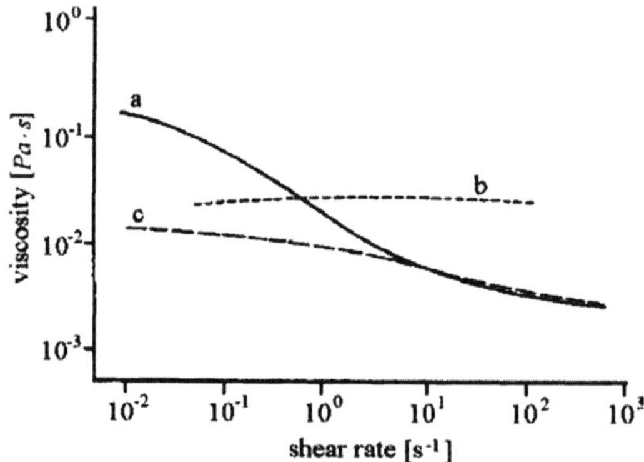

Figure 1. Blood viscosity in steady shear: (a) normal blood, (b) blood with hardened RBCs, (c) blood in a Ringer solution (Chien et al., 1970).

The viscoelastic behavior of blood in large and medium-sized arteries is mainly caused by the deformability of the single RBCs, and the time constant is the relaxation time of the RBCs which is about 0.06 s (Cokelet, 1980). Non-Newtonian behavior is of greater importance for blood flow in small vessels. In large and medium-sized vessels blood may be considered as an isotropic, homogeneous, incompressible Newtonian or non-Newtonian inelastic fluid. Newtonian simplification applies an appropriate constant reference viscosity. The non-Newtonian inelastic description uses macroscopic shear thinning relations. The influence of shear thinning in larger arteries is quantitatively, qualitatively the velocity patterns remains unchanged (Gijsen et al., 1999, Perktold et al., 1991, Perktold and Hofer, 1999). At the occurring shear rates in these vessels, the rheologic properties can be approximated as Newtonian with a viscosity evaluated at a characteristic shear rate of the flow. More accurate studies use appropriate shear thinning relations as, e.g., the modified Cross model (Perktold et al., 1991) or the Carreau-Yasuda model (Gijsen, 1998), where the viscosity is a nonlinear function of the shear rate at physiologic values of the hematocrite. Casson's relation often used for blood flow description at low shear rates in small vessels includes the dependance on the hematocrite. Figure 1 demonstrates the shear thinning property of normal human blood and compares the behavior with hardened RBCs (no deformation) and RBCs in Ringer solution (no aggregation).

Viscoelastic properties of blood were investigated by Thurston (1979). He measured a significant elastic component in oscillatory blood flow. Concerning the influence of viscoelastic blood behavior on flow patterns in large and medium-sized arteries there are only few numerical studies available (Leuprecht and Perktold, 2001, Phillips and Deutsch, 1975, Walitza, 1990), while many studies are available on the simulation of viscoelastic flows for low Reynolds numbers, as polymer processing (Keunings, 1989). The listed viscoelastic blood flow studies used an Oldroyd model. Equations of the Oldroyd-B model reproduce the relevant properties of experimentally found material functions of blood approximately. Thixotropy, a property of structured fluids subjected to a transient load, cannot be express in the Oldroyd-B model. Leuprecht and Perktold (2001) showed that the influence of the viscoelastic properties depends strongly on the shape of the flow domain. At domains with highly elongational flow viscoelastic properties influence the flow patterns. In simpler shear flow, however, viscoelastic hardly affects the flow in larger arteries, and the shear thinning effect forms the dominant non-Newtonian influence on the flow patterns.

In small blood vessels (capillaries) multiphase models expressing the structure of the fluid-particle system have to be applied in the mathematical description.

1.2 Constitutive Relations

Basic relations between stress and strain. In an isothermal flow the primary unknowns include the pressure p and the components of the velocity vector $\boldsymbol{u}$ which are functions of the spatial co-ordinates and the time. In problems involving an unknown boundary, such as an interface, the location of this boundary is usually determined by the kinematic equation. The equations for the three momentum components, for the continuity and the kinematic equation can be solved for the velocity components, the pressure and the location of the boundary if the stress components occurring in these equations are expressed in terms of the primary unknowns. The corresponding relations between stress and strain (or rate of strain) are the constitutive relations. Other flow variables, such as the components of the stress tensor, the streamlines, the fluid particle residence time, can be evaluated using the calculated primary unknowns.

In general flow the stress tensor $\boldsymbol{T}$ can be expressed as a functional of the form

$$\boldsymbol{T} = -p\boldsymbol{I} + f(\boldsymbol{A}_1, \boldsymbol{A}_2, .., \boldsymbol{A}_\mathrm{n}) , \tag{1.1}$$

where $\boldsymbol{I}$ is the unit tensor, $\boldsymbol{A}_\mathrm{k}$, k = 1,.., n, are the Rivlin-Ericksen strain tensors, which are directly related to the rate of strain tensor $\boldsymbol{D}$ and its substantial derivatives

$$\boldsymbol{T} = -p\boldsymbol{I} + f(\boldsymbol{D}, \dot{\boldsymbol{D}}, \ddot{\boldsymbol{D}}, ...) , \tag{1.2}$$

the dots indicate differentiation with respect to time. The memory of the fluid (the behavior to remember and return to its undeformed state) determines the number of tensors necessary to approximate the stresses. The memory arises from the elastic properties of the particles or molecules when stretched or compressed. The occurring elastic forces resist deformation and tend to spontaneously return to the undeformed or unstressed state (Tanner, 1985).

First order fluids. These fluids have small, stiff molecules and exhibit no memory. The local deformation determines the local stress excluding strains which incorporates history effects and rate of strain derivatives that violate the localization of the rate of strain. The constitutive relation for first order fluids is

$$\boldsymbol{T} = -\,p\boldsymbol{I} + \alpha_1 \boldsymbol{D}\,, \tag{1.3}$$

where $\alpha_1 = 2\mu$, $\mu > 0$, being the dynamic viscosity of the fluid. For non-Newtonian inelastic fluids μ is a function of the rate of strain, for Newtonian fluids μ is constant.

Second order fluids. Including higher-order derivatives leads to fluids with memory (viscoelastic fluids), the stresses in a memory fluid are described by

$$\boldsymbol{T} = -\,p\boldsymbol{I} + \alpha_1 \boldsymbol{D} + \alpha_2 \boldsymbol{D}^2 + \alpha_3 \ddot{\boldsymbol{D}}\,, \tag{1.4}$$

where the parameters α_i are material constants or functions. As the elasticity and memory of the fluid increases, additional terms are required to approximate the stresses.

1.3 Macroscopic Blood Flow Description

Blood flow in large and medium-sized arteries can be described mathematically using a first order homogeneous incompressible fluid model. The mathematical model applies the system of time-dependent, incompressible Navier-Stokes equations which are the conservation of momentum and mass (the incompressibility constraint)

$$\rho\,(\frac{\partial \boldsymbol{u}}{\partial t} + (\boldsymbol{u}\cdot\nabla)\boldsymbol{u}) - \nabla\cdot\boldsymbol{T} = \rho\boldsymbol{g}\,,\quad \nabla\cdot\boldsymbol{u} = 0 \qquad \text{in } G \subset R^n,\quad n = 2 \text{ or } 3,\ t \in (t_0, T) \tag{1.5}$$

where G is bounded with a sufficiently regular boudary Γ. The convective terms are

$$(\boldsymbol{u}\cdot\nabla)\boldsymbol{u} = \sum_{j=1}^{n} u_j \frac{\partial}{\partial x_j}\boldsymbol{u}\ .$$

In this model $\boldsymbol{u}$ denotes the flow velocity vector, $\boldsymbol{g}$ is the volume body force, and ρ is the constant fluid density. The total stress tensor (Cauchy stress tensor) $\boldsymbol{T}$ is expressed as the combination of an isotropic pressure p and viscous contribution, the extra stresses $\boldsymbol{S}$

$$\boldsymbol{T} = -\,p\boldsymbol{I} + \boldsymbol{S}\ . \tag{1.6}$$

For turbulent flow $\boldsymbol{S}$ is expressed as

$$\boldsymbol{S} = \boldsymbol{S}^l + \boldsymbol{S}^R\,, \tag{1.7}$$

where $\boldsymbol{S}^l$ is the laminar stress tensor corresponding to time-averaged turbulent kinematics, and $\boldsymbol{S}^R$ is the turbulent Reynolds stress tensor which is a function of turbulent velocity fluctuations. The completition of the model requires appropriate constitutive relations for the extra stresses $\boldsymbol{S}$. The system is closed by imposing appropriate initial and boundary conditions.

1.4 Shear Thinning Model

The non-Newtonian inelastic viscous forces can be modeled as functions of the shear-dependent viscosity

$$\boldsymbol{S} = 2\mu(D_{II}; p_1,..,p_m)\boldsymbol{D}\,, \tag{1.8}$$

where μ describes the non-Newtonian essentially shear-dependent dynamic viscosity; $\boldsymbol{D}$ the rate of deformation tensor, is defined as

$$\boldsymbol{D} = \frac{1}{2}(\nabla \boldsymbol{u} + (\nabla \boldsymbol{u})^T)\,. \tag{1.9}$$

D_{II} denotes the second invariant of $\boldsymbol{D}$, a scalar measure of the rate of deformation tensor with suitable invariance properties with respect to the reference co-ordinate system. Using the trace "*tr*" of a matrix, D_{II} can be expressed as

$$D_{II} = \frac{1}{2}(tr\boldsymbol{D}^2 - (tr\boldsymbol{D})^2)\,, \tag{1.10}$$

$p_1,.., p_m$ are specific parameters. Applying the second invariant the constitutive equation is objective, $\mu(D_{II})$ is invariant with respect to a co-ordinate system transformation.

Eqn. 1.6 is a tensor equation equivalent in 3D to nine scalar equations. Substituting the components of $\boldsymbol{D}$

$$d_{ij} = \frac{1}{2}(\frac{\partial u_i}{\partial x_j} + \frac{\partial u_j}{\partial x_i}) \qquad i,j = 1,2,3 \tag{1.11}$$

in Eqn. 1.6 the Cauchy stress components in the cartesian co-ordinate system read for an incompressible and isotropic medium as

$$t_{ij} = -p\delta_{ij} + \mu(D_{II})(\frac{\partial u_i}{\partial x_j} + \frac{\partial u_j}{\partial x_i}) \qquad i,j = 1,2,3\,. \tag{1.12}$$

In incompressible flow $tr\boldsymbol{D} = 0$, and thus, $D_{II} = ½\, tr\boldsymbol{D}^2 = ½\;\; d_{ij}\, d_{ij}$. The viscous stress tensor is symmetric, i.e., $s_{ij} = s_{ji}$, therefore, there are only six independent stress components. In fully developed (Poiseuille-) flow the occurring extra stress is defined by $s = \mu\gamma$, where $\gamma = du/dr$ is called the shear rate and is the velocity gradient in the direction perpendicular to the flow axis.

Applying parallel shear flow the relation between the scalar shear rate γ and the second invariant D_{II} can be shown. Assuming the velocity field $\boldsymbol{u} = u(z)\boldsymbol{e}_x$, $(u' := \frac{\partial u}{\partial z})$ then

$$\frac{u'^2}{4} = D_{II} \rightarrow |u'| = 2\sqrt{D_{II}}$$

$$\gamma = 2\sqrt{D_{II}}\,. \tag{1.13}$$

The non-Newtionan, inelastic stresses (Eqn. 1.12) can be substituted into the equations of motion resulting in the generalized Navier-Stokes equations in stress-divergence form

$$\rho(\frac{\partial \boldsymbol{u}}{\partial t} + (\boldsymbol{u}\cdot\nabla)\boldsymbol{u}) - \nabla\cdot(\mu(\nabla\boldsymbol{u} + (\nabla\boldsymbol{u})^T)) + \nabla p = \boldsymbol{g}$$
$$\nabla\cdot\boldsymbol{u} = 0 \qquad \text{in } G\subset R^n, \quad n=2 \text{ or } 3, \; t\in(t_0,T). \tag{1.14}$$

Applying Einstein's summation convention (which means that summation is indicated if a subscript repeatedly occurs in a single term) the equivalent form in cartesian components n=3 reads

$$\rho(\frac{\partial u_i}{\partial t} + u_j\frac{\partial u_i}{\partial x_j}) - \frac{\partial}{\partial x_j}(\mu(\frac{\partial u_i}{\partial x_j} + \frac{\partial u_j}{\partial x_i})) + \frac{\partial p}{\partial x_i} = g_i$$
$$\frac{\partial u_j}{\partial x_j} = 0 \qquad i, j = 1,2,3. \tag{1.15}$$

These forms of the Navier-Stokes equations are referred to as the primitive equations since they are written in terms of the primitive variables velocity components and pressure. In addition to the standard convective term in Eqn. 1.14 alternative forms $\nabla\cdot(\boldsymbol{uu}) = \boldsymbol{u}\cdot\nabla\boldsymbol{u} + \boldsymbol{u}(\nabla\cdot\boldsymbol{u})$ or $\boldsymbol{u}\cdot\nabla\boldsymbol{u}$ or $\boldsymbol{u}\cdot\nabla\boldsymbol{u} + 1/2\,\boldsymbol{u}(\nabla\cdot\boldsymbol{u})$ exist.

Shear dependent blood viscosity. The shear-dependent viscosity in large and medium-sized blood vessels can be described using appropriate relations. The Carreau-Yasuda model

$$\mu(\dot\gamma) = \mu_0 + (\mu - \mu_\infty)(1 + (\lambda\dot\gamma)^a)^{\frac{n-1}{a}} \tag{1.16}$$

was used by Gjisen (1998), it contains five positive parameter. Eqn. 1.16 expresses the behavior at low (μ_0) and high shear rate (μ_∞); λ is a time constant, n is the flow index, the parameter a controls the transition to the lower Newtonian range. The modified Cross model

$$\mu(\dot\gamma) = \mu_\infty + (\mu_0 - \mu_\infty)\frac{1}{(1 + (\lambda\dot\gamma)^b)^a} \tag{1.17}$$

was used by Perktold et al. (1991). In the description of pulsatile blood flow frequency-dependent dynamic viscosity data have to be applied. The parameters occurring in Eqn. 1.17 are determined by numerical fitting of experimental data measured by Liepsch et al. (1991) in oscillatory blood flow of two Hertz. The resulting values are $\mu_\infty = 0.03$ Poise, $\mu_0 = 0.1315$ Poise, $\lambda = 0.5s$, $a = 0.3$, $b = 1.7$.

Newtonian fluids. Constant viscosity characterizes Newtonian fluids. Applying the incompressible mass conservation in the equation of motion the divergence of the viscous stresses can be simplified: $\mu(\nabla^2\boldsymbol{u} + \nabla(\nabla\cdot\boldsymbol{u})) = \mu\nabla^2\boldsymbol{u}$, and the classical Navier-Stokes equations for Newtonian fluids in the variables velocity and pressure may be expressed in vector form as

$$\rho(\frac{\partial \boldsymbol{u}}{\partial t}+(\boldsymbol{u}\cdot\nabla)\boldsymbol{u})-\mu\nabla^2\boldsymbol{u}+\nabla p=\boldsymbol{g}$$
$$\nabla\cdot\boldsymbol{u}=0 \qquad \text{in } G\subset R^n,\quad n=2 \text{ or } 3,\quad t\in(t_0,T). \tag{1.18}$$

Initial and boundary conditions. The Navier-Stokes equations can be solved for the variables velocity components and pressure, provided that the viscosity function and appropriate initial and boundary conditions are given. The initial condition requires the specification of the flow velocity at the initial time t_0, e.g., $t_0 = 0$

$$\boldsymbol{u}(\boldsymbol{x},0)=\boldsymbol{u}_0 \qquad \text{in } G, \tag{1.19}$$

where the given initial velocity field is divergence free.

For the flow type as treated in this study the boundary of the flow domain partitions into two non-overlapping sections Γ_u and Γ_t with corresponding types of boundary conditions. With respect to appropriate boundary conditions it is necessary to prescribe either the velocity (Dirichlet or essential boundary condition) or the surface traction force (Neumann or natural boundary condition). At the inflow boundary and at the vessel wall Γ_u the time-dependent flow velocity is prescribed

$$\boldsymbol{u}=\boldsymbol{f} \qquad \text{on } \Gamma_u,\quad t\in(0,T). \tag{1.20}$$

In cases where physiological correct inflow data are not available often fully developed velocity profiles (Womersley solutions) corresponding to the prescribed velocity pulse waveform can be used. These profiles can be calculated as long straight tube profiles. This idealization of the inflow condition is acceptable in relatively long straight artery segments. At the vessel wall the no-slip condition is appropriate, which expresses that the velocity at the vessel wall boundary is the wall velocity ($\boldsymbol{u}$ = 0 for rigid vessels and $\boldsymbol{u} \neq 0$ for distensible vessels).

At the remaining boundary (outflow boundary Γ_t) a condition describing surface traction force can be applied $\boldsymbol{T}\cdot\boldsymbol{n} = \boldsymbol{h}$. The traction condition appropriate for the generalized Navier-Stokes equations in stress-divergence form Eqn. 1.14 is

$$-p\boldsymbol{n}+\mu\,(\nabla\boldsymbol{u}+(\nabla\boldsymbol{u})^T)\cdot\boldsymbol{n}=\boldsymbol{h} \qquad \text{on } \Gamma_t,\quad t\in(0,T), \tag{1.21}$$

where $\boldsymbol{n}$ is the outward pointing normal unit vector at the outflow boundary section. In components n=3 the condition reads as

$$\sum_{j=1}^{3}(-p\delta_{ij}+\mu(\frac{\partial u_i}{\partial x_j}+\frac{\partial u_j}{\partial x_i}))n_j=h_i \qquad i=1,2,3 \tag{1.22}$$

Eqn. 1.22 corresponds to a Neumann condition for the operator

$$\sum_{j=1}^{3}\frac{\partial}{\partial x_j}(-p\delta_{ij}+\mu(\frac{\partial u_i}{\partial x_j}+\frac{\partial u_j}{\partial x_i}))n_j \qquad i=1,2,3$$

If incompressible mass is included in the equation of motion (Eqn. 1.18) the corresponding natural "pseudo-traction" condition

$$-p\boldsymbol{n} + \mu\nabla\boldsymbol{u}\cdot\boldsymbol{n} = \hat{\boldsymbol{h}} \qquad \text{on} \quad \Gamma_t\,, \quad t\in(0,T)\,, \tag{1.23}$$

does not represent the physical forces, in case of zero surface traction force the conditions Eqn. 1.21 and Eqn. 1.23 are equivalent. The condition $-p\boldsymbol{n} + \mu\nabla\boldsymbol{u}\cdot\boldsymbol{n} = \boldsymbol{0}$ is helpful as outflow condition.

Flow analysis through arterial bifurcations is characterized by two outlets and prescribed varying flow partition during the pulse cycle. In case of two outflow boundaries a calculation procedure suggested consists of two steps. In the first step auxiliary velocity profiles at one outlet are assumed. These profiles only approximate the correct profiles but guarantee the presribed flow division ratio. At the second outlet zero surface traction forces are prescribed. This calculation step yields correct velocities at the second outlet. These correct outflow data are applied as velocity boundary condition in the second step where the zero force condition is prescribed at the first outflow boundary.

In the compliant model the outlet traction forces are

$$h_n = -p(t) \qquad h_t = 0, \quad t\in(0,T)\,, \tag{1.24}$$

where h_n and h_t are the normal and tangential force, respectively.

Generally the specification of appropriate outflow boundary conditions is difficult for finite flow domains. Force conditions (traction free outflow $\boldsymbol{h} = 0$ in case of rigid walls) are useful at artificial boundaries (the exit of a tube) which are located downstream the domain under consideration to guarantee that there is no influence of the conditions on the flow variables in the flow region of interest. An improvement is the application of coupling methods for the Navier-Stokes equations in the artery segement of interest and of reduced equations in the remaining artery. Developments in this direction of multiscale methods are carried out by Quarteroni et al. (2001), Formaggia et al. (2001).

The system of the Navier-Stokes equations (Eqn. 1.14 or Eqn. 1.18) describes (in a domain $G\subset R^n$, n = 2 or 3, and for a time intervall (t_0,T)) the velocity $\boldsymbol{u}(x_1,..,x_n,t)$, with the components $(u_1,..,u_n)$, and the pressure $p(x_1,..,x_n,t)$ of an incompressible fluid, for given physical properties (dynamic viscosity μ, density ρ) and prescribed initial and boundary conditions.

Flow parameters. Through a scaling transformation the Navier-Stokes equations are often made dimensionless with the Reynolds number $Re = U_0 L_0/\nu$ as the characteristic parameter, where U_0 is a characteristic velocity (typically, the mean or maximum inflow velocity), L_0 is a characteristic length (typically, a vessel diameter) and $\nu = \mu/\rho$ is the kinematic viscosity (an appropriate reference viscosity in case of non-Newtonian inelastic fluids). The non-dimensional Navier-Stokes equations and the Dirichlet and Neumann boundary conditions read

$$\frac{\partial \boldsymbol{u}^*}{\partial t^*} + (\boldsymbol{u}^* \cdot \nabla)\boldsymbol{u}^* - \frac{1}{Re}\nabla^2 \boldsymbol{u}^* + \nabla p^* = \boldsymbol{g}^*$$

$$\nabla \cdot \boldsymbol{u}^* = 0 \tag{1.25}$$

$$\boldsymbol{u}^* = \boldsymbol{f}^* \quad \text{on} \quad \Gamma_u, \quad -p^*\boldsymbol{n} + \frac{1}{Re}\frac{\partial \boldsymbol{u}^*}{\partial \boldsymbol{n}} = \boldsymbol{h}^*_{\Gamma_t} \quad \text{on} \quad \Gamma_t$$

where the variables are

$$\boldsymbol{x}^* = \frac{\boldsymbol{x}}{L_0}, \quad \boldsymbol{u}^* = \frac{\boldsymbol{u}}{U_0}, \quad t^* = t\frac{U_0}{L_0}, \quad p^* = \frac{p}{\rho U_0^2}. \tag{1.26}$$

In pulsatile flow a dimensionless frequency parameter, the Stokes number $\eta = 2L_0\sqrt{\pi f / \nu}$ (or alternatively the Womersley parameter $\alpha = L_0\sqrt{\pi f / 2\nu}$), where f is frequency of the pulsation, can be defined. Using the Reynolds number and the Stokes number the unsteady Navier-Stokes equations and the continuity equation for a Newtonian incompressible fluid can be written in dimensionless form

$$\frac{\eta^2}{2Re}\left(\frac{\partial \boldsymbol{u}^*}{\partial t^*} + (\boldsymbol{u}^* \cdot \nabla)\boldsymbol{u}^*\right) - \frac{1}{Re}\nabla^2 \boldsymbol{u}^* + \nabla p^* = \boldsymbol{g}^*$$

$$\nabla \cdot \boldsymbol{u}^* = 0 \tag{1.27}$$

(the indication of the dimensionless variables is omitted in the subsequent sections).

Time dependent flow domain. An essential feature of arterial flow analysis is the time-dependent flow domain due to vessel dynamics and vessel compliance. The description of the flow in a time-dependent domain requires an appropriate modification of the Navier-Stokes equations. A successful technique to consider moving boundaries associated with vessel dynamics and distensible vessel wall is the arbitrary Lagrangian Eulerian (ALE) method which is based on arbitrary movement of the reference frame (Hughes et al., 1991). The equation of motion in the ALE formulation can be written as

$$\rho\left(\frac{\partial \boldsymbol{u}}{\partial t} + ((\boldsymbol{u} - \hat{\boldsymbol{u}}) \cdot \nabla)\boldsymbol{u}\right) - \nabla \cdot (\mu(\nabla \boldsymbol{u} + (\nabla \boldsymbol{u})^T)) + \nabla p = \boldsymbol{g}, \tag{1.28}$$

where $\hat{\boldsymbol{u}}$ denotes the velocity of the points of the time-dependent flow domain. A further modification concerns the boundary conditions. The condition at the vessel wall uses the velocity obtained from the wall movement corresponding to the vessel deformation due to vessel dynamics or time-dependent (pulsatile) pressure load.

1.5 Viscoelastic Model

The stresses can be described applying the Oldroyd model

$$\boldsymbol{S} + \lambda_1 \overset{\nabla}{\boldsymbol{S}} = 2\mu_0(\boldsymbol{D} + \lambda_2 \overset{\nabla}{\boldsymbol{D}}) \tag{1.29}$$

where λ_1 is the stress relaxation time, λ_2 is the retardation time, μ_0 shear viscosity; $^{\nabla}$ indicates upper convected Oldroyd time-derivative of a tensor field

$$\overset{\nabla}{\boldsymbol{S}} = \frac{\partial \boldsymbol{S}}{\partial t} + \boldsymbol{u} \cdot \nabla \boldsymbol{S} - \boldsymbol{S} \cdot \nabla \boldsymbol{u} - (\nabla \boldsymbol{u})^T \cdot \boldsymbol{S} . \tag{1.30}$$

The computational approach makes use of a stress decomposition in the non-Newtonian (particle) stress $\boldsymbol{S}_1$ and the Newtonian (solvent) stress $\boldsymbol{S}_2$ (Leuprecht and Perktold, 2001)

$$\boldsymbol{S} = \boldsymbol{S}_1 + \boldsymbol{S}_2 . \tag{1.31}$$

The corresponding stress relations are

$$\boldsymbol{S}_1 + \lambda_1 \overset{\nabla}{\boldsymbol{S}_1} = 2\mu_1 \boldsymbol{D} \tag{1.32}$$

$$\boldsymbol{S}_2 = 2\mu_2 \boldsymbol{D} , \tag{1.33}$$

where μ_1 is the elastic viscosity and μ_2 the Newtonian viscosity. It can be shown that

$$\mu_0 = \mu_1 + \mu_2 \quad \text{and} \quad \lambda_2 = \mu_2 \lambda_1 / \mu_0 . \tag{1.34}$$

If $\lambda_2 = 0$ the model reduces to an upper convected Maxwell fluid, and if $\lambda_1 = \lambda_2$ to a purely viscous Newtonian fluid with viscosity μ_0.

By substituting Eqn. 1.31 and Eqn. 1.33 into Eqn. 1.29 and into the equation of motion (Eqn. 1.5), the momentum, continuity and constitutive equations of the Oldroyd-B model can be written as

$$\begin{gathered} \rho(\frac{\partial \boldsymbol{u}}{\partial t} + (\boldsymbol{u} \cdot \nabla)\boldsymbol{u}) = \nabla \cdot \boldsymbol{S}_1 + \mu_2 \nabla^2 \boldsymbol{u} - \nabla p \qquad \nabla \cdot \boldsymbol{u} = 0 \\ \boldsymbol{S}_1 + \lambda_1 (\frac{\partial \boldsymbol{S}_1}{\partial t} + \boldsymbol{u} \cdot \nabla \boldsymbol{S}_1 - \boldsymbol{S}_1 \cdot \nabla \boldsymbol{u} - \nabla \boldsymbol{u}^T \cdot \boldsymbol{S}_1) = \mu_1 (\nabla \boldsymbol{u} + \nabla \boldsymbol{u}^T) \\ \text{in } G \subset R^n, n = 2 \text{ or } 3 \quad t \in (0,T). \end{gathered} \tag{1.35}$$

The Oldroyd-B model is of mixed parabolic-hyperbolic type. The model has particle trajectories as characteristic curves representing the fluid memory. The extra stresses behave hyperbolic which means that they are only determined by past time. For the Oldroyd-B model the boundary conditions are the same as for the Navier-Stokes equations supplemented by the specification of all the stress components representing the fluid memory at the inlet boundary (Joseph, 1990). The governing equations can be written in non-dimensional form using the reference velocity U_0 the reference length L_0 and the reference viscosity μ_0. Applying U_0, L_0 and $\mu_0 U_0 / L_0$ (for the stresses) leads to the Reynolds number and the Weissenberg number $We = \lambda_1 U_0 / L_0$. Viscoelastic models and computational problems have been discussed by Böhme and Rubart (1993).

3D-Oldroyd-B equations. The cartesian components of the three-dimensional Oldroyd-B equations applying decomposed stresses (Eqn. 1.35) in non-dimensional variables velocity $\boldsymbol{u}^T = (u, v, w)$, pressure p and viscoelastic stresses $\tau_1 = \begin{bmatrix} \tau_{xx} & \tau_{xy} & \tau_{xz} \\ \tau_{yx} & \tau_{yy} & \tau_{yz} \\ \tau_{zx} & \tau_{zy} & \tau_{zz} \end{bmatrix} = \begin{bmatrix} \sigma & \tau & \alpha \\ \tau & \gamma & \theta \\ \alpha & \theta & \beta \end{bmatrix}$ read

$$\frac{\partial u}{\partial t} + u\frac{\partial u}{\partial x} + v\frac{\partial u}{\partial y} + w\frac{\partial u}{\partial z} = -\frac{\partial p}{\partial x} + \frac{1}{\mathrm{Re}}(\frac{\mu_2}{\mu_0}(\frac{\partial^2 u}{\partial x^2} + \frac{\partial^2 u}{\partial y^2} + \frac{\partial^2 u}{\partial z^2}) + (\frac{\partial \sigma}{\partial x} + \frac{\partial \tau}{\partial y} + \frac{\partial \alpha}{\partial z}))$$

$$\frac{\partial v}{\partial t} + u\frac{\partial v}{\partial x} + v\frac{\partial v}{\partial y} + w\frac{\partial v}{\partial z} = -\frac{\partial p}{\partial y} + \frac{1}{\mathrm{Re}}(\frac{\mu_2}{\mu_0}(\frac{\partial^2 v}{\partial x^2} + \frac{\partial^2 v}{\partial y^2} + \frac{\partial^2 v}{\partial z^2}) + (\frac{\partial \tau}{\partial x} + \frac{\partial \gamma}{\partial y} + \frac{\partial \theta}{\partial z}))$$

$$\frac{\partial w}{\partial t} + u\frac{\partial w}{\partial x} + v\frac{\partial w}{\partial y} + w\frac{\partial w}{\partial z} = -\frac{\partial p}{\partial z} + \frac{1}{\mathrm{Re}}(\frac{\mu_2}{\mu_0}(\frac{\partial^2 w}{\partial x^2} + \frac{\partial^2 w}{\partial y^2} + \frac{\partial^2 w}{\partial z^2}) + (\frac{\partial \alpha}{\partial x} + \frac{\partial \theta}{\partial y} + \frac{\partial \beta}{\partial z}))$$

$$\frac{\partial u}{\partial x} + \frac{\partial v}{\partial y} + \frac{\partial w}{\partial z} = 0$$

$$\sigma + We(\frac{\partial \sigma}{\partial t} + u\frac{\partial \sigma}{\partial x} + v\frac{\partial \sigma}{\partial y} + w\frac{\partial \sigma}{\partial z} - 2\frac{\partial u}{\partial x}\sigma - 2\frac{\partial u}{\partial y}\tau - 2\frac{\partial u}{\partial z}\alpha) = 2\frac{\mu_1}{\mu_0}\frac{\partial u}{\partial x}$$

$$\gamma + We(\frac{\partial \gamma}{\partial t} + u\frac{\partial \gamma}{\partial x} + v\frac{\partial \gamma}{\partial y} + w\frac{\partial \gamma}{\partial z} - 2\frac{\partial v}{\partial x}\tau - 2\frac{\partial v}{\partial y}\gamma - 2\frac{\partial v}{\partial z}\theta) = 2\frac{\mu_1}{\mu_0}\frac{\partial v}{\partial y}$$

$$\beta + We(\frac{\partial \beta}{\partial t} + u\frac{\partial \beta}{\partial x} + v\frac{\partial \beta}{\partial y} + w\frac{\partial \beta}{\partial z} - 2\frac{\partial w}{\partial x}\alpha - 2\frac{\partial w}{\partial y}\theta - 2\frac{\partial w}{\partial z}\beta) = 2\frac{\mu_1}{\mu_0}\frac{\partial w}{\partial z}$$

$$\tau + We(\frac{\partial \tau}{\partial t} + u\frac{\partial \tau}{\partial x} + v\frac{\partial \tau}{\partial y} + w\frac{\partial \tau}{\partial z} - (\frac{\partial u}{\partial x} + \frac{\partial v}{\partial y})\tau - \frac{\partial v}{\partial x}\sigma - \frac{\partial v}{\partial z}\alpha - \frac{\partial u}{\partial y}\gamma - \frac{\partial u}{\partial z}\theta) = \frac{\mu_1}{\mu_0}(\frac{\partial u}{\partial y} + \frac{\partial v}{\partial x})$$

$$\alpha + We(\frac{\partial \alpha}{\partial t} + u\frac{\partial \alpha}{\partial x} + v\frac{\partial \alpha}{\partial y} + w\frac{\partial \alpha}{\partial z} - (\frac{\partial u}{\partial x} + \frac{\partial w}{\partial z})\alpha - \frac{\partial w}{\partial x}\sigma - \frac{\partial w}{\partial y}\tau - \frac{\partial u}{\partial y}\theta - \frac{\partial u}{\partial z}\beta) = \frac{\mu_1}{\mu_0}(\frac{\partial u}{\partial z} + \frac{\partial w}{\partial x})$$

$$\theta + We(\frac{\partial \theta}{\partial t} + u\frac{\partial \theta}{\partial x} + v\frac{\partial \theta}{\partial y} + w\frac{\partial \theta}{\partial z} - (\frac{\partial v}{\partial y} + \frac{\partial w}{\partial z})\theta - \frac{\partial w}{\partial x}\tau - \frac{\partial v}{\partial x}\alpha - \frac{\partial w}{\partial y}\gamma - \frac{\partial v}{\partial z}\beta) = \frac{\mu_1}{\mu_0}(\frac{\partial v}{\partial z} + \frac{\partial w}{\partial y}).$$

1.6 Microscopic Blood Flow Model

In order to get basic insight into the interactions between the RBCs as the most essential constituent of blood and the plasma the flow of two-phase solid-fluid suspensions is discussed applying a multiphase flow model. In the homogeneous flow model, especially suited for large particle concentrations, the suspension is viewed as an effective continuum Newtonian fluid with effective properties dependent on the local particle volume fraction (Hofer and Perktold,

1997). The flow dynamics are described by the time-dependent Navier-Stokes equations for incompressible fluids and the variation of the particle concentration is governed by a modified transport equation accounting for shear-induced particle migration effects. In the more complex but also more accurate two-fluid model the particle and the fluid phase are modeled separately as two interpenetrating continua which interact with each other through the exchange of momentum (Nunziato, 1983).

Most of the flow mechanical properties of blood as the shear-thinning effect or the viscoelastic flow behavior have already been correlated to its multiphase nature as a suspension of RBCs and the plasma as the carrier fluid. The large size of the particles (7μm) excludes a solely convection-diffusion approach to discribe the motions of the particles and suggests effects of the microstructural particle transport on the suspension flow. Although there is a large amount of experimental evidence for specific multiphase effects occurring in blood in microcirculation, like the formation of a particleless marginal layer (Fahraeus and Lindquist, 1931) or inertia dominated lateral RBC motions (Goldsmith, 1979, Segre and Silberberg, 1962) up to now relatively few numerical studies concering multiphase models have been carried out.

Mathematical model. In the two-fluid model the flow of the dispersed suspension is described by the momentum conservation equations and by the continuity equations for the solid and the fluid phase, respectively, where incompressible properties are assumed both for the particles and the carrier fluid. The momentum conservation equations take the form

$$\begin{aligned}
&\rho_f(1-\Phi_s)\left(\frac{\partial \boldsymbol{u}_f}{\partial t}+(\boldsymbol{u}_f\cdot\nabla)\boldsymbol{u}_f\right)= \\
&-(1-\Phi_s)\nabla p_f+\mu_f\nabla\cdot((1-\Phi_s)(\nabla\boldsymbol{u}_f+\nabla\boldsymbol{u}_f{}^T))-\rho_f(1-\Phi_s)\boldsymbol{g}-\boldsymbol{m}-(\pi-p_f)\nabla\Phi_s \\
&\rho_f\Phi_s\left(\frac{\partial \boldsymbol{u}_s}{\partial t}+(\boldsymbol{u}_s\cdot\nabla)\boldsymbol{u}_s\right)= \\
&-\Phi_s\nabla p_s+\mu_s\nabla\cdot(\Phi_s(\nabla\boldsymbol{u}_s+\nabla\boldsymbol{u}_s{}^T))-\rho_s\Phi_s\boldsymbol{g}+\boldsymbol{m}+(\pi-p_s)\nabla\Phi_s,
\end{aligned} \tag{1.36}$$

where f and s denote the variables of the fluid and the solid phase, respectively, $\boldsymbol{u}_i, i\in\{f,s\}$ is the velocity vector, $\Phi_i, i\in\{f,s\}$ are the local volume fractions $\Phi_f+\Phi_s=1$, $p_i, i\in\{f,s\}$ is the pressure in the phase i and π is the interface pressure at the solid-fluid interface. The interactions between the phases are represented by the momentum transfer term $\boldsymbol{m}$; ρ_f, ρ_s are the constant densities, μ_f, μ_s are apparent viscosities and $\boldsymbol{g}$ is the vector of gravitation and acceleration. The conservation of the total mass is expressed in the continuity equations

$$\begin{aligned}
&\frac{\partial \Phi_f}{\partial t}+(\boldsymbol{u}_f\cdot\nabla)\Phi_f+\Phi_f\nabla\cdot\boldsymbol{u}_f=0 \\
&\frac{\partial \Phi_s}{\partial t}+(\boldsymbol{u}_s\cdot\nabla)\Phi_s+\Phi_s\nabla\cdot\boldsymbol{u}_s=0
\end{aligned} \tag{1.37}$$

In the case of a solid-fluid suspension with rigid spherical particles at low particle concentration the momentum transfer term can be modeled as

$$\boldsymbol{m} = \Phi_s \boldsymbol{f}_D + \Phi_s \boldsymbol{f}_L \tag{1.38}$$

where $\boldsymbol{f}_D$ and $\boldsymbol{f}_L$ are the hydrodynamical drag and lift forces per unit volume acting on a single particle (Hofer, 1998).

2 Navier-Stokes equations and finite element methods

The numerical solution of the system of Navier-Stokes equations involves several typical difficulties: the incompressibility constraint requires implicit solution techniques, dominant nonlinear effects cause stability problems, complicated flow structure requires fine mesh, convection dominant flow (large Reynolds number flow) requires locally refined and anisotropic meshes in boundary layers. Requirements for an efficient flow simulation are the development and usage of fast, flexible and accurate numerical techniques; simulations on appropriate workstations are advantageous.

The most important numerical methods for the approximate solution of partial differential equation problems are the finite difference method, finite element method, finite volume method and spectral method. The discussion here is restricted on the approximation of the Navier-Stokes equations using the finite element method (FEM). Several corresponding techniques were published by different authors (e.g., Cuvelier et al., 1986, Galdi et al., 2000, Girault and Raviart, 1986, Gresho and Sani, 2000, Quarteroni and Valli, 1994, Temam, 1984). However, the development of accurate, stable and efficient algorithms is still an important mathematical research topic. If the Navier-Stokes equations are solved in primitive variable form, the determination of the pressure is part of the solution procedure and is available when the solution is achieved. Investigations show that the accuracy of the primitive variable solutions is very sensitive to the convergence tolerance used for the pressure solver.

One of the significant advantages of the FEM is the high flexibility, and thus, this method is one of the most efficient approaches to deal with irregular shaped boundaries. The approximations for velocity and pressure are determined in finite dimensional subspaces consisting of piecewise polynomial functions. The FEM provides a high computational variability, and is increasingly attractive in Computational Fluid Dynamics (CFD).

2.1 Weak Navier-Stokes Problem

For the numerical solution of the time-dependent Navier-Stokes equations (Eqn. 1.18) with boundary conditions (Eqn. 1.20) and (Eqn. 1.23) and initial condition (Eqn. 1.19) the Galerkin method will be applied. Thereto the equation of motion is multiplied by arbitrary time-independent vector-valued test functions $\boldsymbol{v}$ and the continuity equation by scalar test functions q. Integration over the domain results in

$$\begin{aligned} &\int_G \frac{\partial \boldsymbol{u}}{\partial t} \cdot \boldsymbol{v} dG + \int_G ((\boldsymbol{u} \cdot \nabla)\boldsymbol{u} - \frac{1}{Re} \nabla^2 \boldsymbol{u} + \nabla p) \cdot \boldsymbol{v} dG = \int_G \boldsymbol{g} \cdot \boldsymbol{v} dG \\ &\int_G (\nabla \cdot \boldsymbol{u}) q dG = 0 \end{aligned} \tag{2.1}$$

Applying the Gauss-theorem the second velocity derivatives and the pressure derivatives can be eliminated

$$\int_G \frac{\partial \boldsymbol{u}}{\partial t}\cdot \boldsymbol{v} dG + \int_G ((\boldsymbol{u}\cdot\nabla)\boldsymbol{u}\cdot\boldsymbol{v} + \frac{1}{Re}\nabla\boldsymbol{u}:\nabla\boldsymbol{v})dG - \int_G p\nabla\cdot\boldsymbol{v} dG =$$
$$\int_G \boldsymbol{g}\cdot\boldsymbol{v} dG + \frac{1}{Re}\int_\Gamma (\nabla\boldsymbol{u}\cdot\boldsymbol{n})\cdot\boldsymbol{v} d\Gamma - \int_\Gamma p\boldsymbol{n}\cdot\boldsymbol{v} d\Gamma \qquad (2.2)$$
$$\int_G (\nabla\cdot\boldsymbol{u}) q dG = 0.$$

In Eqn. 2.2 the definitions $\nabla\boldsymbol{u}:\nabla\boldsymbol{v} = \sum_{i,j=1}^{n} \frac{\partial u_i}{\partial x_j}\frac{\partial v_i}{\partial x_j}$ and $(\boldsymbol{u}\cdot\nabla)\boldsymbol{u}\cdot\boldsymbol{v} = \sum_{i.j=1}^{n} u_j \frac{\partial u_i}{\partial x_j} v_i$ are used.

The occurring boundary integral represents the stresses at the boundary

$$\int_\Gamma (-p\boldsymbol{n}\cdot\boldsymbol{v} + \frac{1}{Re}\frac{\partial \boldsymbol{u}}{\partial \boldsymbol{n}}\cdot\boldsymbol{v})d\Gamma = \int_\Gamma \boldsymbol{h}\cdot\boldsymbol{v} d\Gamma = \int_{\Gamma_t} \boldsymbol{h}_\Gamma\cdot\boldsymbol{v} d\Gamma_t \,. \qquad (2.3)$$

If the test functions are $\boldsymbol{v} = 0$ on Γ_u the boundary integral reduces to the part of the boundary Γ_t where the stress vector $\boldsymbol{t}$ is prescribed ($\Gamma = \Gamma_u \cup \Gamma_t, \Gamma_u \cap \Gamma_t = \{0\}$).

For the final weak formulation let $\mathbf{V}_0$ (in 3D: $\mathbf{V}_0 = \mathrm{V}_0\times\mathrm{V}_0\times\mathrm{V}_0$) be the space of vector functions $\boldsymbol{v} = 0$ at the velocity boundary Γ_u and $\mathbf{V}_f$ the space of vector functions satisfying the prescribed velocity conditions at Γ_u. Then the weak formulation reads :
Find $\boldsymbol{u}(t) \in \mathbf{V}_f$ and $p(t) \in \mathrm{Q}$ such that

$$\int_G \frac{\partial \boldsymbol{u}}{\partial t}\cdot \boldsymbol{v} dG + \int_G (\boldsymbol{u}\cdot\nabla\boldsymbol{u}\cdot\boldsymbol{v} + \frac{1}{Re}\nabla\boldsymbol{u}:\nabla\boldsymbol{v})dG - \int_G p\nabla\cdot\boldsymbol{v} dG =$$
$$\int_G \boldsymbol{g}\cdot\boldsymbol{v} dG + \int_{\Gamma_t} \boldsymbol{h}_\Gamma\cdot\boldsymbol{v} d\Gamma_t \qquad (2.4)$$
$$\int_G (\nabla\cdot\boldsymbol{u}) q dG = 0$$

for all $\boldsymbol{v} \in \mathbf{V}_0$ and all $q \in \mathrm{Q}$. For surface traction free outflow $\boldsymbol{h}_\Gamma = 0$ the boundary integral vanishes.

Galerkin method. The weak form will be approximated on finite-dimensional subspaces. The approximation of the velocity $\boldsymbol{u}_h$ and the pressure p_h are choosen as linear combinations

$$\boldsymbol{u}_h(\boldsymbol{x},t) = \boldsymbol{u}_b(\boldsymbol{x},t) + \sum_{i=1}^{N} \boldsymbol{u}_i(t)\varphi_i(\boldsymbol{x})$$
$$p_h(\boldsymbol{x},t) = \sum_{i=1}^{M} p_i(t)\psi_i(\boldsymbol{x}) \qquad (2.5)$$

where $\boldsymbol{u}_b$ fulfills the velocity boundary conditions in a sense that $\boldsymbol{u}_b|_{\Gamma_u} = \boldsymbol{f}$. Insertion of the basis functions $\varphi_i \in V_0$ ($i = 1,.., N$) and $\psi_i \in Q$ ($i = 1,.., M$) as the test functions leads to the system of Galerkin equations

$$\int_G \frac{\partial u_{1h}}{\partial t}\varphi_i dG + \int_G ((u_{1h}\frac{\partial u_{1h}}{\partial x_1} + u_{2h}\frac{\partial u_{1h}}{\partial x_2} + u_{3h}\frac{\partial u_{1h}}{\partial x_3})\varphi_i + \frac{1}{Re}(\nabla u_{1h}\cdot\nabla\varphi_i))dG - \\ - \int_G p_h \frac{\partial \varphi_i}{\partial x_1} dG = \int_G \hat{g}_1\varphi_i dG + \int_{\Gamma_t} h_1\varphi_i ds \qquad i = 1,\ldots,N \tag{2.6}$$

$$\int_G \frac{\partial u_{2h}}{\partial t}\varphi_i dG + \int_G ((u_{1h}\frac{\partial u_{2h}}{\partial x_1} + u_{2h}\frac{\partial u_{2h}}{\partial x_2} + u_{3h}\frac{\partial u_{2h}}{\partial x_3})\varphi_i + \frac{1}{Re}(\nabla u_{2h}\cdot\nabla\varphi_i))dG - \\ - \int_G p_h \frac{\partial \varphi_i}{\partial x_2} dG = \int_G \hat{g}_2\varphi_i dG + \int_{\Gamma_t} h_2\varphi_i ds \qquad i = 1,\ldots,N \tag{2.7}$$

$$\int_G \frac{\partial u_{3h}}{\partial t}\varphi_i dG + \int_G ((u_{1h}\frac{\partial u_{3h}}{\partial x_1} + u_{2h}\frac{\partial u_{3h}}{\partial x_2} + u_{3h}\frac{\partial u_{3h}}{\partial x_3})\varphi_i + \frac{1}{Re}(\nabla u_{3h}\cdot\nabla\varphi_i))dG - \\ - \int_G p_h \frac{\partial \varphi_i}{\partial x_3} dG = \int_G \hat{g}_3\varphi_i dG + \int_{\Gamma_t} h_3\varphi_i ds \qquad i = 1,\ldots,N \tag{2.8}$$

$$\int_G (\frac{\partial u_{1h}}{\partial x_1} + \frac{\partial u_{2h}}{\partial x_2} + \frac{\partial u_{3h}}{\partial x_3})\psi_i dG = 0 \qquad i = 1,\ldots,M\,. \tag{2.9}$$

Prescribed velocity boundary conditions $\boldsymbol{u}_b$ are included in the right hand side G-integral.

2.2 Concept of the Finite Element Method

The basic idea of the finite element method is to subdivide the domain into a finite number of subregions (elements) and to approximate the solution with a piecewise polynomial in the space variables on each element. The elements are typically triangles or quadrilaterals in two-dimensional problems, tetrahedrons or hexahedrons in three dimensions. An essential aspect of the discretisation is the definition of the nodal points and of the set of corresponding basis functions (interpolation functions) $\{\varphi_k\}_{k=1,\ldots,N_u}$ and $\{\psi_k\}_{k=1,\ldots,N_p}$. In the elements the velocity components $u_1(x,y,z)$, $u_2(x,y,z)$, $u_3(x,y,z)$ are expressed as linear combinations of the local interpolation functions $N_i(\xi,\eta,\zeta)$, which are defined on a reference element (master element) with the property

$$N_i(\xi,\eta,\zeta) = \begin{cases} 1 & \text{at the local element node } i \in (1,\ldots,n^e) \\ 0 & \text{at the remaining nodes} \end{cases}$$

$$u_1^e(x,y,z) = \sum_{i=1}^{n^e} u_{1,i}^e N_i(\xi,\eta,\zeta)$$
$$u_2^e(x,y,z) = \sum_{i=1}^{n^e} u_{2,i}^e N_i(\xi,\eta,\zeta) \tag{2.10}$$
$$u_3^e(x,y,z) = \sum_{i=1}^{n^e} u_{3,i}^e N_i(\xi,\eta,\zeta)$$
$$\xi = \xi(x,y,z), \quad \eta = \eta(x,y,z), \quad \zeta = \zeta(x,y,z)$$

where $u_{1,i}^e, u_{2,i}^e, u_{3,i}^e$ $(i = 1,..,n^e)$, denote the element node values of the velocity components.

Isoparametric element representation. The mapping of the master element onto the element *e* of the physical domain uses the co-ordinate transformation

$$x = \sum_{i=1}^{n^e} x_i^e N_i(\xi,\eta,\zeta), \quad y = \sum_{i=1}^{n^e} y_i^e N_i(\xi,\eta,\zeta), \quad z = \sum_{i=1}^{n^e} z_i^e N_i(\xi,\eta,\zeta) \tag{2.11}$$

x_i^e, y_i^e, z_i^e, denote the co-ordinates of the velocity nodes of the element *e* in the physical system. The fact that the co-ordinate transformation of an element uses the same interpolation functions as the approximation of the dependent variables characterizes the so-called isoparametric elements. The terms under the integrals of the Galerkin equations (Eqns. 2.6 to 2.9) are functions of the physical (global) co-ordinates (x,y,z). Applying the transformation (Eqn. 2.11) the occurring element functions φ^e_i (x,y,z) and the derivatives with respect to (x,y,z) have to be expressed in each element in terms of the local co-ordinates (ξ,η,ζ).

The resulting global representation of the velocity field (under consideration of the co-ordinate transformation) is

$$u_1(x,y,z) = \sum_{k=1}^{N_u} u_{1,k}\varphi_k(x,y,z)$$
$$u_2(x,y,z) = \sum_{k=1}^{N_u} u_{2,k}\varphi_k(x,y,z) \tag{2.12}$$
$$u_3(x,y,z) = \sum_{k=1}^{N_u} u_{3,k}\varphi_k(x,y,z)$$

where the functions φ_k are the global shape functions satisfying the properties: φ_k takes the value 1 at the global node *k* and vanishes at all other nodal points; $\varphi_k \neq 0$ only in the elements to which the node *k* belongs (N_u is the global number of velocity nodes). This property leads to the fact that the coefficient matrix of the resulting algebraic equation system is banded. The global functions have to be continuous over the finite element mesh and continuously differentiable in each element (conforming elements). The co-ordinate transformation is explained in detail in the finite element literature (e.g., Cuvelier et al., 1986, Reddy and Gartling, 1994).

Simultaneously to the velocity field the pressure field has to be calculated, and the pressure nodes as interpolation points for the pressure approximation are to be specified (N_p is the global number of pressure nodes and ψ_k are the global pressure shape functions)

$$p(x,y,z) = \sum_{k=1}^{N_p} p_k \psi_k(x,y,z)\,. \tag{2.13}$$

In order that the approximation of the weak Navier-Stokes equations is stable in an asymptotical sense ($h \to 0$) it is crucial that the spaces for the velocity and pressure approximation satisfy a compatibility condition, the so-called “inf-sup” or “divergence-stability” or “Brezzi-Babuska” condition (Brezzi and Fortin, 1991). This condition expresses that the choice of the space for the velocity approximation $\mathbf{V}_h$ and for the pressure approximation Q_h is not completely free, but should undergo a certain restriction condition. Stable finite element discretization can be found for mixed approximation using polynomial functions of different order for $\boldsymbol{u}$ and p. Elements with linear pressure und quadratic velocity interpolation are successful. The corresponding basis functions ψ_k are either discontinuous at the element boundary and non-trivial only in one element, or continuous and the support ranges over more elements. In both cases the approximation functions of velocity and pressure fulfill the requirements of the weak formulation. The problem of spurious pressure modes and of elements satifying the divergence-stability condition have been discussed by Gunzberger (1989).

Structure of the Navier-Stokes finite element equation system. Substituting the finite element test and approximation functions in the Galerkin system (Eqns. 2.6 to 2.9) the resulting system of ordinary differential equations becomes in matrix-vector notation ($\dot{\boldsymbol{u}} := \frac{\partial \boldsymbol{u}}{\partial t}$)

$$\begin{aligned} \mathbf{M}\dot{\boldsymbol{u}} + \mathbf{K1}(\boldsymbol{u})\boldsymbol{u} + \mathbf{K2}\,\boldsymbol{u} - \mathbf{Q}^T p &= \mathbf{g} \\ \mathbf{Q}\boldsymbol{u} &= \mathbf{0} \end{aligned} \tag{2.14}$$

where this semi-discrete finite element equation system for the vector of nodal velocity components u_1, u_2, u_3 and for the nodal pressure p has the form

$$\begin{aligned} \boldsymbol{M}\dot{u}_1 + \boldsymbol{K1}(\boldsymbol{u})u_1 + \boldsymbol{K2}u_1 - \boldsymbol{Q}_{x_1}^T p &= \mathrm{g}_1 \\ \boldsymbol{M}\dot{u}_2 + \boldsymbol{K1}(\boldsymbol{u})u_2 + \boldsymbol{K2}u_2 - \boldsymbol{Q}_{x_2}^T p &= \mathrm{g}_2 \\ \boldsymbol{M}\dot{u}_3 + \boldsymbol{K1}(\boldsymbol{u})u_3 + \boldsymbol{K2}u_3 - \boldsymbol{Q}_{x_3}^T p &= \mathrm{g}_3 \\ \boldsymbol{Q}\boldsymbol{u} &= 0 \end{aligned} \tag{2.15}$$

with initial conditions $\boldsymbol{u}(0) = \boldsymbol{u}_0$.

Definition of the mass matrix $\boldsymbol{M}$, the convection matrix $\boldsymbol{K1}$, the diffusion matrix $\boldsymbol{K2}$ and the gradient (divergence) matrix $\boldsymbol{Q}_{x_k}^T$ $(\boldsymbol{Q}_{x_k})$, k=1,2,3:

$$
\begin{aligned}
&M(i,j)=\int_G \varphi_i\varphi_j dG\\
&K1(i,j)=\int_G \varphi_i \boldsymbol{u}\cdot\nabla\varphi_j dG\\
&K2(i,j)=\frac{1}{Re}\int_G \nabla\varphi_i\cdot\nabla\varphi_j dG\\
&Q_{x_k}^T(i,j)=\int_G \frac{\partial\varphi_i}{\partial x_k}\psi_j dG = Q_{x_k}(j,i)\\
&\mathrm{g}_k(i)=\int_G \hat{g}_k\varphi_i dG+\int_{\Gamma_t} h_k\varphi_i ds \qquad k=1,2,3.
\end{aligned}
\tag{2.16}
$$

For detailed description of the numerical equations see Cuvelier et al. (1986).

Time discretization. In addition to the discretization with respect to the three space variables the time discretization of the weak problem has to be performed. Due to the fact that the incompressibility constraint should be satisfied at any time the application of explicit time integration methods are unsuitable. An approach often applied uses time-advancing by finite differences

$$
\begin{aligned}
&\mathbf{M}\frac{\boldsymbol{u}^{n+1}-\boldsymbol{u}^n}{\Delta t}+\Theta\mathbf{K1}(\boldsymbol{u}^{n+1})\boldsymbol{u}^{n+1}+(1-\Theta)\mathbf{K1}(\boldsymbol{u}^n)\boldsymbol{u}^n+\Theta\mathbf{K2}\boldsymbol{u}^{n+1}+(1-\Theta)\mathbf{K2}\boldsymbol{u}^n-\\
&-\Theta\mathbf{Q}^T p^{n+1}-(1-\Theta)\mathbf{Q}^T p^n=\Theta\mathbf{g}^{n+1}+(1-\Theta)\mathbf{g}^n\\
&\mathbf{Q}\boldsymbol{u}^{n+1}=\mathbf{0}
\end{aligned}
\tag{2.17}
$$

with Δt is the time step, Θ $(0\le\Theta\le 1)$ denotes the parameter of the Θ-method, $\boldsymbol{u}^{n+1}$, p^{n+1} are the velocity and the pressure at the time level n+1, $\boldsymbol{u}^n$, p^n are the known velocity and the known pressure at the preceding time level n. The Θ-method is second order accurate with respect to the time step if Θ = ½ (Crank-Nicolson scheme), while it is only first order accurate for all other values of Θ (Θ = 0: explicit (forward Euler) scheme, Θ = 1: implicit (backward Euler) scheme).

Linearization techniques. The non-linearity of Eqn. 2.17 requires a linearization technique. Commonly applied are the Picard method or the Newton method. The Picard linearization uses known velocities at the preceding iteration step in the convection matrix.

$$
\begin{aligned}
&\mathbf{M}\boldsymbol{u}^{n+1,m+1}+\Theta\Delta t\mathbf{K1}(\boldsymbol{u}^{n+1,m})\boldsymbol{u}^{n+1,m+1}+\Theta\Delta t\mathbf{K2}\boldsymbol{u}^{n+1,m+1}-\Theta\Delta t\mathbf{Q}^T p^{n+1,m+1}=\\
&\quad=\mathbf{M}\boldsymbol{u}^{n,\overline{m}}-(1-\Theta)\Delta t\mathbf{K1}(\boldsymbol{u}^{n,\overline{m}})\boldsymbol{u}^{n,\overline{m}}-(1-\Theta)\Delta t\mathbf{K2}\boldsymbol{u}^{n,\overline{m}}+\\
&\quad+(1-\Theta)\Delta t\mathbf{Q}^T p^{n,\overline{m}}+\Theta\Delta t\mathbf{g}^{n+1}+(1-\Theta)\Delta t\mathbf{g}^n\\
&\mathbf{Q}\boldsymbol{u}^{n+1,m+1}=\mathbf{0} \qquad m=0,\dots,\overline{m}\text{-}1, \quad \boldsymbol{u}^{n+1,0}=\boldsymbol{u}^{n,\overline{m}}
\end{aligned}
\tag{2.18}
$$

Upwind stabilization. A successful application of the finite element method for the approximation of the Navier-Stokes problem often requires stabilization of the numerical scheme. This aspect is of importance for high Reynolds number flow in complex flow structures, where locally high gradients and flow disturbances as separation and local recirculation occur. To avoid oscillations in the numerical solution the streamlinie-upwind-Petrov-Galerkin method (Brooks and Hughes, 1982) can be applied. This method uses modified velocity-dependent test functions in the weighted residual formulation

$$\widetilde{\varphi} = \varphi + \widetilde{k}_{SUPG} \frac{\boldsymbol{u} \cdot \nabla \varphi}{\|\boldsymbol{u}\|^2}, \tag{2.19}$$

where φ is the Galerkin test function and $\widetilde{k}_{SUPG}$ is the upwind parameter which controls the amount of streamline-upwind weighting. The corresponding modification of the Galerkin equations (Eqns. 2.6–2.8) where the test functions φ are replaced by $\widetilde{\varphi}$ preserve the constistency of the scheme. The summation of the upwind terms in the modified equations results in an integral expression which contains the residual of the original equation.

2.3 Projection Method

The solution of the numerical Navier-Stokes system requires great computational efforts, particularly in three-dimensional problems, and thus, further extensive research was applied to develop more effective solution procedures. An approach particularly attractive for computing time-dependent incompressible flow uses projection schemes based on the Helmholtz decomposition principle of a vector field. According to this theorem any sufficiently regular vector function ($\boldsymbol{u} \in \boldsymbol{L}^2(G)$) can be uniquely represented as $\boldsymbol{u} = \boldsymbol{v} + \nabla \phi$ with $\boldsymbol{v} \in \boldsymbol{L}^2(G), \nabla \cdot \boldsymbol{v} = 0, \boldsymbol{v} \cdot \boldsymbol{n} = 0$ on the boundary Γ and $\phi \in H^1(G)$. The decomposition which defines an orthogonal projection can be applied to the weak formulation of the incompressible Navier-Stokes equations in appropriate spaces (Girault and Raviart, 1986).

Based on this decomposition projection schemes for the incompressible Navier-Stokes equations can be developed (Chorin, 1968, Temam, 1984). The technique results in an sequence of a convection-diffusion problem computing an intermediate non-divergence free velocity field, a Poisson-problem computing the pressure (or pressure correction), and the projection of the intermediate velocity on the divergence free functional space. The operator splitting technique yields a velocity-pressure correction method in which the occurring variables velocity components and pressure are uncoupled. The method results in a Burgers-step for the intermediate velocity field with explicit (known) pressure term. The method was first proposed by Chorin (1968) using finite differences. Later the technique was adapted to the finite element discretization by several authors (e.g., Donea et al., 1981, Gresho et al., 1984, Perktold, 1987). The method is demonstrated here for non-Newtonian inelastic flow and vanishing external forces. Applying backward Euler discretization for the time derivatives the modified equation of motion reads

$$\frac{\rho}{\Delta t}(\tilde{\boldsymbol{u}}^{n+1} - \tilde{\boldsymbol{u}}^{n}) = \hat{\boldsymbol{g}}^{n+1} - \nabla p^{n}$$
$$\tilde{\boldsymbol{u}}^{n+1} = \boldsymbol{u}^{n+1} \quad \text{on } \Gamma_u, \tag{2.20}$$

where the diffusion and convection are contained in the vector $\hat{\boldsymbol{g}}^{n+1}$ at the new time level t^{n+1}; $\Delta t = t^{n+1} - t^{n}$ denotes the time step. The corresponding projection-step is

$$\boldsymbol{u}^{n+1} = P_H \tilde{\boldsymbol{u}}^{n+1} \tag{2.21}$$

In the solution procedure Eqn. 2.21 is replaced using equations for the pressure correction and the velocity correction. At each time step Picard iteration is applied to linearize the non-linear convection and also the diffusion terms in case of non-Newtonian inelastic fluid.

Projection scheme, algorithm. Applying the projection technique to non-Newtonian inelastic flow the procedure at the time step t^{n+1} reads:

1. Calculate an intermediate velocity vector field $\tilde{\boldsymbol{u}}^{n+1,m+1}$ from the linearized system

$$(\mathbf{M} + \Delta t\mathbf{K1}(\boldsymbol{u}^{n+1,m}) + \Delta t\mathbf{K2}(\mu^{n+1,m}))\tilde{\boldsymbol{u}}^{n+1,m+1} = \mathbf{M}\boldsymbol{u}^{n,m} + \Delta t\mathbf{Q}^{T} p^{n+1,m} + \Delta t\mathbf{g}_{\boldsymbol{u},p}^{n+1} \tag{2.22}$$

$\tilde{\boldsymbol{u}}^{n+1,m+1} = \boldsymbol{u}^{n+1}$ at the velocity boundary Γ_u

$\tilde{\boldsymbol{u}}^{n+1,m+1}$ is the vector containing all unspecified velocity node values at the new time level t^{n+1} ($n = 0,..,\bar{n}-1$) at the iteration step m+1 ($m = 0,..,\bar{m}-1$).

This step uses the non-Newtonian viscosity $\mu^{n+1,m} = \mu(D_{II}^{n+1,m})$.

2. Calculate the pressure correction $q^{n+1,m+1}$:

$$\overline{\mathbf{Q}}\mathbf{M}_d^{-1}\overline{\mathbf{Q}}^{T} q^{n+1,m+1} = -\frac{1}{\Delta t}\mathbf{Q}\tilde{\boldsymbol{u}}^{n+1,m+1} \tag{2.23}$$

3. Calculate the divergence-free velocity field $\boldsymbol{u}^{n+1,m+1}$:

$$\boldsymbol{u}^{n+1,m+1} = \tilde{\boldsymbol{u}}^{n+1,m+1} + \Delta t\mathbf{M}_d^{-1}\overline{\mathbf{Q}}^{T} q^{n+1,m+1} \tag{2.24}$$

4. Calculate the updated pressure $p^{n+1,m+1}$:

$$p^{n+1,m+1} = p^{n+1,m} + q^{n+1,m+1} \tag{2.25}$$

In the Eqns. 2.22 to 2.24 for the velocities and the pressure correction $\mathbf{M}$ and $\mathbf{M}_d$ are the consistent and the concentrated mass matrix, $\mathbf{K1}(\boldsymbol{u}^{n+1,m})$ is the linearized convection matrix, $\mathbf{K2}(\mu^{n+1,m})$ is the linearized diffusion matrix, $\mathbf{Q}^T$ and $\mathbf{Q}$ are the gradient and the divergence matrix. $\overline{\mathbf{Q}}^T$ and $\overline{\mathbf{Q}}$ are modified with respect to the velocity boundary conditions, where only the normal component of the velocity is considered. The vector $\mathbf{g}_{u,p}^{n+1}$ contains the prescribed boundary data at the new time level. The validation of this projection scheme has been performed in various Newtonian and non-Newtonian calculations (Perktold et al., 1991, Perktold and Rappitsch, 1994, Leuprecht and Perktold, 2001). The convergence criterion of the Navier-Stokes iteration scheme uses the difference in the nodal values of the velocity components and the pressure successive in the iteration procedure. To verify the grid independency of the solution, a finer grid is empolyed until no observable discrepancy is found in the solution between the two grid densities.

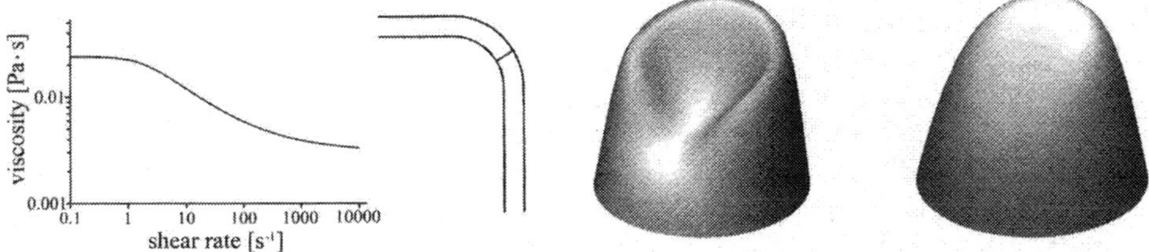

Figure2: Three-dimensional velocity profiles in a curved tube, comparision of shear-thinning flow (left) and Newtonian flow (right).

Time-dependent flow domain. Applying the Reynolds transport theorem

$$\frac{d}{dt}\int_{G(t)} \psi dG = \int_{G(t)} \frac{\partial \psi}{\partial t} dG + \int_{\Gamma(t)} \hat{\boldsymbol{u}} \cdot \boldsymbol{n} \psi d\Gamma \tag{2.26}$$

where ψ is a differentiable function defined in a time-dependent domain $G(t)$ and $\hat{\boldsymbol{u}}$ is the velocity of the moving boundary, the transformation of Eqn. 1.28 into an appropriate variational form leads to the basic formulation for the Galerkin finite element approximation of the Navier-Stokes problem in a time-dependent flow domain

$$\begin{aligned} &\frac{d}{dt}\int_{G(t)} u_i \varphi \, dG + \int_{G(t)} ((u_j - \hat{u}_j)\frac{\partial u_i}{\partial x_j}) \varphi \, dG - \int_{G(t)} \frac{\partial \hat{u}_j}{\partial x_j} u_i \varphi \, dG - \\ &- \int_{G(t)} \tilde{p} \frac{\partial \varphi}{\partial x_i} dG + \int_{G(t)} \nu (\frac{\partial u_i}{\partial x_j} + \frac{\partial u_j}{\partial x_i}) \frac{\partial \varphi}{\partial x_j} dG = \\ &= \int_{\Gamma_t} (-\tilde{p}\delta_{ij} + \nu(\frac{\partial u_i}{\partial x_j} + \frac{\partial u_j}{\partial x_i})) n_j \varphi \, d\Gamma_t \end{aligned} \tag{2.27}$$

$$\int_{G(t)} \frac{\partial u_j}{\partial x_j} \psi \, dG = 0 \tag{2.28}$$

A detailed description and a stability analysis for the ALE technique using finite elements have been published by Formaggia and Nobile, (1999).

2.4 Vessel Wall Model

The arterial wall is a non-homogeneous anisotropic structure with non-linear history-dependent finite deformation responses to dynamic loads. The stretch of the artery at a pressure load depends on its geometry and material properties. Due to the relatively small thickness of the artery wall it can be modeled properly as a shell structure, which is characterized by two curved surface coordinates. The essential feature in shell theory is an appropriate transformation of the three-dimensional equations into a system of two-dimensional equations and the integration over the shell thickness (Koiter and Simmonds, 1973). The behavior of the shell structure is governed by the behavior of the middle surface on which the two-dimensional equations are defined. The mathematics of shells uses tensor notation, and the mathematical formalism is complicated. Assuming the physical characteristics of the wall material, the distribution of the applied forces and the boundary conditions, the shell problem is to determine the deformation consisting of stretching and changing of curvature of the middle surface. From the deformation the strains and the stresses occurring throughout the shell space can be calculated.

The equilibrium equations representing the total potential energy V (consisting of the potential energy due to loading and of the strain energy function U) must be satisfied with respect to the deformed geometry:

$$V = U - \sum_{S_P} u_i P_i \qquad U = \frac{1}{2} \sum_{i=1}^{n} \sum_{j=1}^{n} k_{ij} u_i u_j, \tag{2.29}$$

where S_P denotes the set of the points of external loading. The displacements at the points of loading are $u_i, i = 1,...,n$, the stiffness influence coefficients are $k_{ij}, i, j = 1,...,n$. Further explanations can be found in Fung (1965). The strain energy in the variational formulation is approximated by a middle surface integral. The integrand consists of a quadratic expression in the components of the middle surface strain tensor and in the tensor of curvature changes. The principle of minimum potential energy is applied in the numerical solution with respect to the finite element subspace. Due to the expected relatively large deformations the calculation uses geometrically non-linear shell theory.

The calculation of the wall deformation uses the commercial finite element program package ABAQUS, where the selected element type for the shell model is S4R. This is a four node doubly curved, shear flexible shell element with reduced integration and hourglass control using five degrees of freedom per node: three displacement components and two in-surface rotation components.

Coupling. The computational domain consists of the artery lumen and the artery wall. The continuity of the velocity and the no-slip property form the interface condition between lumen and wall. More specifically, the flow field and the pressure are calculated in the lumenal region with accounting for the wall velocity as boundary condition. Then the shell equations are solved while accounting for the pulse pressure and the pressure from the Navier-Stokes calculation. From the resulting wall deformation the new shape of the flow domain and the wall velocity are calculated. This procedure results in a weak coupling of fluid and structure, where the shell equations are solved once at a time step. Update the flow domain and finite element mesh is an essential part of the coupling procedure. Fluid structure coupling analysis with application to hemodynamics has been studied by Formaggia et al. (2001), Nobile (2001), Quarteroni and Formaggia (2002).

3 Mass Transport Processes in Arteries

The investigation of arterial mass transport processes is important due to the known relationship to the initiation and development of atherosclerotic diseases. The local mass transfer between the blood and the arterial wall affects the transport of nutrients to the cells, the removal of metabolic wastes from the wall, and the accumulation of potentially atherogenic molecules (Fry, 1987). It has been observed that low density lipoprotein (LDL) accumulation in the intima at zones of low and oscillating wall shear stress is associated with the tendency to intimal thickening (Caro et al., 1971).

The dynamics of dissolved gases (e.g., oxygen or carbon dioxide) and of macromolecules (e.g., lipoprotein or albumin) in arteries and in the artery wall is strongly related to the flow dynamics of blood. In several numerical studies irregular blood flow patterns (flow separation and recirculation, low and oscillating wall shear stress), and highly disturbed mass transfer have been described (e.g., Ma et al., 1997, Rappitsch et al., 1997, Rappitsch and Perktold, 1996).

3.1 Mass Transport in the Arterial Lumen

The basis of the mathematical description of arterial mass transport processes is the transient convection-diffusion-reaction-equation

$$\frac{\partial c}{\partial t} + \boldsymbol{u} \cdot \nabla c - \nabla \cdot (D \nabla c) - Q = 0 \quad \text{in} \quad G_l \subset R^n, n = 2 \text{ or } 3, \quad t \in (t_0, T) \tag{3.1}$$

where c denotes the concentration of the substance of interest in the arterial lumen G_l, and Q is a term describing metabolic processes (e.g., 1st order reaction, $Q(c) = kc$, for irreversible chemical reaction). The vector $\boldsymbol{u}$ denotes the velocity of the blood which couples primarily the transport problem to the Navier-Stokes problem in the domain G_l. A further coupling of the concentration field to the flow field can occur through the diffusivity tensor (or diffusivity coefficient) D, if it is a function of the deformations.

For constant diffusion coefficient of the species in plasma (using the normalizations Eqn. 1.26 and $c^* = c/C_0$ with respect to the reference bulk concentration at the entrance C_0) the transport equation (Eqn. 3.1) can be transformed into the dimensionless form

$$\frac{\partial c^*}{\partial t^*} + \boldsymbol{u}^* \cdot \nabla c^* - \frac{1}{Pe} \Delta c^* - Q^* = 0, \tag{3.2}$$

where Pe is the Peclet number, defined as

$$Pe = \frac{U_0 L_0}{D} . \tag{3.3}$$

The parameter Pe characterizes the transport by relating the convective transport to the diffusion. Mass transport processes in medium-sized and large arteries are generally highly convection dominated which is reflected in very large Peclet numbers.

The description of the mass transport in a time-dependent flow domain requires an ALE-modification of the transport equation. The dimensionless convection-diffusion-reaction equation in the ALE formulation can be written as

$$\frac{\partial c^*}{\partial t^*}+(\boldsymbol{u}^*-\hat{\boldsymbol{u}}^*)\cdot\nabla c^*-\frac{1}{Pe}\Delta c^*-Q^*=0 \qquad (3.4)$$

where $\hat{\boldsymbol{u}}^*$ is the domain velocity of the time-dependent flow domain.

Initial and boundary conditions. The solution of the time-dependent mass transport problem requires the prescription of an appropriate condition at an initial time t_0, e.g., $t_0 = 0$

$$c(\boldsymbol{x},0)=c_0(\boldsymbol{x}) \quad \text{on} \quad G_l\,. \qquad (3.5)$$

The choice of appropriate boundary conditions is essential for the simulation of various mass transport processes. At the inflow cross-section Γ_{in} in most cases a constant concentration profile is prescribed as Dirichlet boundary condition

$$c=C_0, \quad \text{on} \quad \Gamma_{in}, \quad t\in(0,T)\,. \qquad (3.6)$$

At the outflow boundary zero diffusive flux can be assumed

$$\frac{\partial c}{\partial \boldsymbol{n}}=0, \quad \text{on} \quad \Gamma_{out}, \quad t\in(0,T)\,. \qquad (3.7)$$

The boundary condition at the inner surface of the artery wall (lumen-endothelium interface) depends on the considered molecules. The transfer of dissolved gases to and into the wall is diffusion boundary layer controlled, due to the fact that the endothelium is not an essential barrier to these relatively small molecules. Therefore, the assumption of a concentration boundary condition at the inner surface is justified. In the case of arterial oxygen transport Back et al. (1977) suggests a wall concentration for the free oxygen in plasma of

$$c=C_0/3, \quad \text{on} \quad \Gamma_w, \quad t\in(0,T)\,. \qquad (3.8)$$

The main resistance to the transfer of macromolecules from blood into the artery wall is the endothelial layer. The flux across the endothelium into the inner layers of the arterial wall is determined by the endothelial permeability and by the concentration differential across the layer. Therefore, the permeability boundary condition at the blood-endothelium surface (which is of Neumann type)

$$\boldsymbol{u}_{filt}\cdot\boldsymbol{n}c-D\frac{\partial c}{\partial \boldsymbol{n}}=P(c-c_i), \quad \text{on} \quad \Gamma_w, \quad t\in(0,T)\,, \qquad (3.9)$$

is applied to the macromolecule transport model, where P is the endothelial permeability, c_i is the concentration in the subendothelial intima, and $\boldsymbol{u}_{filt}$ is the filtration velocity of plasma at the lumenal surface. The model requires prescribed data for the subendothelial concentration c_i. Expressing the fact that the endothelium is not a passive barrier to macromolecules the permeability depends on the local shear stress at the endothelium $P = P(|\tau_w|)$ which is determined as $\tau_w = \boldsymbol{t} \cdot \boldsymbol{T} \cdot \boldsymbol{n}$, where $\boldsymbol{T}$ is the local viscous stress tensor, $\boldsymbol{n}$ and $\boldsymbol{t}$ are the normal and the tangential unit vectors at the wall, respectively. Eqn. 3.9 is a mass balance of macromolecules at the blood-wall boundary, and it states that the amount of macromolecules infiltrating into the vessel wall is the difference between the amount carried to the blood-endothelium boundary by filtration flow and the amount which diffuses back to the mainstream.

Numerical method. The convection-diffusion equation is solved numerically applying a streamline-upwind-Petrov-Galerkin (SUPG) finite element method and the technique of 'discontinuity-capturing' which circumvent spurious oscillations typical for convection dominated (high Peclet number) transport (Brooks and Hughes, 1982, Hughes et al., 1986).

The SUPG method uses modified velocity-dependent test functions in the weighted residual formulation. In order to preclude overshooting around sharp layers an additional stabilization, the 'discontinuity-capturing' term which acts in the direction of the solution gradient is included in the test functions. The resulting modified test functions are

$$\tilde{\varphi} = \varphi + \tilde{k}_{SUPG} \frac{\boldsymbol{u} \cdot \nabla \varphi}{\|\boldsymbol{u}\|^2} + \tilde{k}_{DC} \frac{\boldsymbol{u}_{\|} \cdot \nabla \varphi}{\|\boldsymbol{u}_{\|}\|^2}, \tag{3.10}$$

where φ is the Galerkin test function and $\tilde{k}_{SUPG}$, $\tilde{k}_{DC}$ are parameters which control the amount of streamline-upwind weighting and discontinuity capturing. The velocity $\boldsymbol{u}_{\|}$ is the projection of $\boldsymbol{u}$ onto the direction of the concentration gradient vector. The resulting method is non-linear due to the dependence of $\boldsymbol{u}_{\|}$ on the concentration. In the iterative solution procedure $\boldsymbol{u}_{\|}$ is calculated using the concentration of the previous iteration-step

$$\boldsymbol{u}_{\|} = \begin{cases} \dfrac{(\boldsymbol{u} \cdot \nabla c^i)}{\|\nabla c^i\|^2} \nabla c^i & (\nabla c^i \neq \boldsymbol{0}) \\ 0 & (\nabla c^i = \boldsymbol{0}) \end{cases}. \tag{3.11}$$

3.2 Transport in the Lumen and in the Artery Wall

The transport model as described in section 3.1 is wall free in the sense that the solute transport was considered only in the artery lumen, where appropriate boundary conditions are prescribed. An improved model considers also the artery wall. The domain is composed by the artery lumen G_l and the vessel wall G_w. The endothelial surface Γ_w is the interface between the two domains. In this improved model the subendothelial concentration c_i (occurring in Eqn. 3.9) representing the wall concentration c_w is an additional unknown. The concentration fields in the lumen c_l and in the wall c_w are directly coupled, while the velocity field $\boldsymbol{u}_l$ in the lumen (Navier-Stokes solution) and in the wall $\boldsymbol{u}_w$ (Darcy velocity, filtration velocity) are coupled by the influence of the pressure at the wall on the filtration velocity, which is the velocity boundary condition at the blood-wall surface in the Navier-Stokes calculation.

The solute transport in the fluid-wall model for the concentration field in the lumen c_l and for the concentration field in the wall c_w can be expressed as

$$\frac{\partial c_l}{\partial t} + \boldsymbol{u}_l \cdot \nabla c_l - \nabla \cdot (D_l \nabla c_l) - Q_l = 0 \quad \text{in} \quad G_l \subset R^n, n = 2 \text{ or } 3, \quad t \in (0,T) \tag{3.12}$$

$$c_l = C_{l0}, \quad \text{on} \quad \Gamma_{l,in}, \quad t \in (0,T) \tag{3.13}$$

$$\frac{\partial c_l}{\partial \boldsymbol{n}_l} = 0, \quad \text{on} \quad \Gamma_{l,out}, \quad t \in (0,T) \tag{3.14}$$

$$\frac{\partial c_w}{\partial t} + \boldsymbol{u}_w \cdot \nabla c_w - \nabla \cdot (D_w \nabla c_w) - Q_w = 0 \quad \text{in} \quad G_w \subset R^n, n = 2 \text{ or } 3, \quad t \in (0,T) \tag{3.15}$$

$$\frac{\partial c_w}{\partial \boldsymbol{n}_w} = 0, \quad \text{on} \quad \Gamma_{w,out}, \quad t \in (0,T) \tag{3.16}$$

where $\boldsymbol{u}_l$ and $\boldsymbol{u}_w$ are the convective velocity in the lumen and in the wall, and $\Gamma_{w,out}$ is the outer wall surface (media-adventitia boundary).

Appropriate matching conditions at the interface Γ_w uses the continuity of the convective-diffusive flux

$$\boldsymbol{u}_{filt,l} \cdot \boldsymbol{n}_l c_l - D_l \frac{\partial c_l}{\partial \boldsymbol{n}_l} = P(c_l - c_w) \quad on \quad \Gamma_w \tag{3.17}$$

$$\boldsymbol{u}_{filt,w} \cdot \boldsymbol{n}_w c_w - D_w \frac{\partial c_w}{\partial \boldsymbol{n}_w} = P(c_w - c_l) \quad on \quad \Gamma_w, \tag{3.18}$$

where $\boldsymbol{n}_l$ is the outward normal unit vector at the interface boundary with respect to G_l ; the vector with respect to G_w, $\boldsymbol{n}_w = -\boldsymbol{n}_l$. P is the permeability of the endothelium to the macromolecules considered (e.g., LDL). The terms on the left hand side of Eqn. 3.17 represent the transport to the vessel wall by transmural filtration and the diffusion back into the lumen, the right hand side expresses the rate of molecules passing into the wall. Condition Eqn. 3.17 is formally similar to condition Eqn. 3.9, while in Eqn. 3.9 the subendothelial concentration is known, the wall concentration c_w in Eqn. 3.17 is unknown. Condition Eqn. 3.18 expresses that at the lumen-wall interface the influx of macromolecules from the lumen must be equal to the convective-diffusive flux on the intimal side of the interface. c_l is the concentration of the molecules in the plasma.

Filtration velocities in the porous artery wall. The mathematical description of the pressure driven flow velocity (filtration velocity) within the porous artery wall applies Darcy's law

$$\boldsymbol{u}_{filt} = -\frac{K_p}{\mu_V} \nabla p \qquad \nabla \cdot \boldsymbol{u}_{filt} = 0, \tag{3.19}$$

where K_p denotes the Darcy permeability which results from the structure of the porous medium, and μ_V is the constant viscosity of the fluid (plasma). The ability of Darcy's law to describe very slow filtration flow has been demonstrated in experimental studies. Furthermore, theoretical studies showed that Darcy's law can be directly derived from the Navier-Stokes equations in the case of very slow flow applying a volume-averaging procedure within the porous medium (Shyy et al., 1997). The Darcy permeability is determined from an appropriate fiber matrix model of the particular wall layer (Curry, 1984, Huang et al., 1992, Huang and Tarbell, 1997).

Due to the incompressibility of the fluid the pressure within the porous medium must meet the Laplacean equation

$$\nabla^2 p = 0 . \tag{3.20}$$

The pressure values at the blood-wall boundary and at the media-adventitia boundary (which is seen as the outer wall surface) are the boundary conditions for the Laplacean equation. The pressure at the media-adventitia surface can be prescribed, while the pressure at the blood-wall surface couples the Navier-Stokes problem in the lumen and the Darcy problem in the wall.

Fluid-wall coupling. In the first step the flow velocity in the lumen is calculated applying constant prescribed filtration velocity at the blood-wall boundary. The filtration velocity is updated solving the boundary value problem for the pressure and applying Darcy's law (Eqn. 3.19). With the new filtration velocity as boundary condition at the wall the Navier-Stokes equations are solved again.

The concentration field is first calculated in the lumen with prescribed subendothelial concentration. The second step is the calculation of the concentration field within the wall applying the concentration field in the lumen through the boundary condition at the lumen-wall interface. In the next step the concentration in the lumen is updated with the wall concentration (subendothelial concentration). The iterative process is applied until convergence is fulfilled. Coupling analysis and properly posedness studies have been carried out by Quarteroni et al. (2002).

3.3 LDL Lumenal Surface Concentration Polarization

The concentration polarization effect of LDL at the blood-endothelium boundary and the influence of flow separation have been studied in an idealized axisymmetric stenosed tube with 75% area occlusion. The essential LDL transport parameters are the diffusivity in plasma D = $2.87 \cdot 10^{-11}$ $m^2 \cdot s^{-1}$, the endothelial permeability to LDL $P_{end} = 2.2 \cdot 10^{-10}$ $m \cdot s^{-1}$, the Peclet number $Pe = 3.6 \cdot 10^7$, the Reynolds number $Re = 300$, the filtration velocity $u_{filt} = 4 \cdot 10^{-8}$ $m \cdot s^{-1}$.

The development of lumenal surface concentration and the effect of flow separation on the local mass transfer to and into the wall in the downstream region of the stenosis is demonstrated in Figure 3. The figure demonstrates LDL accumulation near the tube wall. The equilibrium concentration at the fluid-endothelium boundary is higher than the concentration in the bulk stream. This concentration polarization effect at the surface occurs due to the plasma-permeability of the wall. The figure shows the steep concentration gradient across the separating streamline and a concentration increase of about two percent in the separation region compared to the bulk concentration. At the wall region directly downstream the stenosis a decrease of surface concentration polarization of about 35 percent compared to the uniform tube can be observed (Figure 3-A). The results on LDL polarization at the wall of a stenosed tube correspond qualitatively to the results published by Wada and Karino (2000).

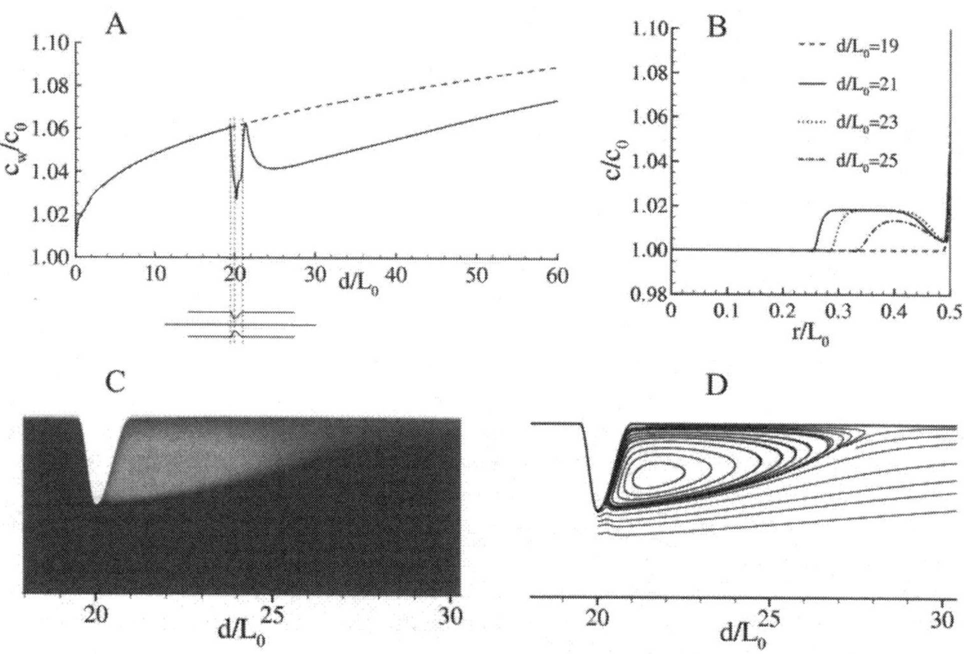

Figure 3: (A) Lumenal surface LDL concentration development over the tube length, (B) Concentration profiles downstream the stenosis, (C) Concentration contours, (D) Streamlines in the separation zone.

The results of the mass transport study support the conclusion that the filtration process at the wall causes a lumenal surface concentration polarization of LDL which is a dominant effect of mass transport processes including plasma-permeable walls. Flow separation downstream the axisymmetric stenosis influences the concentration boundary layer resulting in a decrease of surface concentration. The equilibrium concentration at the lumenal surface depends on the local convective transport near the surface in the lumen, on the filtration velocity of plasma into the wall and on the diffusive processes of the molecules.

3.4 Concentration of Platelet Active Substances in the Vicinity of Mural Microthrombi

Platelet thrombi usually grow over the surface of an implanted artificial organ, at sites of intravascular injury and at the positions of atherosclerotic lesions. The first stage in thrombogenesis is the initiation stage, where platelets adhere to the artificial surface or the injured blood vessel wall and form a mural aggregate. This process is governed by the transport of platelets to the vicinity of the surface by convection and reaction of the platelets with the surface. The onset of the initiation stage depends on the surface and on the conditions in the blood stream. Platelets do not adhere to undamaged endothelium but adhere preferentially to sites of injured blood vessel wall and to implanted artificial surfaces.

In the vicinity of the mural aggregate a region of high concentration of thrombotic substances is generated. This leads to the second stage in thrombogenesis: the aggregation stage, in which other platelets flowing nearby the mural aggregate become activated by the liberated substances. These platelets react with the already adherent platelets, forming a mural thrombus. The level of the local concentrations of the thrombotic substances depends on the rate of generation and on the diffusive and convective transport. The most significant thrombotic substances which are able to activate platelets are thrombin, adenosine diphosphate (ADP) and thromboxane A_2 (TxA_2).

In the study by Karner and Perktold (1998) a three-dimensional model is used to analyse the influence of the flow field on the concentration of thrombogenic substances. The mural microthrombus is modeled as a semisphere attached to a plane surface. For each of the substances of interest the stationary concentration field is calculated for different thrombus diameters and for different inlet flow profiles representing physiologically relevant wall shear rates.

Mathematical model. The mathematical description of blood flow uses the three-dimensional incompressible Navier-Stokes equations for Newtonian fluids. The flow domain is chosen as a cuboid with the artery wall as its base. The microthrombus, idealized as a semisphere (diameter *d*) is attached to the base of the cuboid. In the scale of microthrombi the approximation of the artery wall as a plane surface and the assumption of the near wall inlet velocity as linear function $u(\boldsymbol{x}) = \gamma y$ ($v(\boldsymbol{x}) = w(\boldsymbol{x}) = 0$,) are acceptable. γ denotes the shear rate and y is the normal distance from the artery wall.

The mass transport and reaction processes of platelet active substances are modeled by a system of coupled stationary convection-diffusion-reaction equations published by McIntire and Tay (1989)

Prothrombin (A): $$\boldsymbol{u}\cdot\nabla c_A - \nabla\cdot(D_A \nabla c_A) = 0 \tag{3.21}$$

Thrombin (B): $$\boldsymbol{u}\cdot\nabla c_B - \nabla\cdot(D_B \nabla c_B) + k'' c_C c_B = 0 \tag{3.22}$$

Antithrombin III (C): $$\boldsymbol{u}\cdot\nabla c_C - \nabla\cdot(D_C \nabla c_C) + k'' c_B c_C = 0 \tag{3.23}$$

Adenosine diphosphate (D): $$\boldsymbol{u}\cdot\nabla c_D - \nabla\cdot(D_D \nabla c_D) = 0 \tag{3.24}$$

Thromboxane A_2 (E): $$\boldsymbol{u}\cdot\nabla c_E - \nabla\cdot(D_E \nabla c_E) + k' c_E = 0 \tag{3.25}$$

where c_A, c_B, c_C, c_D, c_E are the concentrations of the substances of interest, $\boldsymbol{u}$ denotes the velocity of blood and D_A, D_B, D_C, D_D, D_E are the diffusion coefficients. The constant coefficient of disintegration of thromboxane A_2 is k' and k'' is the coefficient describing the thrombin inactivation caused by antithrombin III. By adding an antithrombotic agent (e.g., heparin) the inactivation of thrombin is increased. This fact is expressed by a corresponding increase of the coefficient k''.

The boundary conditions for the mass transport and reaction model are constant concentrations at the inlet

$$c_A = c_{A_0},\ c_B = 0,\ c_C = c_{C_0},\ c_D = 0,\ c_E = 0 \tag{3.26}$$

and zero concentration gradients at the outlet

$$\frac{\partial c_i}{\partial n} = 0\ ,\quad i{=}A,..,E. \tag{3.27}$$

The model used considers the thrombin production from prothrombin at the thrombus surface by a fast reaction process

$$c_A = 0,\quad \frac{\partial c_B}{\partial n} = -\frac{\partial c_A}{\partial n} \tag{3.28}$$

and a constant emission flux of TxA_2 and ADP at the thrombus surface

$$D_D \frac{\partial c_D}{\partial n} = \alpha,\quad D_E \frac{\partial c_E}{\partial n} = \beta\ , \tag{3.29}$$

where $\boldsymbol{n}$ is the outward pointing unit normal vector at the thrombus surface, while α and β are the constant emission fluxes at the thrombus surface of ADP and TxA_2, respectively.

Numerical Method. The solution of the Navier-Stokes equations is performed applying the velocity-pressure correction method as discussed in section 2.3. The transport equations of thrombin (Eqn. 3.22) and antithrombin III (Eqn. 3.23) are solved iteratively. To linearize the reaction terms in each iteration step i+1 known concentration values of the previous step i are used

$$\boldsymbol{u} \cdot \nabla c_B^{i+1} - \nabla \cdot (D_B \nabla c_B^{i+1}) + k'' c_C^i c_B^{i+1} = 0 \tag{3.30}$$

$$\boldsymbol{u} \cdot \nabla c_C^{i+1} - \nabla \cdot (D_C \nabla c_C^{i+1}) + k'' c_B^i c_C^{i+1} = 0\,. \tag{3.31}$$

Applying the finite element Galerkin method the system of coupled convection-diffusion equations for the unspecified concentration node values of the substances A,…,E is

$$(\boldsymbol{K}1(\boldsymbol{u}) + \frac{1}{Pe_A}\boldsymbol{K}2)\boldsymbol{c}_A = \boldsymbol{0} \tag{3.32}$$

$$(\boldsymbol{K}1(\boldsymbol{u}) + \frac{1}{Pe_B}\boldsymbol{K}2 + k''\boldsymbol{C}_B(\boldsymbol{c}_C^i))\boldsymbol{c}_B^{i+1} = \boldsymbol{0} \tag{3.33}$$

$$(\boldsymbol{K}1(\boldsymbol{u}) + \frac{1}{Pe_C}\boldsymbol{K}2 + k''\boldsymbol{C}_C(\boldsymbol{c}_B^i))\boldsymbol{c}_C^{i+1} = \boldsymbol{0} \tag{3.34}$$

$$(\boldsymbol{K}1(\boldsymbol{u}) + \frac{1}{Pe_D}\boldsymbol{K}2)\boldsymbol{c}_D = \boldsymbol{0} \tag{3.35}$$

$$(\boldsymbol{K}1(\boldsymbol{u}) + \frac{1}{Pe_E}\boldsymbol{K}2 + k'\boldsymbol{C}_E)\boldsymbol{c}_E = \boldsymbol{0} \tag{3.36}$$

where $\boldsymbol{c}_A, \boldsymbol{c}_B^{i+1}, \boldsymbol{c}_C^{i+1}, \boldsymbol{c}_D, \boldsymbol{c}_E$ are the vectors containing all unspecified node values. The matrices $\boldsymbol{K}1(\boldsymbol{u})$ and $\boldsymbol{K}2$ are the convection matrix and the diffusion matrix, $\boldsymbol{C}_B(\boldsymbol{c}_C^i)$, $\boldsymbol{C}_C(\boldsymbol{c}_B^i)$ and $\boldsymbol{C}_E$ are reaction matrices. $Pe_A,\ldots,Pe_E$ denote the Peclet numbers for the substances A,…,E.

Numerical Results. The results of the flow calculation show regions of flow recirculation directly upstream and downstream of the mural aggregate. At the top of the mural aggregate a local maximum of the wall shear stress occurs. A brief outline of thrombin concentration results is presented here. Detailed information about parameters and further results have been published by Karner and Perktold (1998).

Thrombin is produced on the thrombus surface from prothrombin which is present in blood and arrives by convection and diffusion. The inactivation of thrombin by the protein antithrombin III, which is also present in blood, strongly depends on the heparin concentration in blood. In case of higher shear rates more prothrombin arrives at the thrombus surface, and therefore, more thrombin is produced, but due to the higher shear rate the produced thrombin molecules are transported downstream more rapidly. The effect of faster molecular transport dominates the effect of higher thrombin production and the maximum concentration value generally decreases or remains the same if the shear rate increases. In the downstream direction the faster transport of the produced thrombin causes a higher concentration in the region downstream of the thrombus. Furthermore, two local concentration maxima in the regions of flow recirculation directly upstream and (strongly pronounced) downstream the thrombus occur (Figure 4).

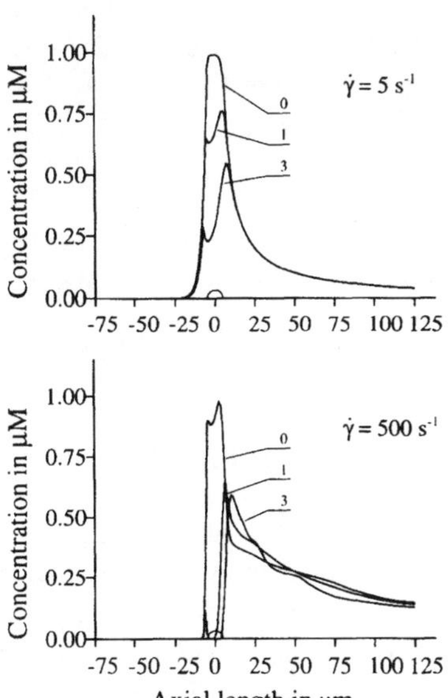

Figure 4. Thrombin concentration profiles for thrombus diameter 10 μm at shear rates $\gamma = 5\ s^{-1}$ and $\gamma = 500\ s^{-1}$ at radial distances from the surface of 0 μm, 1 μm and 3 μm.

The maximum concentration generally increases with increasing thrombus size and decreases with increasing shear rate. The thrombin concentration required to stimulate platelet activity is approximately 0.001μM to 0.003μM (McIntire and Tay, 1989). The results show that the thrombin concentration in the vicinity of mural microthrombi is generally much higher than the required level, and therefore, thrombin plays a very important role in the process of thrombogenesis.

In the case of heparin presence in blood the inactivation of thrombin by antithrombin III is much stronger, and the resulting thrombin concentration is much lower (Figure 5).

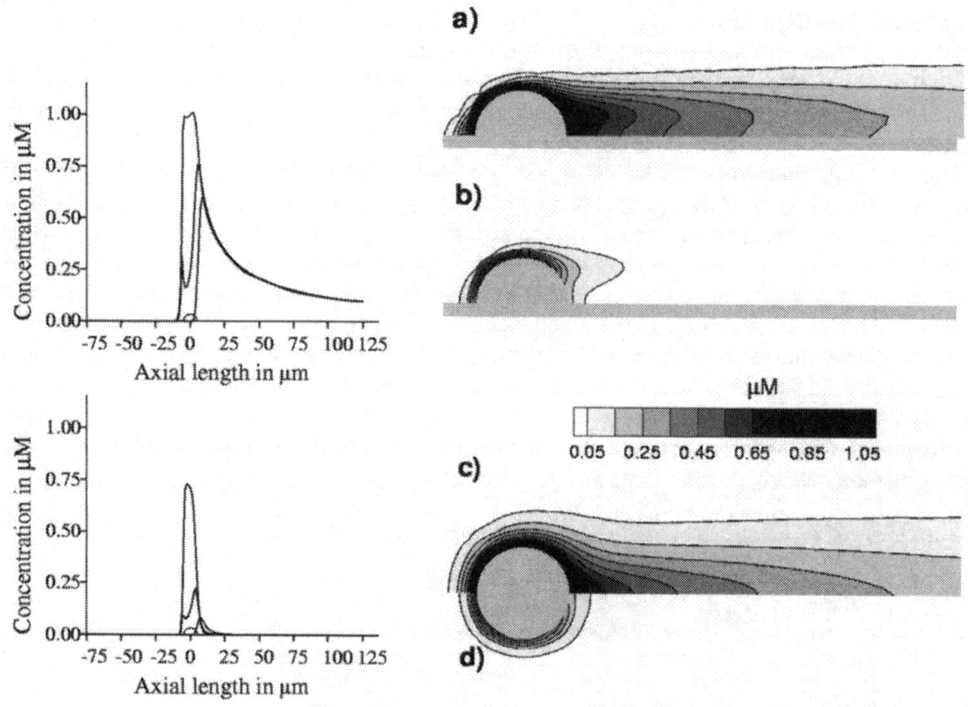

Figure 5. Thrombin concentration profiles and contours for thrombus diameter 10 µm at a shear rate of $\gamma = 100\ s^{-1}$. a) on the plane of symmetry , b) on the plane of symmetry with heparin, c) at a distance from the artery wall of 1 µm and d) at a distance from the artery wall of 1 µm with heparin.

4 Macromolecule Transport in the Multilayered Arterial Wall

Early atherosclerotic lesions are associated with abnormally high macromolecule accumulations within the arterial intima. From this fact it follows that mass transport processes and chemical reactions within the wall layers are important factors in atherogenesis. It has been observed that sites of an increased endothelial permeability to macromolecules are preferred locations of early lesions, which leads to the conclusion that the endothelial barrier plays the most important role in the initiation and development of atherosclerotic diseases. The endothelial resistance to macromolecule entrance into the arterial wall has been investigated in various studies (Friedman and Fry, 1993; Tarbell, 1993). In addition to experimental studies on transport properties of the wall layers (Meyer et al., 1996; Tarbell et al., 1988) different mathematical models have been developed in recent years to investigate the mass transport processes within the artery wall (Fry, 1985; Penn et al., 1994; Huang et al., 1997).

In the studies by Karner and Perktold (2000) and by Karner et al. (2001) a mathematical model has been developed for the description of the mass transport processes in the arterial wall coupled with the mass transport in the arterial lumen. The model considers essential processes like the convective transport across the endothelium and internal elastic lamina (IEL), the interaction between the lumenal and transmural transport processes, and important effects like osmotic forces, sieving of large molecules and restricted convection and diffusion, which are determined by the structure of the wall layers. The mathematical description of the lumenal mass transport uses the convection-diffusion equation. The porous wall layers (consisting of a fluid phase and a solid phase) are macroscopically treated as a homogenous media. The mass transport in the intima and media is described applying a macroscopic volume-averaged convection-diffusion-reaction equation, the calculation of the filtration velocity in the intima and media applies Darcy's law. The transport equations corresponding to the arterial lumen, the intima and the media are coupled by the Kedem-Katchalsky equations which describe the convective and diffusive flux across the endothelium and IEL. The transport parameters of the intima and media (effective diffusivity, Darcy permeability and porosity) are obtained from fiber matrix models (Curry, 1984; Huang and Tarbell, 1997). The parameters of the two semipermeable membranes the endothelium and IEL (permeability, hydraulic conductivity and reflection coefficients) are calculated from the equations of pore theory (Crone and Levitt, 1984; Curry, 1984).

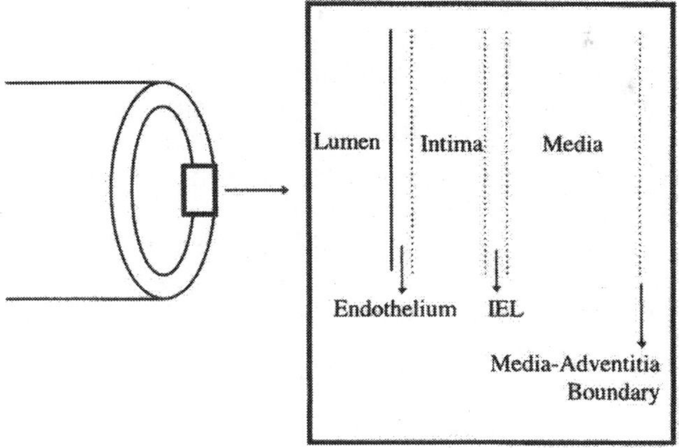

Figure 6. Schematic illustration of geometric artery wall model.

4.1 Mathematical Model

The mathematical description of the blood flow in the arterial lumen uses the incompressible Navier-Stokes equations for Newtonian fluids

$$(\boldsymbol{u}_l \cdot \nabla)\boldsymbol{u}_l - \nu \cdot \nabla^2 \boldsymbol{u}_l + \frac{1}{\rho}\nabla p_l = 0 \qquad \nabla \cdot \boldsymbol{u}_l = 0 \tag{4.1}$$

with the velocity vector field $\boldsymbol{u}_l$ and the pressure p_l.

The filtration velocity in the intima and media is calculated from Darcy's law

$$-\boldsymbol{u}_{filt} = \frac{K_p}{\mu_V} \nabla p , \tag{4.2}$$

where K_p is the Darcy permeability of the wall layer and μ_V is the dynamic viscosity of the fluid (plasma) within the artery wall. The pressure in each wall layer is determined through the Laplacean equation

$$\nabla^2 p = 0 . \tag{4.3}$$

The coupling the pressure across the wall layers uses constant volume flux through the layers.

The mass transport in the arterial lumen is described applying the convection-diffusion equation

$$\boldsymbol{u}_l \cdot \nabla c_l - \nabla \cdot (D_l \nabla c_l) = 0 , \tag{4.4}$$

where c_l is the concentration of the substance of interest and D_l is the diffusion coefficient which is assumed to be constant.

The description of the mass transport processes in the intima and media (microscopically heterogenous, but macroscopically approximated as homogenous media) uses the volume-averaged convection-diffusion-reaction equation

$$K\boldsymbol{u}_{filt} \cdot \nabla c_w - \nabla \cdot (D_w \nabla c_w) + k c_w = 0 , \tag{4.5}$$

where c_w denotes the volume-averaged concentration of the substance in the particular wall layer. The volume averaging procedure of the transport equation is described in (Shyy et al., 1997). The effective diffusivity in the wall layer is D_w, the rate constant for chemical reaction is k, and K is the effective lag coefficient for convective transport in a porous medium. According to Fry (1985) and Huang and Tarbell (1997) metabolic degradation can be appropriately modeled by an irreversible first order reaction term.

The transport processes in the lumen, intima and media are coupled by the flux across the endothelium and IEL which is modeled using the Kedem-Katchalsky equations

$$\begin{aligned} J_V &= L_p (\Delta p - \sigma_d \Delta \Pi) \\ J_S &= P \Delta c_w + J_V (1 - \sigma_f) \bar{c}_w , \end{aligned} \tag{4.6}$$

where J_V and J_S denote the volume flux and the solute flux, respectively. The permeability coefficient of the wall layer is P, the hydraulic conductivity is L_p, the osmotic and solvent drag reflection coefficients are σ_d and σ_f, respectively. The concentration differential across the layer is Δc_w, and $\bar{c}_w$ denotes the average concentration within the layer. The osmotic pressure differential across the wall layer is $\Delta\Pi$, which is calculated using van't Hoff's relation

$$\Pi = R_g T c , \tag{4.7}$$

where R_g is the universal gas constant and T is the absolute temperature.

Boundary conditions. The boundary conditions for the flow calculation in the lumen are fully developed velocity profiles at the inlet of the lumen and the filtration velocity $\boldsymbol{u}_{filt}$ at the lumen-endothelium interface. At the lumenal outflow boundary, zero surface traction force is applied.

The boundary conditions for the lumenal mass transport are constant concentration C_0 at the flow entrance and zero diffusive flux at the outlet.

At the inflow and outflow cross sections of the arterial wall the boundary conditions for the mass transport use zero diffusive flux.

At the media-adventitia interface zero diffusive flux or constant concentration can be assumed. The zero diffusive flux condition corresponds to the assumption, that solute is supplied to the intima and media by the lumenal blood, and to the adventitia by the vasa vasorum. At the interfaces lumen-endothelium-intima and intima-IEL-media continuity of convective-diffusive flux

$$\begin{aligned} -D_l \frac{\partial c_l}{\partial \boldsymbol{n}_l} + \boldsymbol{u}_{filt,l} \cdot \boldsymbol{n}_l c_l = J_{S,end} &= -(-D_i \frac{\partial c_i}{\partial \boldsymbol{n}_i} + \boldsymbol{u}_{filt,i} \cdot \boldsymbol{n}_i c_i) \\ -D_i \frac{\partial c_i}{\partial \boldsymbol{n}_i} + \boldsymbol{u}_{filt,i} \cdot \boldsymbol{n}_i c_i = J_{S,IEL} &= -(-D_m \frac{\partial c_m}{\partial \boldsymbol{n}_m} + \boldsymbol{u}_{filt,m} \cdot \boldsymbol{n}_m c_m), \end{aligned} \tag{4.8}$$

has to be assumed. l, i, m indicate the lumen, intima and media, respectively, $\boldsymbol{n}_l$, $\boldsymbol{n}_i$, $\boldsymbol{n}_m$ denote the outward pointing normal vectors at the endothelium and at the IEL.

4.2 Models of the Wall Layers

Endothelium. The endothelium is assumed to be a layer of constant thickness. Exchange of water and solutes between the arterial lumen and the intima takes place through clefts which occur between the endothelial cells. The study considers normal endothelial clefts which are modeled as cylindrical pores and leaky junctions which are approximated as pores with a ring-shaped cross-section surrounding the leaky cells (cells which are either dying or in mitosis). The resistance of fibers within these junctions is neglected.

The transport parameters of the leaky clefts ($L_{p,lj}$, P_{lj}, Φ_{lj}, σ_{lj}) and normal junctions ($L_{p,nj}$, P_{nj}, σ_{nj}) are calculated from pore theory (Anderson and Malone, 1974, Crone and Levitt, 1984, Curry, 1984). The values of the averaged hydraulic conductivity, permeability and reflection coefficients of the endothelium containing normal clefts and leaky clefts are

$$\begin{aligned} L_{p,e} &= L_{p,nj} + L_{p,lj}\varepsilon_{lj} \\ P_e &= P_{nj} + P_{lj}\varepsilon_{lj}\Phi_{lj} \\ \sigma &= \frac{L_{p,nj}\sigma_{nj} + L_{p,lj}\varepsilon_{lj}\sigma_{lj}}{L_{p,e}} \end{aligned} \tag{4.9}$$

(Huang et al., 1992), where ε_{lj} is the area of leaky clefts per unit area on the endothelial surface.

Intima. The subendothelial intima is modeled as an extracellular matrix of randomly distributed fibers (Huang et al., 1992). The fiber matrix is characterized by the intima thickness H_i, the fiber radius r_f, the average spacing D between the fibers, and by the total length of fibers l_t within the unit volume. In the case of a thickened intima smooth muscle cells are included with a volume fraction $\varepsilon_{SMC,i}$. The fractional void volume ε_i of the fiber matrix and the partition coefficient Φ_i (space available to the solute relative to the space available to water) of the fiber matrix have been published by Curry (1984).

The restricted diffusivity coefficient in the intimal extracellular fiber matrix is calculated from the equation

$$D_i = D \cdot \exp(-(1-\varepsilon_i)^{\frac{1}{2}}(1+\frac{r_{mol}}{r_f})) \ , \tag{4.10}$$

published by Ogston et al. (1973).
The Darcy permeability $K_{p,i}$ is given (Huang et al., 1992) as

$$K_{p,i} = \frac{r_f^2 \varepsilon_i^3}{4G(1-\varepsilon_i)^2} \ , \tag{4.11}$$

where G is the Kozeny constant. The equations for the calculation of the Kozeny constant of the used fiber matrix model have been published by Huang et al. (1992). According to Huang and Tarbell (1997) the lag coefficient for convective transport in the fiber matrix is obtained from

$$K_{Cf,i} = 2 - \Phi_i \ . \tag{4.12}$$

Assuming the existence of smooth muscle cells in the intima the total porosity of the intima is $\varepsilon_{i,eff} = \varepsilon_i(1-\varepsilon_{SMC,i})$. In this case the intimal transport parameters (diffusivity coefficient, Darcy permeability and lag coefficient) have to be transformed into 'effective' parameters (Huang and Tarbell, 1997). For numerical calculation purposes it is assumed that, on a macroscopic scale, the intima can be represented by a homogeneous continuum, even though it is microscopically heterogeneous.

Internal elastic lamina. The IEL is an elastin layer which is assumed to be of constant thickness H_{IEL}. It contains fenestral pores through which transport between the intima and the media takes place. According to Huang et al. (1997) and Lever (1995) the fenestrae are approximated as cylindrical pores with radius r_{fen}. The hydraulic conductivity of the IEL, neglecting the resistance of fibers within the fenestral pores, can be calculated from Poiseuille's law as

$$L_{p,IEL} = \varepsilon_{fen} \frac{r_{fen}^2}{8\mu H_{IEL}} \ , \tag{4.13}$$

where $\varepsilon_{fen} = \pi r_{fen}^2 \rho_{fen}$ is the average fraction of the IEL surface area occupied by fenestral pathways and ρ_{fen} is the average fenestral density in the IEL. The permeability of the IEL is

$$P_{IEL} = \varepsilon_{fen} \frac{D_{fen}}{H_{IEL}}, \tag{4.14}$$

where D_{fen} denotes the restricted diffusivity coefficient in the fenestral pores. This pore diffusivity coefficient is defined as

$$D_{fen} = DF(\alpha_{fen}), \tag{4.15}$$

where $\alpha_{fen} = r_{mol} / r_{fen}$ is the ratio of molecule radius to the radius of the fenestral pores, and the function of restricted pore diffusivity $F(\alpha_{fen})$ is described in Curry (1984). The reflection coefficients of the fenestral pores are calculated in the same way as the reflection coefficients of the endothelial normal junctions.

Media. Equivalent to the intima the media is modeled as a medium composed of smooth muscle cells with a volume fraction $\varepsilon_{SMC,m}$ and of an extracellular fluid phase with fibers (porosity ε_{fl}). The total porosity of the media is $\varepsilon_m = \varepsilon_{fl}(1-\varepsilon_{SMC,m})$. The calculation of the transport parameters corresponds to the calculation of the intimal transport parameters.

4.3 Numerical Method

The approximation of the axisymmetric Navier-Stokes equations applies quadrilateral isoparametric elements with biquadratic interpolation for the velocity field and bilinear interpolation for the pressure. The finite element approximation of the lumenal concentration uses bilinear interpolation functions on a 4×4-subgrid of the biquadratic velocity elements. The approximation of the volume averaged concentration in the intima and in the media also uses bilinear interpolation functions. The finite element grids for the solution of the mass transport equations in the lumen, intima and media are generated in such a way that for each node at the lumen-endothelium interface a corresponding node at the endothelium-intima interface exists (similar for the intima-IEL and IEL-media interfaces). This grid structure is necessary for the coupling of the transport processes in the lumen, intima and media using the Kedem-Katchalsky equations. The coupling of the mass transport problems in the three domains in the finite element approach yields one equation system where the unknowns are all node values of the lumenal concentration and of the volume-averaged wall concentration. In the following the coupling of the intimal transport process with the mass transport in the media using the flux across the IEL is demonstrated. The variational formulation of the intimal transport equation is

$$\begin{aligned} & K_{Cf,i} \int_{\Omega} \boldsymbol{u}_{filt,i} \cdot \nabla c_i \, \varphi d\Omega + D_i \int_{\Omega} \nabla c_i \cdot \nabla \varphi d\Omega + \\ & + k_i \int_{\Omega} c_i \, \varphi d\Omega - D_i \int_{\Gamma_{Int-End}} \frac{\partial c_i}{\partial \boldsymbol{n}} \varphi_\Gamma d\Gamma - D_i \int_{\Gamma_{Int-IEL}} \frac{\partial c_i}{\partial \boldsymbol{n}} \varphi_\Gamma d\Gamma = 0, \end{aligned} \tag{4.16}$$

where φ is a Galerkin test function, φ_Γ is a test function at the boundary surface and $\boldsymbol{n}$ is the outward pointing unit normal vector at the boundary surface. Due to mass conservation the mass flux from the intima into the IEL must be equal to the mass flux across the IEL

$$-D_i \frac{\partial c_i}{\partial \boldsymbol{n}}\bigg|_{\Gamma_{Int-IEL}} + K_{Cf,i} u_{filt,i} c_i\big|_{\Gamma_{Int-IEL}} = J_{S,IEL} \, , \tag{4.17}$$

where $u_{filt,i}$ is the intimal filtration velocity component normal to the intima-IEL boundary. Using this flux condition and the Kedem-Katchalsky equation for $J_{S,IEL}$ in the variational formulation yields the formulation for the integral term at the intima-IEL boundary

$$\int_{\Gamma_{Int-IEL}} \frac{\partial c_i}{\partial \boldsymbol{n}} \varphi_\Gamma \, d\Gamma = \alpha \int_{\Gamma_{Int-IEL}} c_i \, \varphi_\Gamma \, d\Gamma + \beta \int_{\Gamma_{Med-IEL}} c_m \varphi_\Gamma \, d\Gamma \tag{4.18}$$

with the coefficients

$$\begin{aligned} \alpha &= \frac{1}{D_i \varepsilon_i} \left(\varepsilon_i K_{Cf,i} v_i - P_{IEL} - L_{p,IEL} (\Delta p - \sigma_{d,IEL} \Delta \Pi) \frac{(1-\sigma_{f,IEL})}{2} \right) \\ \beta &= \frac{1}{D_i \varepsilon_m} \left(P_{IEL} - L_{p,IEL} (\Delta p - \sigma_{d,IEL} \Delta \Pi) \frac{(1-\sigma_{f,IEL})}{2} \right), \end{aligned} \tag{4.19}$$

which couples the concentrations at the intima-IEL interface and at the IEL-media interface. Applying this method at all appearing wall layer interfaces yields an equation system where the unknowns are all lumenal and mural concentration values.

Due to the dependence of the osmotic pressure Π on the concentration, the equation system is non-linear. For this reason the equation system is solved iteratively where the osmotic pressure is calculated using the concentration values of the previous iteration-step. The iteration procedure is performed until the stopping criterion is fulfilled.

Due to the low diffusivities of macromolecules the transport process is convection dominated and an upwind stabilization has to be applied to the solution procedure. The large sparse linear equation systems for the velocity components and the concentration are solved using a modified bi-conjugate gradient algorithm with a preconditioning technique (incomplete LU-factorization) and a compact storage scheme (Hofer and Perktold, 1995). The solution of the symmetric and positive-definite pressure equation system is performed applying the preconditioned conjugate gradient method.

4.4 Numerical Results

The results concerning concentration distribution have been calculated under the assumptions of steady flow and pressure in a straight tube. The sets of parameters and data applied for albumin and LDL transport in the artery lumen and in the wall layers have been described in detail in Karner et al. (2001).

Figure 7 shows the concentration profiles in the artery wall layers for (a) albumin and (b) LDL applying physiologically realistic wall data. The cross-section where the results are presented is located 11cm downstream the inflow (indicated at the top of the figure). The results correspond to the known fact, that the normal (healthy) endothelium provides the main resistance to the transport of large molecules from blood into the arterial wall. This can be concluded from the resulting steep concentration drop across the endothelial layer for large molecules like LDL. The IEL is not an essential barrier for mass transport (compared to the endothelium), because the radius of the fenestral pores is between one and two orders of magnitude larger than the radius of the considered molecules. The main reason for the difference in the intimal and medial concentrations is the difference in the porosity values of these two layers. The normalized albumin concentration values in the media between 0.01 and 0.02 correspond with the values measured by Lever and Coleman (1995) and by Meyer et al. (1996).

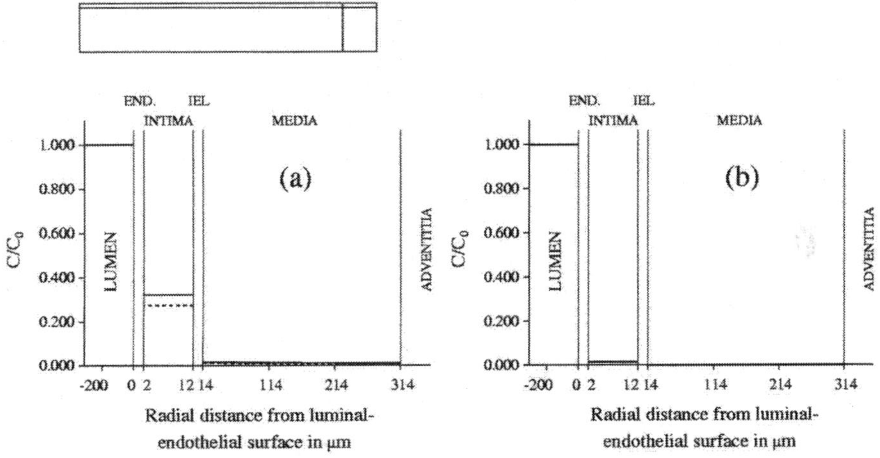

Figure 7. Volume averaged concentration profiles of (a) albumin and (b) LDL.

In the case of a thickened intima the occurrence of smooth muscle cells (SMC) has to be considered which influences the mass transport within the wall by decreasing the intimal porosity and the effective transport parameters (Darcy permeability and diffusivity) and by additional chemical reactions (metabolization processes at the SMC surface).

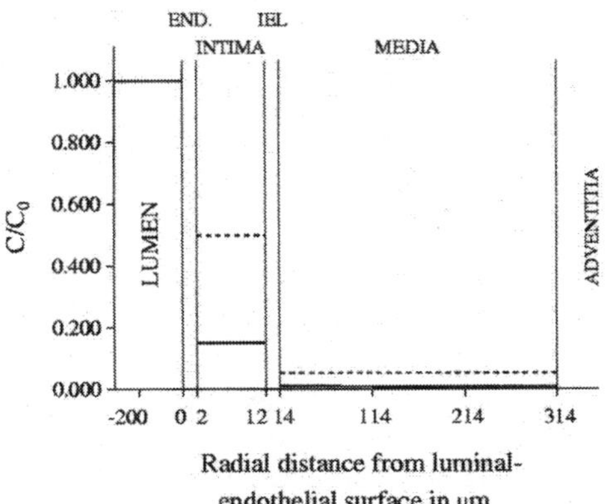

Figure 8. Volume-averaged albumin concentration profiles assuming SMC in the intima with (solid line) and without chemical reactions (dashed line) at the surface of SMC.

Figure 8 shows the albumin concentration if SMC (volume fraction of 40%) are included in the intima and chemical reaction occurs (solid line). The intimal concentration is reduced by 53% compared with the result without SMC in the intima (Figure 7(a)). The figure also demonstrates the effect of ignoring the chemical reactions at the SMC surface on the albumin concentration (dashed line). Without chemical reactions the concentration increase is about 230% in the intima.

Figure 9 demonstrates the effect of endothelial injury on the albumin concentration. Endothelial injury is modeled by increasing the percentage of leaky endothelial cells. The result shows that endothelial injury causes an increase of albumin concentration in the wall layers. This concentration increase is caused by a higher mass flux across the endothelium into the wall due to the reduced transport resistance of the injured endothelium. For 1% leaky endothelial cells the increase in intimal concentration is about 16%, for 5% leaky cells the concentration increase is about 70%.

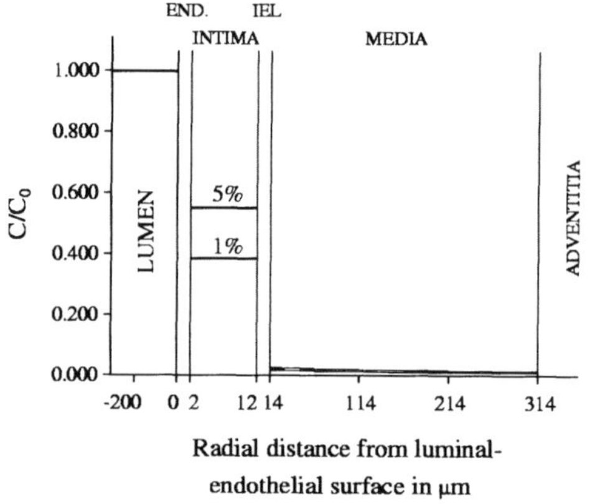

Figure 9. Volume-averaged albumin concentration profiles for different degrees of endothelial injury (1% leaky cells, 5% leaky cells).

5 Computer Simulation of Flow and Transport in Arterial Bifurcation Models

5.1 Human Carotid Bifurcation Model

It has been found that vascular diseases preferentially develop at sites where the flow is complex, particularly at bifurcations of large arteries. This is an indirect proof of the significant role of blood flow dynamics in the initiation and development of diseases. Therefore, detailed fluid dynamic studies are required to investigate the regions of disturbed blood flow in arterial segments. The most interesting flow phenomena include boundary layer separation, flow recirculation and stagnation and low and oscillating wall shear stress. The tendency of intimal thickening as a consequence of low-density lipoprotein (LDL) accumulation in the intima at sites where the wall shear stress is low and changes direction during the pulse cycle has been observed (Caro et al., 1971, Friedman et al., 1987). The locally disturbed mass transfer into the artery wall due to endothelial permeability variations at sites of low and oscillating wall shear stress seems to be an important factor in the process of atherogenesis. The flow characteristics depend on local vessel geometry which may vary considerably within the population. Therefore, the influence of the vascular geometry in atherogenesis has been studied by several authors (Nerem and Cornhill, 1980, Friedman et al., 1983, Ku et al., 1985). The computer simulation under anatomically realistic conditions is enabled by the development of efficient numerical methods and by the application of modern computer technology. Geometrically realistic modeling is an essential aspect since the flow field can be significantly affected by slight alterations in the geometry of the flow domain.

In the study published by Karner et al. (1999) pulsatile blood flow in an anatomically realistic compliant model of the human carotid artery bifurcation is analyzed numerically. The major points of concern are the investigation of flow characteristics as axial velocity, secondary motion and wall shear stress. To demonstrate the influence of wall compliance the results have been compared with results of a corresponding rigid wall model. The study applies an anatomically correct cast which was fabricated and optically digitized by D. Liepsch, Fachhochschule, Munich (Karner et al., 1999). The digitized surface data were used to generate the computational flow domain. The mathematical description of the blood flow applies the time-dependent, three-dimensional, incompressible ALE-modified Navier-Stokes equations for non-Newtonian inelastic fluids. The complex rheological behavior of blood is approximated using a shear thinning model.

Geometrical model and flow conditions. The measured surface data were smoothed using weighted least squares B-splines to get the computational model surface demonstrated in Figure 10. The figure shows the slight non-planarity of the branching. The finite element grid using 3D-brick elements was created applying our developed mesh generator based on local optimization of geometric properties such as smoothness and orthogonality of the grid, which are controlled by a corresponding function. Further details have been presented by Perktold et al. (1998). The finite element discretization of the flow domain yields 50.568 brick elements with 223.233 degrees of freedom for the three velocity components and the pressure.

The measured flow pulse waveform at the common carotid artery inlet and at the external carotid outlet and the pressure pulse waveform at the internal outlet applied in this study (Figure 10) have been published by Osenberg (1991). One pulse cycle (pulse period $t_p = 0.75\ \mathrm{s}$) is discretized into 112 non-equidistant time intervals, where the chosen time steps are smaller for the pulse cycle intervals of high flow acceleration and deceleration. To reach periodicity in the numerical solution the calculations are repeated over three to four pulse periods. The mean flow rate in the common carotid artery over the pulse cycle is $5.5\ \mathrm{ml \cdot s^{-1}}$. Applying the diastolic common carotid artery diameter of $0.67 \cdot 10^{-2}$ m as a reference length and using the reference kinematic viscosity $\nu_{ref} = 0.35 \cdot 10^{-5}\ \mathrm{m^2 \cdot s^{-1}}$ (constant fluid density $\rho = 1.05 \cdot 10^3\ \mathrm{kg \cdot m^{-3}}$) the resulting mean Reynolds number is $\mathrm{Re}_{ref} = 300$ and the Womersley number is $\alpha = 5.18$.

Mechanical wall model. In the present study the complex mechanical artery wall properties are described using hyperelastic behavior. The corresponding material parameters in the region of the carotid artery bifurcation have been published by Delfino et al. (1997). Incrementally linearly elastic behavior were applied by Karner et al. (1999), where in the range of the pulse pressure from 78 mmHg to 102 mmHg the assumed elastic modulus is $E = 2.14 \cdot 10^5\ \mathrm{N \cdot m^{-2}}$. The comparison of the deformation results with experimental results as published by Van Merode et al. (1988) justify both assumptions.

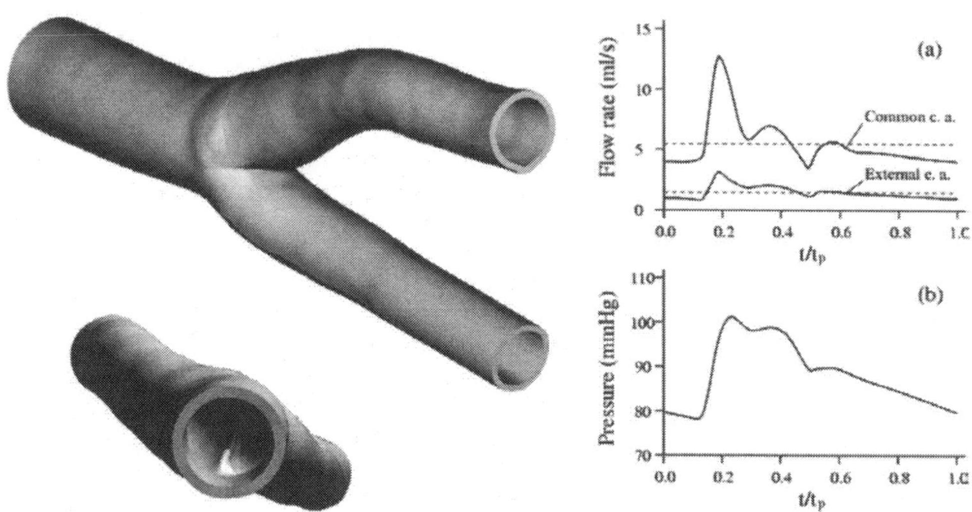

Figure 10. Computational carotid artery bifurcation model. Flow pulse waveform at the common carotid artery inlet and at the external carotid outlet (a) and pressure pulse waveform at the internal outlet (b).

In the wall calculation the diastolic reference geometry (at 78 mmHg) is used as initial configuration. The calculation of the wall displacements is performed employing the finite element program package ABAQUS.

Wall shear stress. In viscous flow a stress component parallel to the flow vector occurs. This shear stress is defined as the force per unit area between two adjacent layers of fluid in relative motion parallel to each other. In undisturbed tube flow the shear stress is proportional to the slope of the velocity profiles (shear rate), and thus, the shear stress is maximum at the wall. The proportionality factor is the viscosity coefficient of the fluid. In incompressible flow (in non-uniform geometry, such as the region of branches and bifurcations) the shear stress is determined by the extra stress tensor $\tau_w = \boldsymbol{t} \cdot \boldsymbol{T} \cdot \boldsymbol{n}$. The wall shear stress vector in this flow type (assuming rigid vessel wall and no-slip condition) is given by the normal derivative of the tangential velocity vector at the wall and by the viscosity of the fluid. The wall shear stress is calculated from the finite element velocity field near the wall.

Numerical results. The flow patterns in the internal carotid artery where plaques often develop are of main interest. A three-dimensional representation of the axial velocity during systolic flow deceleration and at the diastolic flow minimum at the proximal and mid sinus location is shown in Figure 11. The figure demonstrates the skewing of the profiles where high velocity gradients occur at the flow divider wall and separation tendency and locally reversed flow at the outer sinus wall. Although the branching of the bifurcation is non-planar, the axial velocity profiles are relatively symmetric with respect to the side walls.

Results concerning the axial flow velocity in the “bifurcation plane” during systolic flow deceleration and at the diastolic flow minimum as well as inplane velocities during flow deceleration at proximal and mid sinus location are displayed in Figure 12. High velocity flow can be seen on the carina side of each branch. At the outer walls of the branching region flow separation and recirculation can be observed at the considered pulse phases.

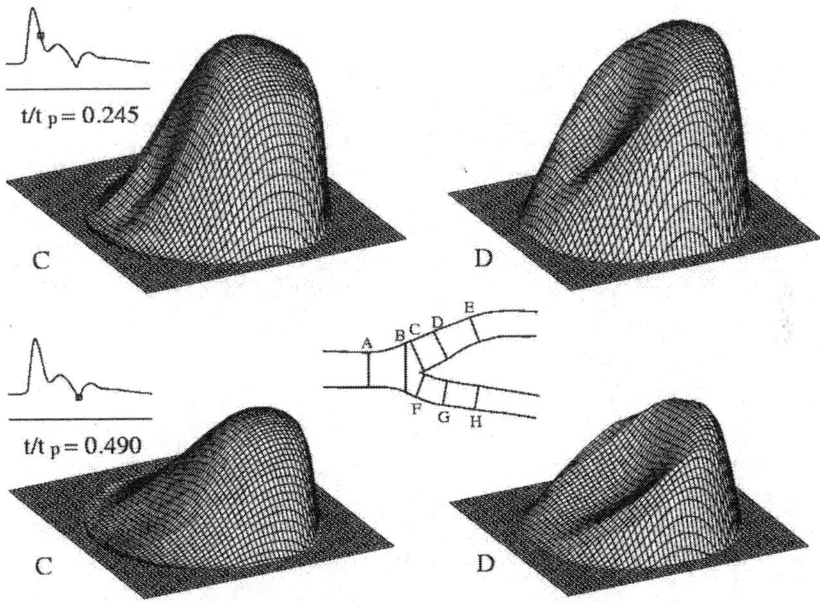

Figure 11. Three-dimensional representation of the axil flow velocity during systolic flow deceleration phase and at the diastolic flow minimum.

The quantitative influence of the wall distensibility on the axial velocity is demonstrated by comparing the profiles with the axial velocity profiles in a corresponding rigid wall model (dashed lines), where the geometry of the rigid model agrees with the diastolic geometry of the distensible model. The comparison indicates decreased flow separation in the realistic distensible model. At peak pressure the hyperelastic expansion of the outer sinus wall is about 6 % of the diameter. This expansion corresponds to an older subject (Reneman et al., 1985). Flow motion lateral to the main flow direction occurs in curved 3D flow. This inplane velocity results from the interaction between inertial (centripetal) and viscous forces. Figure 12 shows significant inplane motion directed from the outer wall towards the flow divider wall in the center of the tube. The stream back to the outer wall occurs along the side walls where this back stream is more pronounced along the lower side wall (left side of the cross sections). The inplane velocity increases the flow separation or the separation tendency at the outer wall of the carotid sinus. Contrary to the axial velocity, the slightly non-planar branching of the model results in a significant asymmetry of the inplane velocity field.

The axial wall shear stress distribution at the inner surface of the vessel wall during systolic deceleration is presented in Figure 13. The results show regions of reversed flow (negative axial shear stress) at the outer and side walls of the bifurcation. High axial shear stress occurs at the inner wall (flow divider wall) of the internal carotid artery.

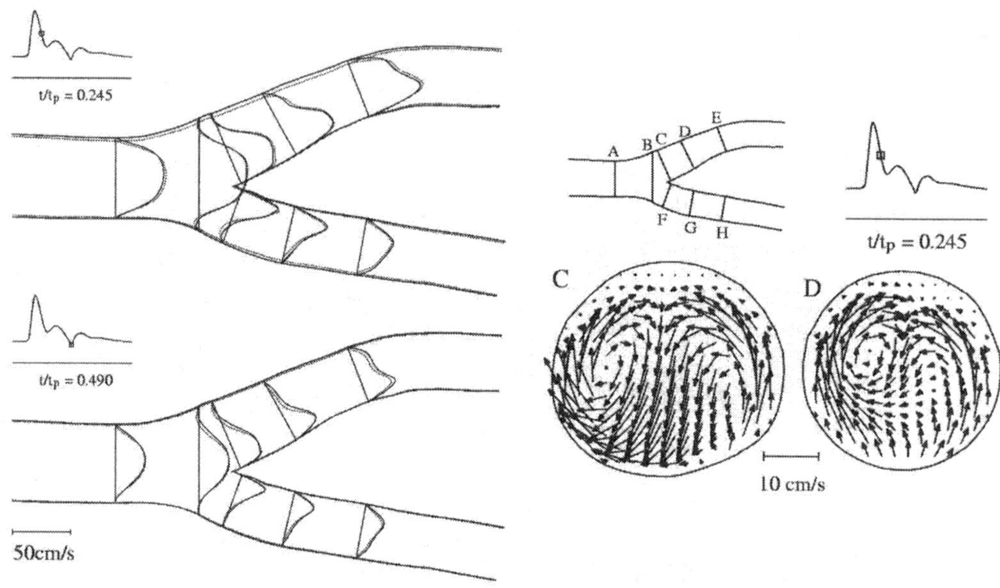

Figure 12. Axial flow velocity profiles in the bifurcation plane at the pulse phases of systolic flow maximum, flow deceleration and at minimum flow rate (solid line: distensible wall model; dashed line: rigid wall model). Inplane motion at specified cross sections (as indicated) during systolic flow deceleration.

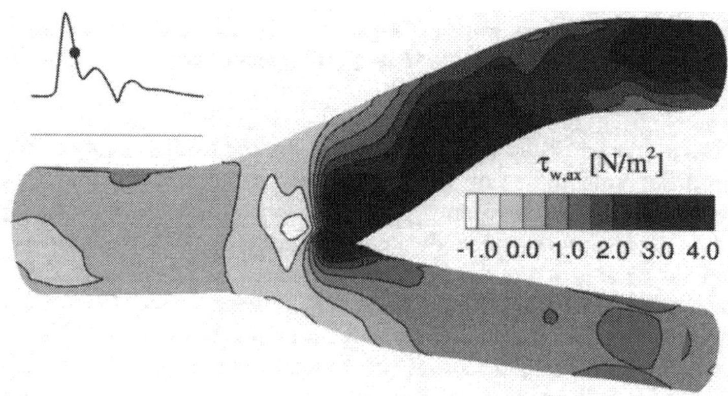

Figure 13. Axial wall shear stress contours at the inner surface of the vessel wall during systolic flow deceleration.

Figure 14 displays the axial wall shear stress distribution along the outer common-internal artery wall and at different sinus locations. The upper panel shows the wall shear stress during flow deceleration (a) and the time-averaged axial wall shear stress (b). The diagram (a) demonstrates the extension of the flow recirculation zone during flow deceleration. The positions of the separation point and the reattachment point (zero axial wall shear stress) can be seen. Flow separation occurs from the very beginning of the internal carotid artery up to the mid sinus location. The time-averaged axial shear stress (diagram (b)) demonstrates minimum axial shear stress, and thus, tendency of flow separartion at the beginning of the internal carotid. To demonstrate the influence of wall distensibility on the wall shear stress the figure illustrates the results of the realistic distensible wall model (solid line) and of the corresponding rigid wall model (dashed line). The comparison of the time-averaged wall shear stresses shows a slight decrease at the outer internal wall in the distensible model. Figure 14 (lower panel) shows the axial wall shear stress during the pulse cycle at selected locations at the internal divider wall (C_i, D_i) and at the outer sinus wall (C_o, D_o). At the divider wall high shear stress occurs, whereas at the selected locations at the outer sinus wall low and oscillating shear stress can be observed. Negative axial wall shear stresses indicate that flow separation exists at the proximal outer sinus region (C_o) throughout the systolic flow deceleration and at diastolic minimum flow rate.

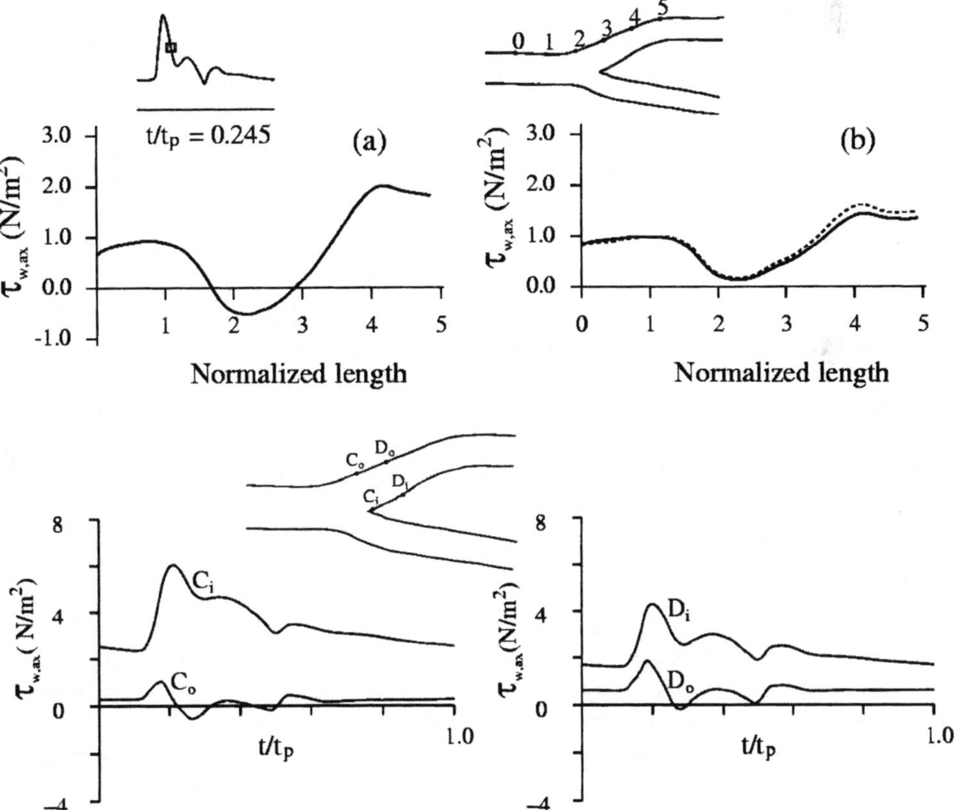

Figure 14. Axial wall shear stress distribution at the outer common-internal artery wall during systolic flow deceleration (a), and time-averaged axial wall shear stress distribution (b) (solid line, distensible wall model; dashed line, rigid wall model). Lower panel: axial wall shear stress distribution at the internal flow divider wall (locations C_i and D_i) and at the outer sinus wall (locations C_o and D_o) during the pulse cycle.

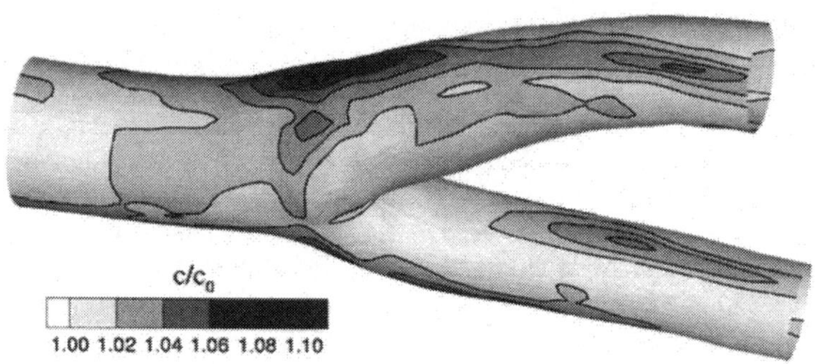

Figure 15. LDL concentration at the blood-endothelium boundary.

Figure 15 demonstrates an increase of lumenal LDL concentration at the blood-endothelium boundary. The maximum concentration increase of eight to ten percent of the bulk flow concentration occurs at the outer sinus region. The zone of maximum increase is correlated to the flow separation zone. This local accumulation of LDL at the lumenal surface in zones of low velocity (and low shear stress) causes basically an increased infiltration of LDL into the wall. Further LDL accumulation in the intima could be explained with concentration-dependent and shear-dependent permeability of the endothelium in the model applied.

5.2 Stenosed Carotid Artery Model

Plaque formation in the human carotid bifurcation tends to localize in regions of low velocity and low wall shear stress and, more specifically, in the vicinity of the outer wall of the carotid sinus. Stenoses develop from the outer sinus wall toward the interior lumen. In many advanced cases, the only open area is a small orifice located on the carina side. In this zone high-velocity flow can be observed. Clinical non-invasive diagnosis of stenoses is often based on detection of axial velocity disturbances induced by these stenoses. In order to gain insight into the influence of a stenosis, present in the carotid sinus, on the flow field and on related quantities two diseased models have been analysed fluid dynamically. The stenosed models under consideration differ in the height of the plaque, i.e., in the degree of the constriction. The computer simulation is performed at the area constrictions of 30% (mild stenosis) and 60% (moderate stenosis). The shape of both stenoses is relatively smooth and downstream directed non-symmetric. The location and extention of the stenoses are displayed in Figure 16.

Axial velocity profiles and secondary motion in the moderately stenosed model during systolic deceleration phase downstream of the stenosis are demonstrated in Figure 16 (right side). Due to the influence of strong secondary flow the profiles are asymmetric with respect to the side wall. At location J the profile shows flow separation which is twisted from the outer wall towards the side wall. The secondary flow field is much more pronounced than in the normal bifurcation and shows a strong single vortex. At the considered cross-section the maximum secondary velocity is about 28% of the maximum axial velocity. The basic characteristics high axial velocity and steep velocity gradients at the divider side of the sinus are much more pronounced in the stenosed model.

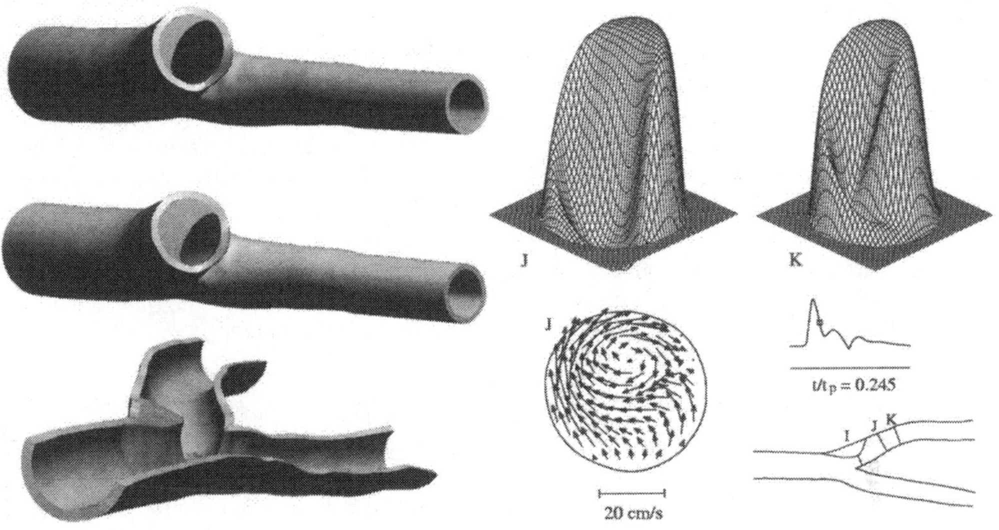

Figure 16. Left: Computational models of the 30% stenosed (mild stenosis) and of the 60% stenosed (moderate stenosis) carotid artery bifurcation. Right: Axial velocity profiles and secondary velocity vectors in the moderately stenosed bifurcation model during flow deceleration phase.

The time-averaged axial wall shear stresses in the two diseased models along the common-internal wall are displayed in Figure 17. The main characteristics are the shear stress peak at the top of the stenosis, the steep descent and the flow separation tendency or flow separation directly downstream of the stenosis.

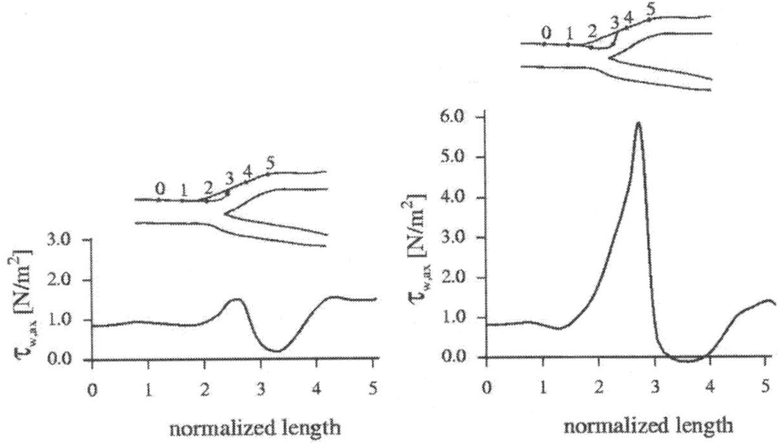

Figure 17. Time-averaged axial wall shear stress along the outer common-internal wall and time-dependent wall shear stress.

While the peak in the mild stenosis is of minor significance, strong elevation of wall shear stress can be observed in the moderate stenosis model, furthermore, the time-averaged wall shear stress indicates flow separation in this model. In physiologically realistic flow through a carotid bifurcation model essentially two types of near wall flow zones which are important in pathogenetic processes can be observed. These zones are characterized by flow separation and flow recirculation where low wall shear stresses occur, and by high velocity gradients and high wall shear stresses. The most important zone where flow separates is the region at the outer sinus wall, the highest shear stress can be observed at the flow divider wall near the apex. The flow dynamical analysis of two carotid artery bifurcation models with a stenosed carotid sinus shows primarily significantly altered velocity and shear stress patterns at the outer sinus wall region. Additional results have been published in Perktold and Karner (2001).

5.3 Curvature Effect on Bifurcating Coronary Artery Flow

The coronary arteries undergo large dynamic variations during each cardiac cycle due to their position on the beating heart. The local artery curvature varies significantly during myocardial contraction. The amount of change in curvature depends on the particular artery and can be very large. The influence of the time-dependent movement and deformation on flow patterns and wall shear stresses has been rarely studied until now (Santamaria et al., 1998, Weydahl and Moore, 2001). The aim of the present study is to analyse the effects of dynamic curvature on coronary artery hemodynamics and to compare these with the effects due to the vessel branching (Perktold et al., 2001). The pulsatile flow field is simulated in an anatomically realistic model of the bifurcation of the left anterior descending coronary artery (LAD) and its first diagonal branch (D1) including time-dependent curvature of the arteries (Figure 18).

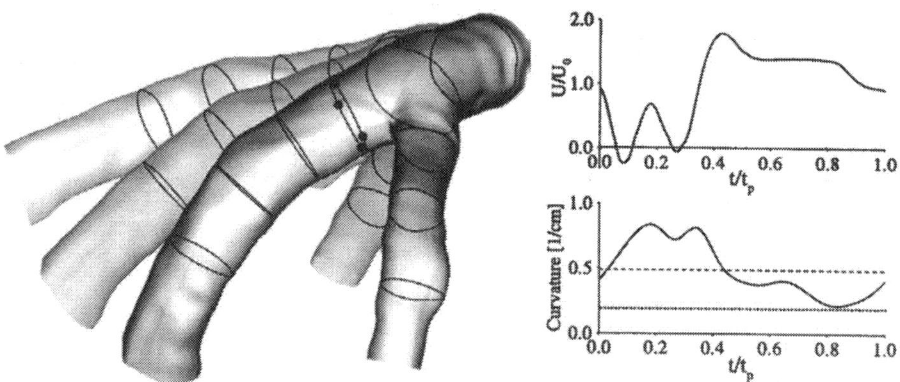

Figure 18. Computational geometric model of the LAD-D1 bifurcation. The models are related to minimum curvature, average curvature and to maximum curvature. Velocity pulse waveform in the LAD (top right) and representative curvature pulse waveform for the bifurcation segment at the branching site (bottom right).

The computational model of the LAD-D1 bifurcation was developed from a digitized arterial cast produced by M.H. Friedman, Duke University, Durham, NC. The generation procedure of the corresponding computational model has been explained in Perktold et al. (1998).

Experimental curvature data of the heart surface measured by Z. Ding, Yale University, New Haven, CT, at the site of the LAD-D1 branching were used to obtain curvature dynamics of the vessel. The physiologically realistic flow profiles used in the simulation were measured using laser Doppler anemometry by M.H. Friedman (Perktold et al., 1998). The LAD-D1 branching segment is known as an atherosclerotic-prone district. The model allows to separate the effects on the flow field and the shear field, the out-of-plane curvature, and of dynamic curvature. The mathematical model applies the ALE-modified Navier-Stokes equations describing the blood flow in the time-dependent geometry with large externally imposed boundary motion.

Numerical results. To analyse the effects of curvature and time-varying curvature on the velocity field and the wall shear stress distribution pulsatile flow calculations in the branch have been carried out using constant curvature at the minimum of curvature during the cycle, at constant curvature at the average curvature, and applying dynamic curvature according to the curvature pulse waveform (Figure 18). The comparison of the velocity and wall shear stress results demonstrates the influence of curvature and curvature dynamics on bifurcating flow and wall shear stresses using a physiologically correct flow pulse waveform and a representative curvature waveform. The results demonstrate that the skewing of the axial velocity profiles near the branching site are mainly determined by the vessel bifurcation. The influence of the curvature and the curvature dynamics is small at the branching site.

The results demonstrate clearly that the bifurcation effect dominates (Figure 19-B). Near the branching the curvature effect on bifurcating flow at a fixed site is moderate (Figure 19-A).

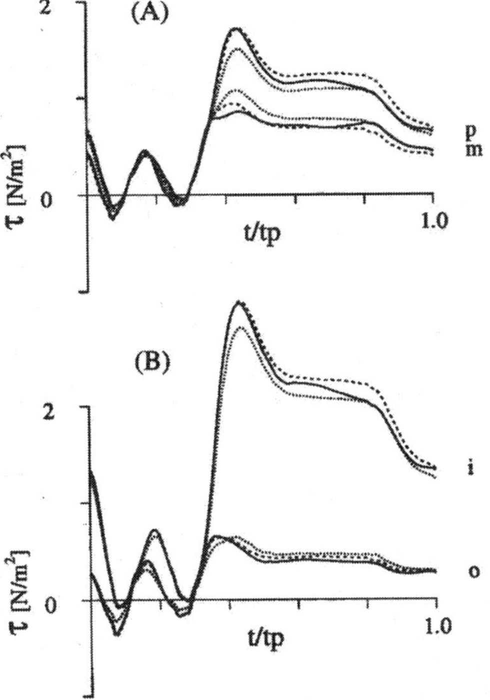

Figure 19. Wall shear stresses at the pericardial (p) and myocardial side (m), and at the inner (i) and outer wall (o) with respect to the bifurcation: Influence of curvature and curvature dynamics in the LAD near the bifurcation (as indicated in Figure 18); constant minimum curvature (dotted line), constant average curvature (dashed line), dynamic curvature (solid line).

Downstream of the bifurcation the pericardial and myocardial wall shear stresses are significantly different. In the range of curvature variation only minor influence of curvature dynamics can be observed. The relation of the flow pulse waveform and of the curvature pulse waveform seems to be important to the influence of curvature and curvature dynamics on flow and wall shear stress patterns occurring in coronary arteries.

6 Computer Simulation of Flow and Wall Mechanics in Peripheral Bypass Anastomoses[1]

At present bypass graft surgery is a routine reconstructive procedure for treating occlusive arterial diseases to restore blood supply to an organ or to the lower limbs. Despite the success of bypass surgery, however, restenosis and long term graft failure often occur. For the improvements of patency rates following peripheral reconstructions in which synthetic prosthesis must be used because of insufficient autologous venous replacement material, several modifications at the distal bypass anastomosis have been suggested. The long term function of such vascular reconstructions is influenced by several factors. Apart from the thrombogenicity of the surface in synthetic vascular prostheses reactive mechanisms resulting from non-physiological flow dynamical and wall mechanical conditions play an important role in the disease process.

The formation of post-operative intimal hyperplasia, a process in which the thickness of the inner wall of a blood vessel increases, is a major cause of long term failure of arterial bypass grafts. The increase preferentially occurs at the distal end of the graft forming an end-to-side anastomosis with the host artery. The most common sites of intimal thickening in an end-to-side vascular anastomosis are the suture line, the graft toe and heel and the artery floor region opposite the junction (Sottiurai et al., 1989). The thickening along the suture line is generally attributed to the normal healing process, but the mismatch between the compliance of the host artery and the graft has also been hypothesized to be a factor. Matching the compliance of graft and artery results in an improvement of patency rates (Trubel et al., 1995). Artery floor hyperplasia is hypothesized to be due to hemodynamics, particularly due to low and oscillating wall shear stress (Bassiouny et al., 1992), and due to high shear gradient (Ethier et al., 1998, Lei et al., 1996, Ojha 1994). Local oxygen concentration influences metabolic processes in the wall and possible degradation of plasma components. Presumably, with adequate knowledge of the hemodynamic, mass transport and wall mechanic consequences of surgical procedures, improved designs and techniques can be developed to minimize flow and stress disturbances, and thus, maximize long term graft function. Cell biological and biochemical influences on intimal hyperplasia have been considered in the studies by Kissin et al. (2000) and by Lemson et al. (2000).

The present study (results have already been published by Leuprecht et al. (2001) and Perktold et al. (2002)) aims to investigate the relationship between flow characteristics and mechanical stresses and post-operative intimal hyperplasia in three established types of distal end-to-side anastomoses using expanded polytetrafluoroethylene (e-PTFE) prostheses. Related to the surgical techniques the models differ in shape and structure of the graft-artery connector

[1] The majority of this section is taken from the authors' article "Fluid dynamics, wall mechanics and oxygen transfer in peripheral bypass anastomoses", *Annals of Biomedical Engineering* , Vol. 30, pp. 447-460, 2002.

region. A significant aspect of the modification is the reduction of the compliance mismatch between the prosthetic graft and the native artery using an interposed vein patch and cuff. Due to the important influence of the geometry on the anastomotic flow field anatomically correct modeling is applied where the geometric features are well replicated.

The computer simulation of anastomotic flow dynamics, mass transport and wall mechanics is part of an interdisciplinary project which includes also in-vivo investigations in a sheep model. In a comparative experimental study in 16 sheep 32 femoro-popliteal vascular prostheses (eight grafts each group) were implanted. The experiments were carried out following the national law for animal experiments and after permission by the local University Ethics Commission.

The flow calculation is based on the Navier-Stokes equations. The dissolved oxygen transport is analysed using the convection-diffusion equation. The influence of the time-dependent flow domain due to vessel compliance is taken into account applying the ALE kinematical description. The calculations of the mechanical stresses in the vessel walls and in the suture were performed using the finite element program package ABAQUS. The wall of the vessels have been modeled as deformable shell structures of elastic isotropic material. Besides the behavior of the graft the mechanical properties of the suture method and material influence the mechanical properties of the connection area. The effect of two different connecting techniques (interrupted suture and running suture) has been included in the analysis.

6.1 Geometric Model and Flow Conditions

The geometrical modeling is based on experimental lumenal casts of the conventional type anastomosis, the Taylor-patch and Miller-cuff anastomoses which were acquired in an in-vivo study in sheep. In the conventional method the e-PTFE prosthesis and the host artery were directly sutured together. The Miller-cuff technique uses an autologuous vein interposed as cuff between prosthesis and artery (Miller et al., 1984). Further anastomotic patches following the techniques proposed by Taylor (Taylor et al., 1992) and Linton (Batson et al., 1984) were included in the study. In the Taylor-patch the prosthetic graft is modified by excision of a form patch from the anastomotic tip. In this free remaining region of the anastomotic toe, a vein patch is sewed with the graft and the artery.

Anatomically realistic computational geometric models of the conventional type anastomosis, the Taylor-patch and Miller-cuff have been developed on the basis of the corresponding experimental lumenal casts taken from femoro-popliteal bypass grafts. The three distal end-to-side anastomosis models were generated from corresponding surface data sets. To generate the computational domains, the experimental surface data were smoothed using weighted least squares B-splines to yield the surface shown in Figure 20. The figure demonstrates the irregularities and non-planarity of the realistic geometries. The simulations are carried out under correct pulsatile flow conditions in the femoral artery of a sheep. The inflow pulse waveform used in the calculations (Figure 20) was measured electromagnetically immediately after the graft implantation. The mean Reynolds numbers in the three models are $Re_{Conv} = 399$, $Re_T = 379$, $Re_M = 384$. The Womersley numbers are $\alpha_{Conv} = 2.94$, $\alpha_T = 3.09$, $\alpha_M = 3.06$. The measured pressure pulse waveform at the distal artery outlet is shown in Figure 20.

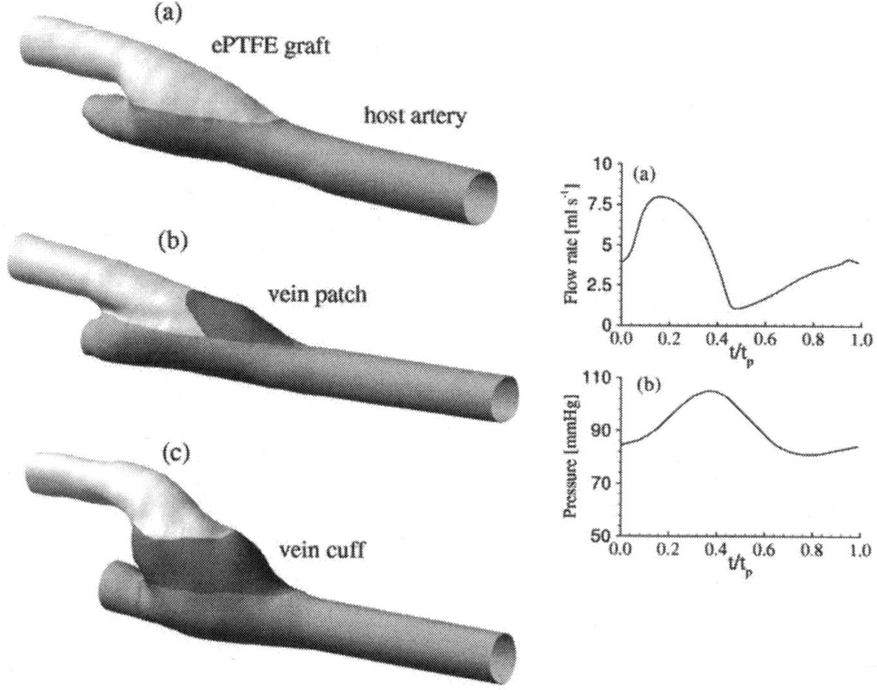

Figure 20. Computational geometric models: (a) Conventional end-to-side anastomosis, (b) Taylor-patch and (c) Miller-cuff. Right side: (a) Flow pulse waveform in the femoral artery of a sheep (b) pressure pulse waveform at the artery outlet.

6.2 Mechanical and Mathematical Models

Vessel wall model. The wall of the host artery, the graft and the vein patch and cuff can be modeled as a deformable shell structure of isotropic material. The analysis uses geometrical non-linear shell theory. The mechanical properties of the wall materials (native artery, e-PTFE graft and vein patch and cuff) are described applying elastic behavior. Corresponding elastic moduli of the materials are determined from experimental deformation data (Trubel et al., 1995). Incremental elastic moduli were calculated by considering the maximum systolic and minimum diastolic values on the corresponding radius-pressure curves. Young's moduli are for the native artery $E_a = 4.1 \cdot 10^5$ N$\cdot$ m^{-2}, the stiff e-PTFE graft $E_g = 7.5 \cdot 10^6$ N$\cdot$ m^{-2} and for the less compliant vein $E_v = 8.2 \cdot 10^5$ N$\cdot$ m^{-2}). The materials are nearly incompressible expressed by the Poisson ratio $\nu = 0.48$. The unstressed wall thickness of the artery and the vein is $h = 5 \cdot 10^{-4}$ m, the thickness of the graft is $h = 3.5 \cdot 10^{-4}$ m.

Suture model. Besides the elastic behavior of the graft material the mechanical properties of the suture technique influence the mechanical properties of the connection region. Stress analysis has been carried out for interrupted sutures and for continuous (running) sutures. The interrupted suture model, assuming that two structures are attached at discrete sites around their junction, idealizes a vascular clips suture. An appropriately fine finite element mesh has been generated in the transition region to model the suture technique resulting in a mesh width of $0.17 \cdot 10^{-3}$ m along the suture line. The clips have a thickness of $0.34 \cdot 10^{-3}$ m and were placed $1.02 \cdot 10^{-3}$ m apart. In this study the clips are modeled as stiff springs which connect the different shell structures at discrete locations. Between the clips graft and artery are not attached, which models the bypass configuration immediately after implantation and prior to any healing. The continuous suture model using polypropylene (6.0 Prolene) suture was applied for the conventional anastomosis and for the Miller-cuff. The diameter of the 6.0 Prolene suture is $0.093 \cdot 10^{-3}$ m. The modulus of elasticity is $1.44 \cdot 10^{9}$ $N \cdot m^{-2}$. The geometric suture model assumes that the needle passes through the wall $0.5 \cdot 10^{-3}$ m from the edge of the sutured structures and at $1 \cdot 10^{-3}$ m distance. This produces an angle of 45° at the outer wall surface and crosses perpendicularly to the contact line. The analysis of the continuous suture requires to represent the vessel wall as three-dimensional structure. To demonstrate basic features of the continuous suture only a part of the junction has been modeled as three-dimensional structure and has been inserted into the shell structure. Corresponding conditions at the boundary resulting from this insertation guarantee the well posedness of the problem. At the remaining part of the suture line the model of the interrupted suture is still applied.

Numerical model. The discretization of each of the three flow domains yields about 80.000 brick elements. For the solution of the mass transport problem each velocity element is subdivided into 32 subelements. The approximation of the concentration field applies about 2.500.000 linear elements. The approximate solution of the shell problem describing the wall mechanics is accomplished applying the commercial finite element program package ABAQUS. The selected element type is S4R, a four-node doubly curved shear flexible element with reduced integration and hourglass control using five degrees of freedom: three displacement components and two in-surface rotation components. To create appropriately fine mesh in the suture line region also a three-node triangular shell element type S3R, which is a degenerated version of the S4R-element, is applied. The discretization of the vessel surface results in about 7.500 elements for the conventional-type anastomosis, in about 11.200 elements for the Taylor-patch and in 10.700 elements for the Miller-cuff. The continuous suture calculation applies also the ABAQUS package. The suture has been modeled using two node truss elements T3D2H with linear interpolation for position and displacement and constant stress. The discretization of the three-dimensional wall patches applies the ABAQUS brick elements C3D8H. In the calculation of the vessel wall deformation the applied pressure load corresponds to the pressure pulse waveform relative to the diastolic reference pressure and the Navier-Stokes pressure.

The computational domain consists of the artery lumen and the artery wall. The continuity of the velocity and the no-slip property form the interface condition between lumen and wall. More specifically, the flow field and the pressure were calculated in the lumenal region with the wall velocity as boundary condition. Then the shell equations were solved while accounting for the pulse pressure and the pressure from the Navier-Stokes calculation. From the resulting wall deformation the new shape of the flow domain and the wall velocity are calculated.

6.3 Results and Discussion

The computer simulation of the complete flow field, the oxygen transport and the mechanical stresses is carried out in the three anastomoses models. Differences in flow structure, local oxygen concentration and mechanical stresses are demonstrated.

Flow velocity and wall shear stress. Figure 21 shows an outline of flow velocity and shear stress results in the (a) conventional type anastomosis, (b) Taylor-patch and (c) Miller-cuff anastomoses during flow deceleration $t/t_p = 0.366$ (which is also phase angle of peak pressure). The three models show common flow features, such as extremely skewed profiles towards the outer wall, time-dependent flow separation on the artery floor opposite the junction and nearly flow stagnation in the anastomotic region. The axial velocity profiles in the host artery distal of the anastomotic region have still a shape very different from parabolic. The profiles are locally more flat, and as a consequence, the velocity gradient at the wall is locally high. The in-plane velocity vector fields reflect the geometric non-planarity of the models. The velocity patterns in the conventional type and in the Taylor-patch are qualitatively similar, but the Taylor-patch connector shows slightly reduced non-uniform flow pattern mostly due to the low bifurcation angle. In the Miller-cuff the characteristics are more pronounced. A secondary vortex occurs in the entire cross-section, this produces a wash out effect which likely protects the artery. The shape of the diastolic geometry is indicated in the flow cross-section panels as dashed lines illustrating the deformation of the compliant models. At peak pressure the Miller-cuff shows significant deformation at the side walls of the anastomotic region (Miller cuff, panel B). In the artery section downstream the toe the normal deformation reaches up to 4%. The flow velocity patterns calculated in the models under rigid wall assumptions are only insignificantly different compared to the distensible wall results. The wall shear stress patterns indicate low and partly reversed shear stress on the artery floor, locally high shear and shear gradients directly proximal to the toe of the anastomoses. In the Miller-cuff type anastomosis a high shear region occurs downstream on the artery bed. The graft to host artery diameter ratio is larger (greater than one) in the Taylor-patch analyzed than in the other models. As a consequence the mean wall shear stress is lower in this model under unchanged inflow rate. Further geometric characteristics of the Taylor-patch are the small anastomotic angle and the relatively great hood length both also reducing the wall shear stress. But the stress pattern still shows the basic characteristics, particularly the wall shear stress concentration directly upstream the toe and the low wall shear stress on the artery floor opposite the graft.

Oxygen transport. Figure 22 illustrates the oxygen transfer into the vessel wall in the conventional type anastomosis. The main question concerns a possible correlation of low wall shear stress and decreased oxygen transfer into the wall. The black bold line and the white bold line indicate the location of the axial and the circumferential wall shear stress stagnation. A correlation of the sites of low shear and of low Sherwood number (normalized flux into the wall) on the artery floor is obvious. The Sherwood number contours show decreased oxygen flux into the vein cuff side wall. There is also a correlation between decreased wall shear stress and decreased mass transfer into the wall.

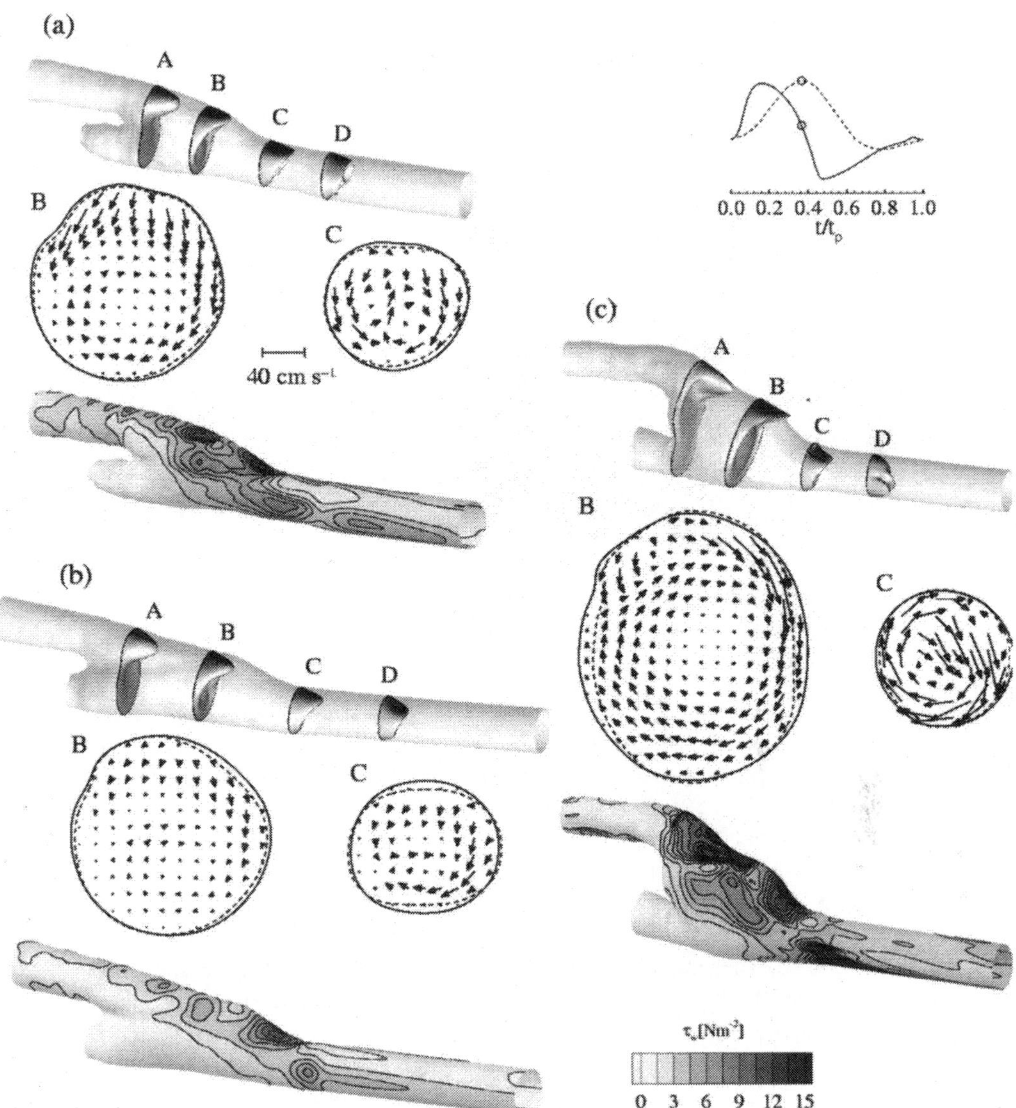

Figure 21. Axial velocity profiles, in-plane velocity vector field (at indicated cross-sections) and wall shear stress contours during systolic flow deceleration phase ($t/t_p = 0.366$): (a) Conventional end-to-side anastomosis, (b) Taylor-patch and (c) Miller-cuff; (dashed lines denote the cross section shape of rigid wall models). Top right: Flow pulse waveform (solid line) and pressure pulse waveform (dashed line).

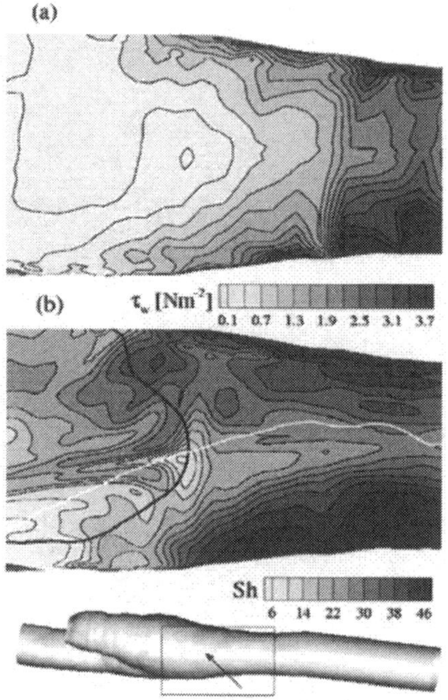

Figure 22. (a) Wall shear stress on the artery floor of the conventional anastomosis and (b) normalized oxygen transfer (Sherwood number) into the wall; Peclet number $Pe = 0.5 \cdot 10^6$.

Principal stresses. The calculation of the intramural stresses in the models uses static pressure inflation to P=1.33·10^4 N·m^{-2} (100 mmHg). The maximum principal stresses at the inner wall surface in the three configurations are displayed in Figure 23. Elevated principal stress can be observed at the anastomotic regions, where the occurring highest stress is a function of the compliance of the materials. The figure demonstrates the stress discontinuity across the suture lines in case of high compliance mismatch where higher stress occurs in the e-PTFE graft as the stiffer material. The stress pattern indicates better matching of compliance between vein and artery as can be observed at the transitions vein patch-artery in the Taylor-patch and vein cuff-artery in the Miller-cuff. The stresses in the host arteries downstream the anastomosis are slightly different, and the mean stress is constant in the three models at about eight times the inflating pressure. The stress in the grafts varies significantly.

Suture mechanics. Besides the elasticity properties of the graft, the suture method influences the mechanical properties of the connection area. Figure 24 and 25 illustrate the local maximum principal stresses and the stress gradients in the applied models of the interrupted sutures (Taylor-patch) and continuous sutures (conventional type and Miller-cuff) at the side walls of the anastomoses. The pressure load applied in the suture stress analysis is P=1.33·10^4 N·m^{-2} (100mmHg). To close the graft-artery contact in the conventional anastomosis and the contacts graft-vein and vein-artery in the Miller-cuff the applied strength of the Prolene suture without a pressure load is 0.02 N. This force draws the edges of the tissue together and causes local tissue compression. In the interrupted suture the stresses are functions of the load, in the continuous suture the stresses depend on both the strength of the suture and the loading placed on it, which are two opposing factors.

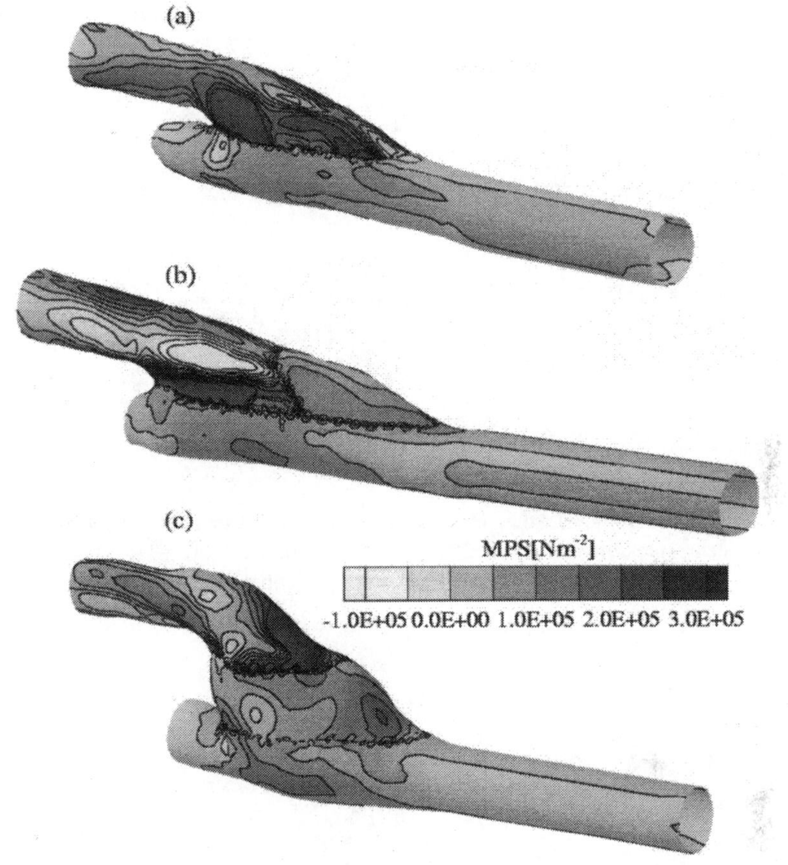

Figure 23. Maximum principal contours at the inner surface of the wall at pressure load $P = 1.33 \cdot 10^4$ N·m^{-2} (100 mmHg): (a) Conventional type anastomosis, (b) Taylor-patch and (c) Miller-cuff.

Stress concentration occurs at the suture attachment sites in the interrupted suture model and at the stitch location in the running suture. The stress concentration strongly depends on the compliance of the structures sutured.

Stress concentration occurs at the suture attachment sites in the interrupted suture model and at the stitch location in the running suture. The stress concentration strongly depends on the compliance of the structures sutured. The maximum principal stresses in the side wall area of the Taylor-patch are shown in Figure 24. The maximum stress occurs at the e-PTFE graft attachment site (Figure 24(a)) where the stress concentration normalized by the average maximum principal stress in the distal host artery is 7.4. Matched compliances (vein-artery) results in lower stresses and stress gradients (Figure 24(c)). The normalized stress concentration at the vein-artery attachment site is 2.0.

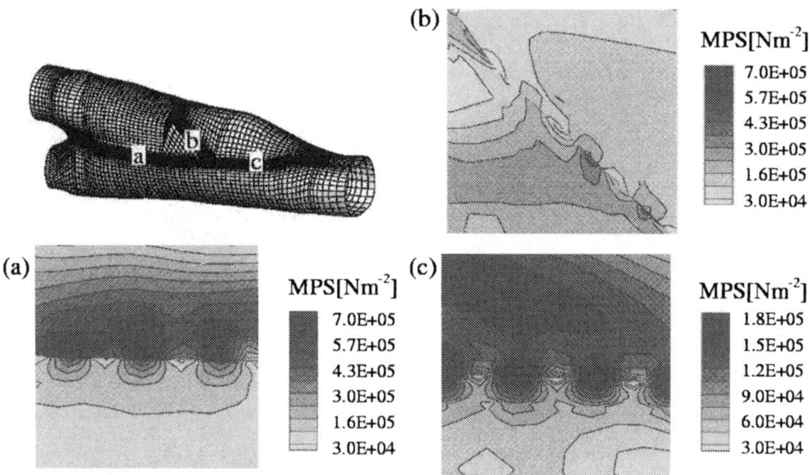

Figure 24. Maximum principal stress contours at the interrupted suture side wall region of the Taylor-patch at pressure load $P = 1.33 \cdot 10^4\ \mathrm{N \cdot m^{-2}}$ (100 mmHg): (a) e-PTFE graft/artery, (b) e-PTFE graft/vein and (c) vein/artery

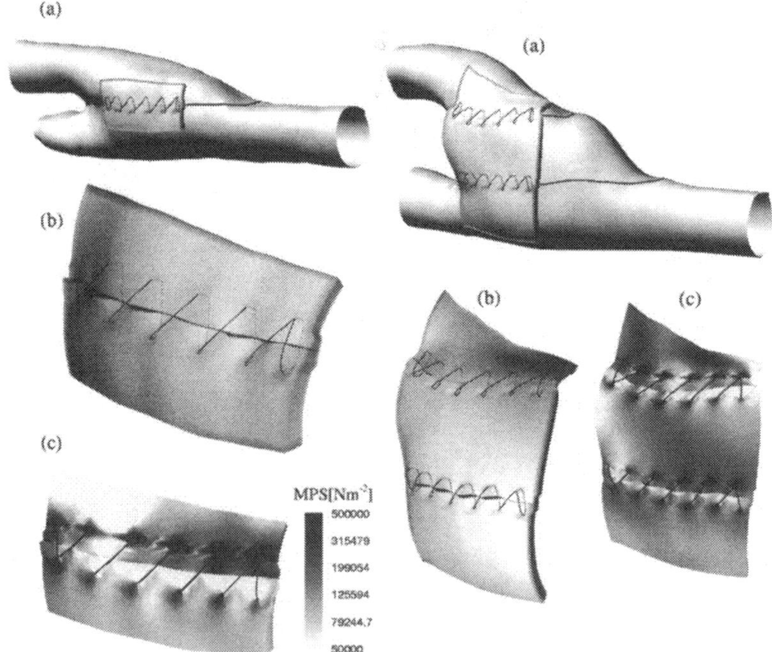

Figure 25. Maximum principal stress contours at the continuous suture side wall area of the conventional anastomosis. Right side: Maximum principal stress contours at the continuous suture side wall area of the Miller-cuff.

Figure 25 illustrates the deformation and the stress patterns of the continuous suture area at the side wall of the conventional anastomosis and of the Miller-cuff. The maximum principal stress contour plots indicate the stress concentration at the stitch locations and the high stress gradients where higher stresses occur in the stiffer part of the two sutured structures. Highest stress concentration occurrs at the e-PTFE graft of the conventional anastomosis. The force obtained in the running Prolene suture caused by the pressure load of 100 mmHg is highest at the vein-artery suture of the Miller-cuff: force is 0.017 N, in the PTFE graft-vein suture the force is 0.012 N. In the conventional PTFE graft-artery suture the force is 0.008 N. The total suture force is the sum of the force caused by the pressure load and the knot strength. The knot strength applied in this analysis is non-realistic low, and therefore, the total suture strength is too low.

The simulation demonstrates the influence of geometrical asymmetries and irregularities on the anastomotic flow field and mass transfer which are factors for intimal hyperplasia. The stress field calculation confirms that more compliant grafts result in lower principal stresses at the suture lines. The in-vivo studies of the project demonstrate correlations to compliance mismatch to some extent, with the graft-artery junctions showing the largest amount of intimal hyperplasia in mean values. However, at detailed locations, the situation may look different, due to geometry and flow field influences. The design of a peripheral vascular bypass configuration may affect the long term patency of the reconstruction. In-vivo studies and correlation aspects concernig intimal hyperplasia and fluid dynamics and wall mechanics (carried out by H. Schima, W. Trubel and M. Czerny, University of Vienna) are included in Perktold et al. (2002).

References

Anderson, J.L., and Malone, D.M. (1974). Mechanism of osmotic flow in porous membranes. *Biophys. J.* 14:957-982.

Back, L.H., Radbill, J.R., and Crawford, D.W. (1977). Analysis of oxygen transport from pulsatile, viscous blood flow to diseased coronary arteries of man. *J. Biomechanics* 10:763-774.

Bassiouny, H.S., White, S., Glagov, S., Choi, E., Giddens, D.P., and Zarins, C.K. (1992). Anastomotic intimal hyperplasia: mechanical injury or flow induced. *J. Vasc. Surg.* 15:708-717.

Batson, R.C., Sottiurai, V.S., and Craighead, C.C. (1984). Linton patch angioplasty: An adjunct to distal bypass with polytetrafluoroethylene grafts. *Ann. Surg.* 199:684-693.

Böhme, G., and Rubart, L. (1993). Einblick in die theoretische Analyse der Stroemungen viskoelastischer Fluessigkeiten. In Mennicken, R., ed., *GAMM Mitteilungen* 16:59-97.

Brezzi, F., and Fortin, M. (1991). *Mixed and Hybrid Finite Elements.* SSCM n. 5, Springer-Verlag, Berlin.

Brooks, A.N., and Hughes, T.J.R. (1982). Streamline upwind/Petrov-Galerkin formulations for convection dominated flows with particular emphasis on the incompressible Navier-Stokes equations. *Comput. Methods Appl. Mech. Eng.* 32:199-259.

Caro, C.G., Fitz-Gerald, J.M., and Schroter, R.C. (1971). Atheroma and arterial wall shear: Observation, correlation and proposal of a shear dependent mass transfer mechanism of atherogenesis. In *Proc. Roy. Soc. Lond.* 177:109-159.

Chien, S., Usami, S., Dellenback, R.J., and Gregersen, M.I. (1970). Shear-dependent deformation of erythrocytes in rheology of human blood. *American Journal of Physiology* 219:136-142.

Chorin A. (1968). Numerical solution of the Navier-Stokes equations. *Math. Comp.* 22:745-762.

Cokelet, G.R. (1980). Rheology and hemodynamics. *Ann. Rev. Physiol.* 42:311-324.

Crone, C., and Levitt, D.G. (1984). Capillary permeability to small solutes. In *Handbook of Physiology. Microcirculation. The Cardiovascular System*. Bethesda, MD: Am. Physiol. Soc., sect. 2, vol. IV, pt. 1, Chapter 8, 411-466.

Curry, F. E. (1984). Mechanics and thermodynamics of transcapillary exchange. In *Handbook of Physiology. Microcirculation. The Cardiovascular System*. Bethesda, MD: Am. Physiol. Soc., sect. 2, vol. IV, pt. 1, Chapter 8, 309-374.

Cuvelier, C., Segal, A., and van Steenhoven, A.A. (1986). *Finite Element Methods and Navier-Stokes Equations*. D. Reidel Pulishing Company, Dordrecht/Boston/Lancaster/Tokyo.

Delfino, A., Stergiopulos, N., Moore, Jr., J.E., and Meister, J.-J. (1997). Residual strain effects on the stress field in a thick wall finite element model of the human carotid bifurcation. *J. Biomechanics* 30:777-786.

Donea, J., Giuliani, S., Laval, H., and Quartapelle, L. (1981). Solution of the unsteady Navier-Stokes equations by a finite element projections method. In Taylor, C., and Morgan, K., eds., *Computational Techniques in Transient and Turbulent Flow*. Pineridge Press, Swansea. 97-132.

Ethier, C.R., Steinman, D.A., Zhang, X., Karpik, S.R., and Ojha, M. (1998). Flow waveform effects on end-to-side anastomotic flow patterns. *J. Biomechanics* 31:609-617.

Fahraeus, R., and Lindquist, T. (1931). The viscosity of the blood in narrow capillary tubes. *Am. J. Physiol.* 96:562-568.

Formaggia, L., Gerbeau, J.-F., Nobile, F., and Quarteroni, A. (2001). On a coupling of 3D and 1D Navier-Stokes equations for flow problems in compliant vessels. *Comp. Methods in Appl. Mech. Engng.* 191:561-582.

Formaggia, L., and Nobile, F. (1999). A stability analysis for the Arbitrary Lagrangian Eulerian formulation with finite elements. *East-West J. Numer. Math.* 7:105-131.

Friedman, M.H., Bargeron, C.B., Deters, O.J., Hutchins, G.M., and Mark, F.F. (1987). Correlation between wall shear and intimal thickness at a coronary artery branch. *Atherosclerosis* 68:27-33.

Friedman, M.H., Deters, O.J., Mark, F.F., Bargeron, C.B., and Hutchins, G.M. (1983). Arterial geometry affects hemodynamics - a potential risk factor for atherosclerosis. *Atherosclerosis* 46:225-231.

Friedman, M.H., and Fry, D.L. (1993). Arterial permeability dynamics and vascular disease. *Atherosclerosis* 104:189-194.

Fry, D.L. (1985). Mathematical models of arterial transmural transport. *Am. J. Physiol.* 248:H240-H263.

Fry, D.L. (1987). Mass transport, atherogenesis, and risk. *Arteriosclerosis* 7:88-100.

Fung, Y.C. (1965). *Foundations of Solid Mechanics*. Prentice-Hall International Series in Dynamics: Prentice-Hall, Inc., Englewood Cliffs, N.J.

Galdi, G.P., Heywood, J.G., and Rannacher, R. (2000). *Fundamental Directions in Mathematical Fluid Mechanics*. Birkhaeuser Verlag, Basel.

Gijsen, F.J.H. (1998). *Modeling of wall shear stress in large arteries*. Thesis, TU-Eindhoven.

Gijsen, F.J.H., Allanic, E., van de Vosse, F.N., and Janssen, J.D. (1999). The influence of the non-Newtoniean properties of blood on the flow in large arteries: unsteady flow in a 90° curved tube. *J. Biomechanics* 32:705-713.

Girault, V., and Raviart, P.-A. (1986). *Finite element methods for Navier-Stokes equations*. Springer-Verlag, Berlin Heidelberg New York Tokyo.

Goldsmith, H.L, and Marlow, J. (1979). Flow behaviour of erythrocytes. II. Particle motions in concentrated suspensions of ghost cells. *Journal of Colloid Interface Science* 71:383-407.

Gresho, P.M., Chan, S.T., Lee, RL., and Upson, C.D. (1984). A modified finite element method for solving the time-dependent incompressible Navier-Stokes equations. *Int. J. Num. Meth. Fluids* 4:557-598.

Gresho, P.M., and Sani, R.L. (2000). *Incompressible Flow and the Finite Element Method*, Vol. 2, John Wiley and Sons, Chichester.

Gunzberger, M.D. (1989). *Finite Element Methods for Viscous Incompressible Flows, A Giude to Theory, Practice and Algorithms*. Academic Press Inc., San Diego.

Hofer, M. (1998). *Numerische Simulation von Multiphasenstroemungen und Anwendungen auf die Blutstroemung*. Dissertation, TU-Graz.

Hofer, M., and Perktold, K. (1995). Vorkonditionierter konjugierter Gradienten Algorithmus für große schlecht konditionierte unsymmetrische Gleichungssysteme. Suppl. Vol. *ZAMM Z. angew. Math. Mech.* 75 SII:641-642.

Hofer, M., and Perktold, K. (1997). Computer simulation of concentrated fluid-particle suspension flows in axisymmetric geometries. *Biorheology* 34:261-279.

Huang, Y., Rumschitzki, D., Chien, S., and Weinbaum, S. (1997). A fiber matrix model for the filtration through fenestral pores in a compressible arterial intima. *Am. J. Physiol.* 272:H2023-H2039.

Huang, Z.J., and Tarbell, J.M. (1997). Numerical simulation of mass transfer in porous media of blood vessel walls. *Am. J. Physiol.* 273:H464-H477.

Huang, Y., Weinbaum, S., Rumschitzki, D., and Chien, S. (1992). A fiber matrix model for the growth of macromolecular leakage spots in the arterial intima. *Advances in Biological Heat and Mass Transfer*, HTD-Vol. 231, ASME 1992:81-92.

Hughes, T.J.R., Liu, W.K., and Zimmermann, T.K. (1991). Lagrangian-Eulerian finite element formulation in incompressible viscous flows. *Comput. Methods Appl. Mech. Engrg*, 29:329-349.

Hughes, T.J.R., Mallet, M., and Mizukami, A. (1986). A new finite element formulation for computational fluid dynamics: II. Beyond SUPG. *Comput. Methods Appl. Mech. Engrg.* 54:341-355.

Joseph, D.D. (1990). *Fluid Dynamics of Viscoelastic Liquids.* Vol. 84 of Applied Mathematical Science. Springer-Verlag, New York, Berlin, Heidelberg, London, Paris, Tokyo, Hong Kong.

Karner, G., and Perktold, K. (1998). The influence of flow on the concentration of platelet active substances in the vicinity of mural microthrombi. *Computer Methods in Biomechanics and Biomedical Engineering* 1:285-301.

Karner, G., and Perktold, K. (2000). Effect of endothelial injury and increased blood pressure on albumin accumulation in the arterial wall: a numerical study. *J. Biomechanics* 33:709-715.

Karner, G., Perktold, K., Hofer, M., and Liepsch, D. (1999). Flow characteristics in an anatomically realistic compliant carotid artery bifurcation model. *Computer Methods in Biomechanics and Biomedical Engineering* 2:171-185.

Karner, G., Perktold, K., and Zehentner, H.P. (2001). Computational modeling of macromolecule transport in the arterial wall. *Computer Methods in Biomechanics and Biomedical Engineering* 3:491-504.

Keunings, R. (1989). Simulation of viscoelastic fluid flow. In Tucker III, C.L., ed., *Computer Modeling for Polymer Processing.* Hanser Publishers, Munich, Vienna, New York. 403-469.

Kissin, M., Kansal, N., Pappas, P.J., DeFouw, D.O., Durán, W.N., and Hobsen, R.W. (2000). Vein interposition cuffs decrease the intimal hyperplastic response of polytetrafluorethylene bypass grafts. *J. Vasc. Surg.* 31:69-83.

Koiter, W.T., and Simmonds, J.C. (1973). Foundations of shell theory. In *Proc. 13th Int. Congress Theor. Appl. Mech.* Berlin: Springer-Verlag, 150-176.

Ku, D., Giddens, D.P., Zarins, C.K., and Glagov, S. (1985). Pulsatile flow and atherosclerosis in the human carotid bifurcation. *Atherosclerosis* 5:293-302.

Lei, M., Kleinstreuer, C., and Archie, Jr., J.P. (1996). Geometric design improvements for femoral graft-artery junctions mitigating restenosis. *J. Biomechanics* 29:1605-1614.

Lemson, M.S., Tordoir, J.H.M., Daemen, M.J.A.P., and Kitslaar, P.J.E.H.M. (2000). Intimal hyperplasia invascular grafts. *Eur. J. Vasc. Endovasc. Surg.* 19:336-350.

Leuprecht, A., and Perktold, K. (2001). Computer simulation of non-Newtonian effects on blood flow in large arteries. *Computer Methods in Biomechanics and Biomedical Engineering* 4:149-163.

Leuprecht, A., Perktold, K., Prosi, M., Berk, T., Trubel, W., and Schima, H. (2001). Numerical study of hemodynamics and wall mechanics in distal end-to-side anastomoses of bypass grafts. *J. Biomechanics* 35:225-236.

Lever, M.J. (1995). Mass transport through the walls of arteries and veins. In Jaffrin, M.Y., and Caro, C.G., eds., *Biological Flow*, Plenum Press: New York, 177-197.

Lever, M.J., and Coleman, P.J. (1995). Fractionation of plasma proteins during their passage through blood vessel walls. In Hochmuth, R.M., Langrana, N.A., and Hefzy, M.S., eds., *Proc. 1995 Bioengineering Conference*, BED-Vol. 29, ASME, New York, 133-134.

Liepsch, D.W., Thurston, G., and Lee, M. (1991). Studies of fluids simulating blood-like rheological properties and applications in models of arterial branches. *Biorheology* 28:39-52.

Ma, P., Li, X., and Ku, D.N. (1997). Convective mass transfer at the carotid bifurcation. *J. Biomechanics* 30:565-571.

McIntire, L.V., and Tran-Son Tay, R. (1989). Concentration of materials released from mural platelet aggregates: flow effects. In Yang, W.J., and Chun, J.L, eds., *Biomedical Engineering*, Hemisphere Publishing Corporation, New York - Washington, 229-245.

Meyer, G., Merval, R., and Tedgui, A. (1996). Effects of pressure-induced stretch and convection on low-density lipoprotein and albumin uptake in the rabbit aortic wall. *Circ. Res.* 79:532-540.

Miller, J.H., Foreman, R.K., Ferguson, L., and Faris, I. (1984). Interposition vein cuff for anastomosis of prosthesis to small artery. *Aust. N.Z. J. Surg.* 54:283-285.

Nerem, R.M., and Cornhill, J.F. (1980). The role of fluid mechanics in atherogenesis. *ASME J. Biomech. Eng.* 102:181-189.

Nobile, F. (2001). *Numerical Approximation of Fluid-Structure Interaction Problems with Application to Hemodynamics*. PhD thesis, École Polytechnique Fédérale de Lausanne (EPFL), Thesis N. 2458.

Nunziato, J.W. (1983). A multiphase mixture theory for fluid-particle flows. In Meyer R.E., ed., *The Theory of Dispersed Multiphase Flow*. Proc. of an Advanced Seminar Conducted by the Mathematics Research Center, University of Wisconsin-Madison. Academic Press, New York. 191-226.

Ogston, A.G., Preston, B.N., and Wells, J.D. (1973). On the transport of compact particles through solutions of chainpolymers. *Proc. R. Soc. London Ser.* A 333:297-316.

Ojha, M. (1994). Wall shear stress temporal gradient and anastomotic intimal hyperplasia. *Circ. Res.* 74:1227-1231.

Osenberg, H.P. (1991). *Simulation des arteriellen Blutflusses - Ein allgemeines Modell mit Anwendung auf das menschliche Hirngefäßsystem*. Dissertation, ETH Zürich 9342, IBT Zürich.

Penn, M.S., Saidel, G.M., and Chisolm, G.M. (1994). Relative significance of endothelium and internal elastic lamina in regulating the entry of macromolecules into arteries in vivo. *Circ. Res.* 74:74-82.

Perktold, K. (1987). On numerical simulation of three-dimensional physiological flow problems. *Ber. Math.-Stat. Sektion, Forschungsges. Johanneum Graz*, Nr. 280, Graz.

Perktold, K., and Hofer, M. (1999). Mathematical modelling of flow effects and transport processes in arterial bifurcation models. In Xu, X.Y., and Collins, M.W., eds., *Haemodynamics of Arterial Organs*. WIT Press. Southampton, Boston. 43-84.

Perktold, K., Hofer, M., Rappitsch, G., Löw, M., Kuban, B.D., and Friedman, M.H. (1998). Validated computation of physiologic flow in a realistic coronary artery branch. *J. Biomechanics* 31:217-228.

Perktold, K., and Karner, G. (2001). Computational principles and models of hemodynamics. In Hennerici, M., and Meairs, S., eds., *Cerebrovascular Ultrasound-Theory, Practice and Future Developments*. Cambridge University Press, 63-76.

Perktold, K., Leuprecht, A., Prosi, M., Berk, T., Czerny, M., Trubel, W., and Schima, H. (2002). Fluid dynamics, wall mechanics and oxygen transfer in peripheral bypass anastomoses. *Annals of Biomedical Engineering* 30:447-460.

Perktold, K., Prosi, M., Leuprecht, A., Ding, Z., and Friedman, M.H. (2001). Curvature effects on bifurcating coronary artery flow. BED-Vol. 50, 2001 Bioengineering Conference, ASME 2001:69-70.

Perktold, K., and Rappitsch, G. (1994). Mathematical modeling of local arterial flow and vessel mechanics. In Crolet, J.M., and Ohayon, R., eds., *Computational methods for fluid-structure interaction.* Pitman Research Notes in Mathematics Series 306, Longman Scientific & Technical, J. Wiley and Sons, New York. 230-245.

Perktold, K., Resch, M., and Florian, H. (1991). Pulsatile non-Newtonian flow characteristics in a three-dimensional human carotid bifurcation model. *Jounal of Biomechanical Engineering* 113:464-475.

Phillips, W., and Deutsch, S. (1975). Towards a constitutive equation for blood. *Biorheology* 12:383-389.

Quarteroni, A., and Formaggia, L. (2002). *Mathematical Modelling and Numerical Simulation of the Cardiovascular System.* Modeling and Scientific Computing, MOX-Report No. 01.

Quarteroni, A., Ragni, S., and Veneziani, A. (2001). Coupling between lumped and distributed models for blood problems. *Computing and Visualisation in Science* 4: 111-124.

Quarteroni, A., and Valli, A. (1994). *Numerical Approximation of Partial Differential Equations.* Springer-Verlag, Berlin, Heidelberg, New York.

Quarteroni, A., Veneziani, A., and Zunino, P. (2002). Mathematical and numerical modelling of solute dynamics in blood flow and arterial walls. SIAM *J. Numer. Anal.* 39:1488-1511.

Rappitsch, G., and Perktold, K. (1996). Computer simulation of convective diffusion processes in large arteries. *J. Biomechanics* 29:207-215.

Rappitsch, G., Perktold, K., and Pernkopf, E. (1997). Numerical modelling of shear-dependent mass transfer in large arteries. *Int. J. Numer. Meth. Fluids* 25:847-857.

Reddy, J.N., and Gartling, D.K. (1994). *The Finite Element Method in Heat Transfer and Fluid Dynamics.* CRC Press, Boca Raton.

Reneman, R.S., van Merode, T., Hick, P.J.J., and Hoeks, A.P.G. (1985). Flow velocity pattern in and distensibility of the carotid artery bulb in subjects of various ages. *Circulation* 71:500-509.

Santamaria, A., Siegel, J.M., and Moore Jr., J.E. (1998). Computational analysis of flow in a curved tube model of the coronary arteries: Effects of time-varying curvature. *Annals of Biomedical Engineering* 26:944-954.

Schmid-Schoenbein, H., Grunau, G., and Braeuer, H. (1980). *Exempla haemorheologica.* Albert-Roussel Pharma GmbH.

Segre, G., and Silberberg, A. (1962). Behaviour of macroscopic rigid sheres in Poiseulle flow, parts 1 and 2. *Journal of Fluid Mechanics* 12:115-157.

Sharp, M.K., Thurston, G.B., and Moore, Jr., J.E. (1996). The effect of blood viscoelasticity on pulsatile flow in stationary and axially moving tubes. *Biorheology* 33:185-208.

Sottiurai, V.S., Yao, J.S.T., Baston, R.C., Sue, S.L., Jones, R., and Nakamura, Y.A. (1989). Distal anastomotic intimal hyperplasia: histological character and biogenesis. *Ann. Vasc. Surg.* 3:26-33.

Shyy, W., Thakur, S., Ouyang, H., Liu, J., and Blosch, E. (1997). *Computational techniques for complex transport phenomena.* Cambridge University Press, Cambridge, UK.

Tanner, R.I. (1985). *Engineering Rheology.* Clarendon Press, Oxford.

Tarbell, J.M. (1993). Bioengineering studies of the endothelial transport barrier. *Bioengineering Science News, BMES Bulletin* 17:35-39.

Tarbell, J.M., Lever, M.J., and Caro, C.G. (1988). The effect of varying albumin concentration on the hydraulic conductivity of the rabbit common carotid artery. *Microvascular Research* 35:204-220.

Taylor, R.S., Loh, A., McFarland, R.J., Cox, M., and Chester, J.F. (1992). Improved technique for polytetrafluoroethylene bypass grafting: long-term results using anastomotic vein patches. *Br. J. Surg.* 79:348-354.

Temam, R. (1984). *Navier-Stokes Equations, Theory and Numerical Analysis.* North Holland, Amsterdam.

Thurston, G.B. (1979). Rheological parameters for the viscosity, viscoelasticity and thixotropy of blood. *Biorheology* 16:149-162.

Trubel, W., Schima, H., Moritz, A., Raderer, F., Windisch, A., Ullrich, R., Windberger, U., Losert, U., and Polterauer, P. (1995). Compliance mismatch and formation of distal anastomotic intimal hyperplasia in externally stiffened and lumen-adapted venous grafts. *Eur. J. Vasc. Endovasc. Surg.* 10:1-9.

Van Merode, T., Hick, P.J.J., Hoeks, A.P.G., Rahn, K.H., and Reneman, R.S. (1988). Carotid artery wall properties in normotensive and borderline hypertensive subjects of various ages. *Ultrasound in Med. & Biol.* 14:563-569.

Wada, S., and Karino, T. (2000). Computational study on LDL transfer from flowing blood to arterial walls. In Yamaguchi, T., ed., *Clinical Application of Computational Mechanics to the Cardiovascular System*. Springer-Verlag, Tokyo. 157-173.

Walitza, E. (1990). *Zum nicht-Newtonschen Flieszverhalten von Blut und einigen damit verbundenen Konsequenzen fuer laminare Stroemungen*. Dissertation, Universitaet Stuttgart.

Weydahl, E.S., and Moore, Jr., J.E. (2001). Dynamic curvature strongly affects wall shear rates in a coronary artery bifurcation model. *J.Biomechanics* 34:1189-1196.

Finite Difference and Finite Volume Techniques for the Solution of Navier-Stokes Equations in Cardiovascular Fluid Mechanics

Sokrates Tsangaris and Theodora Pappou

Fluids Section, Department of Mechanical Engineering,National Technical University of Athens
e-mails: sgt@fluid.mech.ntua.gr, dora@fluid.mech.ntua.gr

Abstract. The present notes summarize the flow equations involved in cardiovascular-fluid mechanics and the finite difference and finite volume numerical methods that are widely used for the modeling of flow through arteries of large or small size. In the case of flow through large size arteries two numerical methods are presented extensively based on a pressure-correction type methodology and a pseudocompressibility methodology for the solution of steady and unsteady flows through domains of deformable-moving walls. In the case of flow though small size arteries, the level set method is treated for the modeling of the multicomponent flow. Representative flow cases for cardiovascular fluid dynamics are also modeled and solved by using the above methodologies.

1 Basic equations involved in cardiovascular fluid mechanics

The problem of simulation of blood flow in arteries has several levels of complexity. For the blood flow in large arteries the system of incompressible, unsteady momentum equations for viscous fluid together with the equation of mass conservation (continuity equation) are considered. The system of the above equations can be coupled with the system of the motion equations for the deformable walls for the case of the largest arteries. The constitutive relations of the blood and the corresponding equations for the behavior of the arterial wall complete the system of equations.
For moderate size arteries the unsteady flow equations are taken into account while the deformation – motion equations of the arterial walls are neglected, because the artery wall can be considered as solid.
For small size arteries (arterioles and capillaries), in the so-called microcirculation, the multicomponent character of the blood should be taken into account. The flow can be described as steady because the flow pulse is damped. Anyway for the flow at least two components (plasma and red blood cells) should be regarded as important, because the size of the red blood cells are comparable with the diameter of the arteries in microcirculation. The last case of the multicomponent flow is treated with the level set method in the last section of this chapter.

1.1 Fundamental equations of blood flow

Incompressible flow equations inside domains with moving-deformable walls are mainly involved in cardiovascular fluid mechanics.
The equations presentation of this section includes both differential and integral form of conservation laws, which are optionally used for the formulation of finite difference or finite volumes and

finite elements numerical techniques respectively. The flow equations will be written for three-dimensional flows and for the quasi one-dimensional case, both taking into account the movement and the elasticity of the computational domain bounds.

For a moving control volume, the basic conservation laws for mass, momentum and energy can be summarized in the general transport equation (in integral form):

$$\underbrace{\frac{\partial}{\partial t}\int_V \rho\phi dV}_{\text{time derivative}} + \underbrace{\int_S \vec{U}\cdot\vec{n}\rho\phi dS}_{\text{advection}} = \underbrace{\int_S \Sigma\vec{n}dS}_{\text{diffusion}} + \underbrace{\int_V S_\phi \rho dV}_{\text{source term}} \tag{1.1}$$

which essentially states a balance of flows caused by different physical phenomena such as advection, diffusion and source effects. ϕ is the transported quantity and $\vec{U}=\vec{\upsilon}-\vec{w}$ is the velocity of fluid relative the velocity of the control – surface $\vec{w}$ and $\vec{n}$ is the outward from the control volume normal vector. The following table shows the specification of ϕ,Σ,S_ϕ for the different conservation equations.

	ϕ	Σ	S_ϕ
Mass balance	1	0	0
Momentum balance	$\vec{\upsilon}$	$\vec{\sigma}^T$	$\vec{g}$
Energy balance	$\frac{1}{2}\upsilon^2+e$	$\vec{\sigma}\cdot\vec{\upsilon}-\vec{q}$	$\vec{g}\cdot\vec{\upsilon}$

$\vec{g}$ is the acceleration of the imposed field forces , $\vec{\sigma}$ is the stress tensor, e is the specific internal energy of blood, which is a quantity proportional to temperature and $\vec{q}$ is the specific heat transfer vector.

In the case of a fixed control volume ($\vec{w}=0$) , by using the theorem of Green-Gauss, the surface integrals are replaced by volume integrals. Since the specific choice of the integration region is arbitrary, the obtained volume integral vanishes if the integrant is zero, and the conservation laws in differential form are obtained:

continuity

$$\frac{\partial\rho}{\partial t}+\nabla\cdot(\rho\vec{\upsilon})=0 \tag{1.2}$$

momentum

$$\frac{\partial(\rho\vec{\upsilon})}{\partial t}+\nabla\cdot\left[\rho(\vec{\upsilon}\otimes\vec{\upsilon})\right]=\nabla\cdot\vec{\sigma}+\rho\vec{g} \tag{1.3}$$

energy

$$\frac{\partial}{\partial t}\left[\rho\left(\frac{1}{2}\upsilon^2+e\right)\right]+\nabla\cdot\left[\rho\left(\frac{1}{2}\upsilon^2+e\right)\vec{\upsilon}\right]=\nabla\cdot(\vec{\sigma}\cdot\vec{\upsilon})-\nabla\cdot\vec{q}+\rho\vec{g}\cdot\vec{\upsilon} \tag{1.4}$$

For blood flows in large arteries the blood behaves macroscopically as homogeneous fluid having the nonlinear rheologic behavior of a non-newtonian fluid, with yield stresses appearing at low values of rate of deformation. The yield stresses are initiated by the aggregation of blood cells and some of its plasma proteins (fibrinogen), while the viscoelasticity mainly by the elasticity of the red blood cell membrane. For the largest size arteries it can be shown that the rheological behavior of blood can be approximated as a Newtonian fluid. The limit value of the blood viscosity for high shear rates is considered as 4-5 times value of the water viscosity at the same temperature. Anyway, the extension of the numerical methodologies for non-newtonian blood behavior is straightforward. For the Newtonian fluid behavior of the blood we use the linear constitutive relation for incompressible fluid:

$$\vec{\sigma} = -pI + 2\mu D \tag{1.5}$$

where p is the static pressure and μ the Newtonian limit viscosity of the whole blood and $\ddot{D}$ is the symmetric tensor of rate of deformation, which can be expressed by the tensor of gradient of the velocity:

$$\ddot{D} = \frac{1}{2}[\nabla \cdot \vec{\upsilon} + (\nabla \cdot \vec{\upsilon})^T] \tag{1.6}$$

Unsteady compressible viscous flow equations are of hyperbolic-parabolic type. Incompressible flow equations are of elliptic-parabolic type because of the elliptic character of the equation for pressure prediction. The pressure prediction equation is constructed because of the absence of density or pressure time derivative from the equation of mass conservation.
The incompressible N-S equations are actually more difficult to solve, due to their mixed character. One has to either iteratively solve the pressure equation at each iteration level, or use the concept of artificial compressibility.
The type of the differential equations that describe the flow problem, determines the numerical scheme that will be used for time integration, spatial discretization and mainly the suitable boundary conditions of the numerical method to be formulated.
For incompressible flows (ρ=const.), as they are the blood flows and the small temperature changes of the living organism, the equations of balance for mass and momentum can be written in their classical conservative form of the incompressible Navier – Stokes:

$$\frac{\partial u}{\partial x} + \frac{\partial w}{\partial z} + \alpha \frac{w}{z} = 0 \tag{1.7}$$

$$\rho\left[\frac{\partial u}{\partial t} + \frac{\partial}{\partial x}\left(u^2 + \frac{p}{\rho}\right) + \frac{\partial}{\partial z}(uw) + \alpha \cdot \frac{uw}{z}\right] = \frac{\partial \tau_{xx}}{\partial x} + \frac{\partial \tau_{xz}}{\partial z} + \alpha \cdot \frac{\tau_{xz}}{z} \tag{1.8}$$

$$\rho\left[\frac{\partial w}{\partial t} + \frac{\partial}{\partial x}(uw) + \frac{\partial}{\partial z}\left(w^2 + \frac{p}{\rho}\right) + \alpha \cdot \frac{w^2}{z}\right] = \frac{\partial \tau_{xz}}{\partial x} + \frac{\partial \tau_{zz}}{\partial z} + \alpha \cdot \frac{\tau_{zz} - \tau_{\phi\phi}}{z} \tag{1.9}$$

where $\vec{\upsilon} = (u, w)$ is the velocity vector in Cartesian Coordinate system and the shear stresses for a Newtonian fluid are expressed as:

$$\tau_{xx} = 2\mu \frac{\partial u}{\partial x},\ \tau_{xz} = \tau_{zx} = \mu\left(\frac{\partial u}{\partial z} + \frac{\partial w}{\partial x}\right),\ \tau_{zz} = 2\mu \frac{\partial w}{\partial z},\ \tau_{\phi\phi} = 2\mu \frac{w}{z} \tag{1.10}$$

The above system of equations (1.7)-(1.9) consists a complete system of four differential equations for $\vec{\upsilon}$, p and can be solved independently from the energy equation.

The parameter α (=0 or 1) in eqs. (1.7)-(1.9) is a switch for the addition of the axisymmetric terms into the two dimensional expressions.

1.2 Foundamental equations for the arterial wall

For modeling the mechanical behavior of the arterial wall there is also a large variety of the deformation behavior. For more details one can mention the review articles of Patel (1973) and Gow (1973). The most of the largest arteries are thin walled. This means that the wall thickness is much smaller than the diameter of the artery. For this reason most authors use the membrane theory for the modeling of arterial wall movement. An often used form of the motion equations of a thin walled, linear elastic, isotropic cylindrical artery is the following proposed by Attinger (1968) in axial (x) and radial (z) directions respectively:

$$\rho_w h \frac{\partial^2 \xi}{\partial t^2} = -\tau_{zx}\Big|_{z=R_o} + \frac{Eh}{1-\sigma^2}\left(\frac{\partial^2 \xi}{\partial x^2} + \frac{\sigma}{R_o}\frac{\partial \zeta}{\partial x}\right) \tag{1.11}$$

$$\rho_w h \frac{\partial^2 \zeta}{\partial t^2} = p\Big|_{z=R_o} - \tau_{zz}\Big|_{r=R_o} - \frac{Eh}{1-\sigma^2}\left(\frac{\zeta}{R_o^{\,2}} + \frac{\sigma}{R_o}\frac{\partial \xi}{\partial x}\right) \tag{1.12}$$

ξ and ζ are the axial and radial components of the artery wall displacement. In the above equations ρ_w denote the density of the artery wall, h the thickness of the thin-walled artery, and R_o the undisturbed internal radius. E is the Young's modulus of elasticity and σ the Poisson ratio. The coupling of the equations of the artery wall-motion with the blood flow equations is succeeded through the pressure and shear stress terms of the right hand of the above equations. In addition the kinematic conditions at the blood-artery wall interface should be satisfied, which means that the velocity of the fluid particles at the interface should be equal to the time derivatives of the wall displacements (non slip boundary condition).

1.3 Fundamental incompressible equations in matrix form and non-dimensional equations.

For the numerical solution of various flow cases, by using the theory of similarity, the flow equations are non-dimensionalized using appropriate reference state for each problem. In our work, the fundamental reference quantities are the reference values of the velocity, the length and the time. The remaining reference quantities for the non-dimensional equations are concluded from the above fundamental quantities. In the final expressions of the flow equations, some specific non-dimensional numbers are formed, whose values are known and representative for the simulated flow field. The most important dimensionless numbers that are involved in cardiovascular fluid mechanics (incompressible, unsteady flows through domains of deformable bounds) are summarized in Table 1 accompanied with their physical meaning.

In general, even if some non-dimensional numbers do not participate in the flow equations, their values are representative for each examined case and are related directly to the evolution of flow phenomena. As an example, we can refer that flow fields with relative small (in relation to the flow case) Womersley number indicate a quasi-steady flow field, and the transition of the flow to its

turbulence region occurs when the Reynolds number of the flow field overcomes some critical value, characteristic for each category of problems.
Some additional dimensionless coefficients are also formed for the macroscopic description of the flow fields and the quantification of the flow dynamic characteristics. As examples, we refer lift and drag coefficients, for external flows either for aerodynamics or bluff bodies. These coefficient are usually in correspondence with other dimensionless numbers of the flow (mainly Reynolds number).
Estimation of the order of the values of the dimensionless parameters as well as the limit upper values for man are indicated in the next table.

	Re	$\hat{Re}/Re$ (pulsatility)	α	M
Ascending Aorta	1500	6.3	21	0.2
Femoralis	200	4.3	4	0.05

The non-dimensional forms of the equations of motion of the artery-wall take the following form:

$$\frac{\rho_w}{\rho}\frac{h}{R_o}\frac{\partial^2\xi}{\partial t^2} = -\frac{1}{Re}\tau_{zx}\Big|_{z=1} + \frac{2}{M^2(1-\sigma^2)}\left(\frac{\partial^2\xi}{\partial x^2} + \sigma\frac{\partial\zeta}{\partial x}\right) \tag{1.14}$$

$$\frac{\rho_w}{\rho}\frac{h}{R_o}\frac{\partial^2\zeta}{\partial t^2} = p\Big|_{z=\alpha} - \frac{1}{Re}\tau_{zz}\Big|_{z=1} - \frac{2}{M^2(1-\sigma^2)}\left(\zeta + \sigma\frac{\partial\xi}{\partial x}\right) \tag{1.15}$$

Table 1: Reference state – Dimensionless numbers

Physical quantity	**Symbol of reference quantity**	**Dimensions**
Length (radius)	R_o	m
Velocity	U_o	m/s
Peak velocity	$\hat{U}$	m/s
Time	R_o/U_o	s
Density	ρ	kg/m^3
Constant fluid viscosity	μ	N s/m^2
Pressure	ρU_o^2	N/m^2
Shear stress	$\mu U_o/R_o$	N/m^2
Axial tension	$\rho U_o^2 R_o$	N/m

Table 1 (continued)

Dimensional number	**Physical meaning**	**Related flow fields**
Reynolds number = $\frac{U_0 R_0}{\nu}$ =Re	Ratio of inertial forces to viscous forces	Viscous flows in general
Peak Reynolds number= $\frac{\hat{U} R_0}{\nu}$	Ratio of unsteady inertial forces to viscous forces	Viscous pulsating flow
Strouhal number = $\frac{\omega R_0}{U_0}$ =St	Ratio of unsteady inertial to steady inertial forces	Unsteady flows
Womersley number = $R_0\sqrt{\frac{\omega}{\nu}} = \alpha$ $=\sqrt{Re \cdot St}$	Ratio of unsteady forces to viscous forces	Unsteady flows
Mach number= $\frac{U_0}{c_0}$ =M $c_0 = \sqrt{\frac{Eh}{2\rho R_0}}$ =Velocity of sound (Moens – Korteweg)	Ratio of inertial forces to elasticity forces	Flow with compli-ant effects

1.4 Quasi 1D equations of Incompressible flows inside tubes with deformable walls

A large number of the fluid-carrying vessels in the human body are compliant. In order to perform a numerical simulation of the inherently unsteady biological flow fields generated within such moving boundaries, the fluid-structure interaction problem must first be modeled and solved.

In many cases of cardiovascular fluid simulations, or biological flows simulation in general, an adequate numerical model can be based on the quasi one-dimensional fluid flow equations inside tubes of deformable walls. In such models, some specific laws-models for the wall deformation in accordance to transmular pressure are also investigated.

1D wave-type equations models simulate unexceptionably the pressure wave propagation inside tubes with moving-deformable walls, like the branches of the arterial system. The quasi 1D problems are also modeled by simple numerical methods, the calculations are extremely faster (on the order of seconds) than corresponding 2D and of course 3D models, while permit the prediction of collapse of the arterial wall under some specific flow conditions.

The time-dependent flow equations for the quasi one-dimensional fluid flow, consist of the continuity equation (1.15), and the momentum equation (1.16)

$$\frac{\partial A}{\partial t} + \frac{\partial (Av)}{\partial x} = 0 \qquad (1.15)$$

$$\frac{\partial v}{\partial t}+\frac{\partial}{\partial x}\left(\frac{v^2}{2}+\frac{p}{\rho}\right)=-F \tag{1.16}$$

where A is the cross-sectional area of the flow domain (1D). The term F represents the viscous friction force. In spite of the unsteady character of the flow, the friction force is assumed to correspond to the Poiseuille resistance for laminar flow in circular tube. This assumption is valid for relative small Womersley numbers. In these circumstances, the additional inertial and frictional losses arising from velocity profile distortion in unsteady flow do not differ greatly from those in steady (Poiseuille) flow conditions. So, the friction force can be expressed as

$$F=\frac{8\pi\nu\, v}{A} \tag{1.20}$$

To solve the above problem, the relationship between the transmural pressure acting on each vessel and its cross-sectional area (elasticity) is required. This is a functional relationship between A and p, such as A=A(p), with the property $dA/dp > 0$, that depends on the elastic properties of the vessel wall and its geometrical data. The elasticity coefficient of a vessel depends on the elastic properties of the wall material (Young's modulus, E), the wall thickness (h) and the vessel diameter, as it is provided from various models.

2 Basic Concepts for the Construction of Numerical Schemes for the Solution of Incompressible Navier-Stokes Equations

Numerical solution methods determine discretization of the solution domain and the model equations and consequent solution of a system of algebraic equations for the prediction of the unknown solution flow vector on the computational points of the discretized flow domain. The computational domain is discretized by defining a numerical grid. Continuous physical quantities are replaced by discrete values on computational points in space only, namely the grid nodes. Spatial discretization of flow equations may be based on finite differences, finite volumes finite elements and spectral methods. Each formulation is related directly to the expression of flow equations (differential of integral form) and the type of problem to be solved (i.e. the type of differential flow equations, hyperbolic, parabolic, elliptic). For each term in the model equations, convective, diffusive, transient or source terms, specific discretization is also applied.

The construction of a numerical method for the solution of flow equation consists in the following basic steps:

- Formulation of the proper system of differential equations that obey the examined category of flows. Completion of the problem equations with required constitutive relation for the specific flow cases that are examined (equations of state, elasticity relations)
- Study of the type of the problem to be solved in accordance to desired time accuracy (steady or unsteady problems)
- Determination of the proper time integration scheme
- Determination of the methods of spatial discretization to be used

- Modeling of the physical problem to be solved mainly by the construction of the proper numerical grid, the imposition of the suitable boundary conditions and the accurate approximation of the real fluid properties
- Considerations about concepts of stability, consistency, convergence, spatial and time accuracy
- Construction of algorithm for the solution of the algebraic discretized system of equations. Valuations of accelerating techniques for the analysis procedure.

2.1 Time Integration –Spatial Discretization

Time Integration. The time derivatives of the Navier-Stokes equations are used either for the construction of iterative methods by using time marching techniques in the case for steady flows, or for the evolution of the flow field to sequence physical time levels in the case of unsteady flows. In accordance to the handling of the numerical integration of N-S equation we can classify numerical methods as follows:

One-step and multistep methods in accordance to the steps between two successive physical time levels. For nonlinear conservation laws the standard high-resolution methods are all one-step methods and we will concentrate on these.

Implicit and explicit methods. An explicit method is one which yields an explicit expression for each value $q^{n+1}(i)$ at time t^{n+1} in terms of nearby values at time t^n. An implicit method couples together values at different grid points at time t^{n+1} and hence an algebraic system of equations must be solved in each time step in order to advance the solution.

For many problems, an explicit method turns out to be unstable unless the time step is orders of magnitude smaller than what seems reasonable based on accuracy considerations.

For a nonlinear differential equation an implicit method will give rise to a nonlinear algebraic system to solve in each time step which may be much more expensive than an explicit method with the same time step. However, in situations where an explicit method would force us to take much smaller time steps than desired, the use of an implicit method usually seems to be much more efficient. Implicit methods have less stringent stability bounds and are usually chosen because we wish to obtain solutions, which require fine grid spacing for numerical resolution and we do not want to limit the time steps by employing a conditionally stable explicit scheme.

Spatial Discretization (Finite Differences, Finite Volumes and Finite Elements)

The domain discretization may be consisted by either an unstructured grid of finite elements, an orthogonal Cartesian grid, or a curvilinear body fitted numerical grid. In addition, if any of the flow domain boundaries are moving or deformable, the arithmetic grid used for the spatial discretization of the flow domain must also be moving.

The discretization of flow equations on the discretized domain can be obtained by many different approaches. They can be mainly grouped into four major categories: finite differences, finite volumes, finite elements methods, Perktold et al. (1995) and spectral methods. In the present notes finite differences and finite volume techniques are incorporated in numerical schemes for the solution of unsteady incompressible flows through domains with deformable walls.

Finite difference methods. The physical domain is covered by a grid of (mostly) structured system of nodal points, arranged along grid-lines, which must be aligned to the direction of differentiation. The differential operator of flow equations is discretized by using finite differences, central or upwind, forming the so-called difference molecule. Finite differences are applied on the differential form of flow equations, that it is not necessary to be expressed in conservative form. This is the main disadvantage of finite difference equations, especially for flow fields with discontinuities that are only preserved if a conservative numerical scheme is applied to conservative expressions of flow equations. This is also crucial in pressure-correction methods, where pressure prediction is related directly to the satisfaction of mass conservation, especially near solid walls, or other boundaries. For the confrontation of this important concept, *staggered* grid (i.e. MAC type grid) were introduced and used, mainly for the most precise satisfaction of continuity equation in discrete form. Finite difference formulations are usually more flexible, especially in the implementation of boundary conditions and in obtaining higher order accuracy.

Finite volume methods. The finite-volume viewpoint is often more valuable in developing methods for conservation laws. In this case, the domain is subdivided into small cells (finite volumes) which may compose a structured or an unstructured numerical grid. The discrete value of variables vector is viewed as an approximation to the average value of this quantity over a grid cell rather than an approximation to a pointwise value. The cell average is simply an integral over the cell divided by its area, so conservation can be maintained by updating this value bases on fluxes though the cell faces. More specifically, the conservation equations in its integral form are written for each finite volume separately. By simply requiring that the outflow of one cell is the inflow into its neighboring cell, a conservative scheme will result collapse of flux terms, resulting identically in the global conservation and certain conservation of mass, momentum and energy. This property is particularly important when computing flows with non-linearities and solving in moving-deforming computational cells.

Although the derivation of such methods may be quite different from that of finite difference methods the resulting formulas may be identical for conservative schemes and canonical grids (rectangular computational domain).

Using the above a discretization scheme, central or upwind representations of the flow variables could be used on the sides of the control volumes, the sides of the finite elements or the computational nodes of a finite difference scheme. Upwind schemes are used mainly for the discretization of convective terms in accordance of the theory of hyperbolic systems of equations, while diffusive terms are discretized by using central schemes.

The choice of the type and order of the spatial differencing is important both in terms of accuracy and stability.

More specifically, upwind operators direction (Steger and Warming, Roe, Harten) should be dictated by characteristic curves inside solution domain. Along the characteristics the information is transmitted in a direction related to the slope (sign) of this characteristic. This is based on theory of hyperbolic systems of equations. When these characteristic direction are not obvious (incompressible flow equations) as a sensor for upwind differentiation is suggested the direction of flow velocity as it is entered into a control volume (as in SOLA algorithm, Hirt et al., 1975). Upwind schemes seems to be much more stable and permit the construction of a fully implicit schemes for the solution of the flow equations, since it loads up the diagonals of the implicit factors of the system of flow equations. They in general can be shown to produce sharp, disturbance free, shocks without added artificial dissipation (they include inherently an amount of internal dissipation).

Upwinding, also gives a physical sense in the discretization process that is related to the physical wave propagation of information inside flow.
When central schemes are used for the discretization of convective terms, explicit and implicit artificial dissipation terms must be added to achieve non-linear analysis.

2.2 Error Analysis and Estimation. Stability, Accuracy Stability and Consistency of the Numerical Scheme

The various types of errors which are unavoidable in the numerical solution of fluid flow problems. More specifically, the user must be able to estimate and eliminate the following types of errors, Ferziger and Peric (1996):

Modeling Errors. The flow equations may be considered as a mathematical model of the problem. We define the modeling error as the difference between the real flow and the exact solution of the mathematical model.

Modeling errors are related directly to the inadequate modeling of each specific flow problem and may concern the numerical grid that discretize the computational domain, the modeling of the physical flow field through the imposition of suitable boundary conditions, the extents of the flow field, the approximation of fluid properties depending strongly on temperature, species concentration or pressure, etc.

Discretization Errors, Stability, Accuracy. We define the discretization error as the difference between the exact solution of the governing equations and the exact solution of the discrete approximation. Discretization errors can only be estimated if solutions on systematically refined grids are compared.

It is much easier to compute the local truncation error (LTE), which is a measure of how well the difference equations model the differential equation locally. This is defined by inserting the true solution into the difference equations at a single arbitrary point in space-time.

Concerning stability, there is typically some restriction on the relation between the time step Δt and the spatial grid size Δx, that is resulted from stability analysis of the numerical scheme to be used. Usually, stability analysis concludes a space varying (local) maximum value for time step in order to a condition like the Courant number restriction to be accomplished (almost uniform Courant number throughout the field).

Iteration Errors. The discretization process normally produces a coupled set of non-linear algebraic equations. These are usually linearized and the linearized equations are also solved by an iterative method since direct solution is usually too expensive (construction of a Newton iterative method can be used for the prediction unknown solution vector by a known time level).Usually, iteration is continued until the levels of residual has been reduced by a particular amount.

We define the iteration error as the difference between the exact and the iterative solutions of the discretized equations.

It is essential to choose an optimum level of iteration error -one that is small enough compared to the other errors (which could not be assessed otherwise) but not smaller (because the cost would be larger than necessary). Knowing when to stop the iteration process is crucial from the point of view of computational efficiency, especially for time accurate algorithms.

As a general rule, should be noted that the iteration errors (sometimes also called convergence errors) should be at least an order of magnitude lower than discretization errors.

2.3 Detection of Errors – Validation of a CFD Code

Once a fluid dynamics code has been written, there are a number of first order checks, which must be passed to assess accuracy and efficiency. Any new code or added feature should undergo systematic analysis with the aim of assessing the discretization errors (both spatial and temporal), of defining convergence criteria in order to assure small iteration errors, and of eliminating as many ''bugs'' as possible.

For this purpose one has to select a set of test cases representative of the range of problems solvable by the code, and for which sufficiently accurate solutions (analytical or numerical) are available.

The major checks, which must be passed to access accuracy and efficiency, are based on the determination of:

- Convergent solution until machine accuracy for some reference cases
- Grid independent solution
- Time step independent solution
- Solution independent on the extents of the computational domain

3 Finite Volumes and Finite Difference Techniques for Blood Flows in Large Arteries. Numerical Techniques for the Solution of Incompressible, Unsteady Flows through Domains with Deformable Walls.

Various time-dependent methods, both implicit and explicit, have been proposed the last three decades for the solution of incompressible flow fields. For the development of implicit iterative procedures for incompressible flows the discretization of the Navier-Stokes equations requires particular consideration since the time derivative of density no longer appears. Hence the time-dependent implicit methods suitable for compressible flows cannot be applied without adaptation.

For time iterative procedures, an approach consists on solving the time-dependent momentum equations in connection with a Poisson equation for the pressure obtained by taking the divergence of the momentum equations and expressing the condition of the divergence free velocity field; it is about the pressure correction methods. The above method of solving Poisson's equation for the pressure was developed firstly by Harlow and Welch (1965).

An arithmetic structure similar to compressible equations can be recovered by adding an artificial compressibility term, under the form of the time derivative of the pressure added to the continuity equation. This is the pseudocompressibility method introduced initially by Chorin (1967), Steger and Kutler (1977), and developed for the solution of steady, unsteady and 3-D flow fields by Peyret and Taylor (1983), Chang and Kwak (1984) , Choi and Merckle (1985), Rizzi and Erikson (1985), Kwak et al. (1986), Merkle and Athavale (1987), Soh and Goodrich (1987) and others.

Representative numerical methods of the above two categories will be presented and tested in biological flows, later in this notes.

3.1 Pressure correction techniques.

Pressure correction methods are commonly used in numerical algorithms for the incompressible Navier-Stokes equations. For an incompressible fluid the density ρ is constant and the continuity equation, becomes the condition of the divergence free velocity field.

$$\nabla \cdot \vec{u} = 0 \tag{3.1}$$

The conservation of momentum equation has the form

$$\vec{u}_t + \vec{u} \cdot \nabla \vec{u} + \nabla p = \frac{1}{Re} \nabla^2 \vec{u} \text{ (non-conservative form)} \tag{3.2}$$

The incompressible Navier-Stokes equations consist of (3.2) together with the constraint (3.1). There is no energy equation and no equation of state. Instead the pressure p must be determined in each time step in such a way that mass conservation of (3.1) remains satisfied. There is no local evolution for p; it must be determined using this global constraint. From the above form of incompressible flow equations, it is evident that there is no way for the construction of an implicit algorithm for the solution of flow field, and a multistage algorithm must be constructed consisted of *prediction-correction* stages.

One common approach is the projection method introduced by Fortin (1971). At each time level, equation (3.2) is integrated in time without the term of ∇p in order to obtain the prediction of velocity field $\vec{u}^*$, i.e. the equation to be solved is

$$\vec{u}_t + (\vec{u}\nabla) \cdot \vec{u} = \frac{1}{Re} \nabla^2 \vec{u} \tag{3.3}$$

For an explicit scheme of first order of accuracy for time integration, the momentum equation yields

$$\frac{\vec{u}^* - \vec{u}^n}{\Delta t} + \left(\vec{u}^n \nabla\right)\vec{u}^n = \frac{1}{Re} \nabla^2 \vec{u}^n \tag{3.4}$$

The resulting velocity field $\vec{u}^*$ is not divergence-free because of the nonlinear terms. From the full momentum equation, is resulted the following equation for the flow field at new time level (n+1)

$$\frac{\vec{u}^{n+1} - \vec{u}^*}{\Delta t} + \nabla p^{n+1} = 0 \tag{3.5}$$

which must be satisfied together with continuity equation at time level (n+1)

$$\nabla \vec{u}^{n+1} = 0 \tag{3.6}$$

By taking the divergence of eq. (3.5), for a divergence-free velocity flow field at (n+1) time level, Poisson equation for the pressure is resulted:

$$\nabla^2 p^{n+1} = \frac{1}{\Delta t} \nabla \vec{u}^* \tag{3.7}$$

The Poisson equation for the pressure is solved with Neumann boundary conditions for the normal pressure gradient obtained by taking the normal component of momentum equations.

After the solution of Poisson equation for pressure, the velocity field is finally corrected using relation (3.5) for the termination of a computational cycle of pressure correction method, until a steady state is achieved (for steady flow problems).

The convergence of Poisson equation for pressure is very crucial in pressure correction methods because it is the sensor for the satisfaction of continuity equation.

An additional condition is essential for the numerical accuracy of the resolution of the pressure equation, namely the compatibility condition that should be identically satisfied by the space discretization. This condition is obtained from Green's theorem, when it is applied to the Poisson equation and results the following boundary condition for normal derivative of pressure

$$\left(\frac{\partial p}{\partial n}\right)_\Gamma^{n+1} = -\frac{1}{\Delta t}\left(\vec{u}_\Gamma^{n+1} - \vec{u}_\Gamma^*\right)\vec{n} \tag{3.8}$$

where Γ is the bound of computational domain, and $\vec{n}$ its outward normal vector.
In accordance to another interpretation of pressure correction methods (introduced by Hirt and Cook, 1972), the first step of prediction of the flow field is of the form:

$$\frac{\vec{u}^* - \vec{u}^n}{\Delta t} + \left(\vec{u}^n \nabla\right)\vec{u}^n = -\nabla p^n + \frac{1}{Re}\nabla^2 \vec{u}^n \tag{3.9}$$

which holds pressure terms of momentum equations. As a consequence, the following velocity correction results

$$\frac{\vec{u}^{n+1} - \vec{u}^*}{\Delta t} + \nabla p' = 0 \tag{3.10}$$

where p' is the pressure correction between two successive time step, such that

$$p^{n+1} = p^n + p' \tag{3.11}$$

The divergence of equation (3.10) yields

$$\nabla^2 p' = \frac{1}{\Delta t}\nabla \vec{u}^* \tag{3.12}$$

for pressure correction with the boundary condition

$$\oint_\Gamma \frac{\partial p'}{\partial n} d\Gamma = \frac{1}{\Delta t}\oint_\Gamma \vec{u}' \cdot d\vec{S} \tag{3.13}$$

for the exactly satisfaction of continuity equation ($\vec{u}^{n+1} = \vec{u}^* + \vec{u}'$).
Instead of Poisson equation for pressure, an algebraic equation may be obtained in accordance to SOLA algorithm (Hirt et al, 1975) when the velocity corrections, of equation (3.10) for example, are discretized and substituted into the also discretized equation of mass conservation, in accordance to the following steps

– velocity corrections relation $\nabla p = -\left(\frac{\delta u}{\delta t}, \frac{\delta v}{\delta t}\right) \Rightarrow \left(\frac{\delta p}{\delta x}, \frac{\delta p}{\delta y}\right) = -\left(\frac{\delta u}{\delta t}, \frac{\delta v}{\delta t}\right)$ (for a two-dimensional problem)

– discrete expressions at nodal points $u_{i,j}^{n+1} = u_{i,j}^* + \frac{\delta t\, \delta p}{\delta x}$, $u_{i-1,j}^{n+1} = u_{i-1,j}^* - \frac{\delta t\, \delta p}{\delta x}$

$v_{i,j}^{n+1} = v_{i,j}^* + \frac{\delta t\, \delta p}{\delta y}$, $v_{i,j-1}^{n+1} = v_{i,j-1}^* - \frac{\delta t\, \delta p}{\delta y}$

– substitution into discretized mass conservation equation

$$\frac{1}{\delta x}\left(u_{i,j}^{n+1} - u_{i-1,j}^{n+1}\right) + \frac{1}{\delta x}\left(v_{i,j}^{n+1} - v_{i,j-1}^{n+1}\right) = 0 \Rightarrow$$

$$\frac{1}{\delta x}\left(\underline{u_{i,j}^*} + \frac{\delta t\, \delta p}{\delta x} - \underline{u_{i-1,j}^*} + \frac{\delta t\, \delta p}{\delta x}\right) + \frac{1}{\delta y}\left(\underline{v_{i,j}^*} + \frac{\delta t\, \delta p}{\delta y} - \underline{v_{i,j-1}^*} + \frac{\delta t\, \delta p}{\delta y}\right) = 0$$

The underscored terms constitute velocity divergence, D, for the values calculated during prediction step. By grouping the terms of pressure corrections, δp, the expression for pressure correction takes the form

$$\delta p = \frac{-D}{2\,\delta t\left(\frac{1}{\delta x^2}+\frac{1}{\delta y^2}\right)} \tag{3.14}$$

which is an algebraic relations requiring much less effort to be solved. By comparing this relation with Poisson equation for pressure or pressure correction, the correspondence of the equations is evident. In both considerations, pressure correction vanishes when the velocity field is divergence-free. In SOLA method the flow equations are discretized on a MAC type grid where scalar quantities (pressure, temperature) are stored at cell center and velocities at east (horizontal component) and north (vertical component) edges of each cell. This type of grid ensures better accuracy for the mass conservation equation discretization and for the imposition of pressure boundary conditions (also for ensuring mass conservation).

The general algorithm for pressure-correction methods consists of the following steps

- Prediction of velocity field from momentum equation with or without pressure terms, using an implicit or explicit scheme
- Solution of a Poisson-type differential or algebraic equations for pressure or pressure correction
- Correction of velocity fields so as to mass conservation be ensured
- Readjustment of boundary conditions for velocity field by taking into account the above corrected distribution
- Evolution to the next time level until convergence to be reached to a steady or an unsteady situation.

Between the second and the fourth step, an iterative procedure for the exchange of more precise data is usually designed for a deeper convergence of continuity equation and acceleration of the whole algorithm convergence.

SOLA method has been extended by the authors (Mathioulakis et al, 1997) in curvilinear coordinate systems for the solution of steady and unsteady flows inside domains of arbitrary shaped walls.

3.2 Pseudo-compressibility Techniques

The incompressible equations after the addition of the pseudocompressibility term, take on an hyperbolic character with pseudo-pressure waves propagating with finite speed. In such type of problems "the information" inside the flow field is transmitted along its characteristic curves. In this sense we can relate the sign of eigenvalues with the upwind representation of the flow variables at the cell faces. The upwinding of the inviscid fluxes gives more freedom in devising implicit algorithms (Steger and Kutler, 1977, and Thomas and Walters, 1985), since it loads up the diagonals of the implicit factors. Upwind differencing (Hartwich et al.,1988, and Rogers and Kwak, 1990), also, alleviates the necessity to add and to tune the numerical dissipation for numerical stability and accuracy as the schemes with central differencing that belong to the family of Beam and Warming Schemes (Beam and Warming, 1976).

The upwind scheme of the hyperbolic problem, in this paper, is based on the extended by the method of pseudocompressibility Flux Vector Splitting method. FVS is a shock-capturing upwind method, well known for solving compressible high speed (transonic and hypersonic) flows. Two of the most known FVS methods are the FVS methods of Steger and Warming (1981) and van Leer (1982).
Here, we extend FVS method of Steger and Warming for solving incompressible flow fields implicitly (Pappou and Tsangaris, 1997) and we are based on the implicit scheme of Drikakis and Tsangaris (1991,1993). In such flow fields the splitting of the convective flux vectors has to change sense because of their non-homogeneous property. This is a very important element of the present study. The values of the flux vectors at the cell faces are approached by upwind schemes up to third order of accuracy, such as the Monotone Upstream Centered for Conservation Law (MUSCL) scheme (Steger and Kutler, 1977) and Hybrid upwind extrapolation (Drikakis and Tsangaris 1993).
The unfactored discretized Navier-Stokes equations are solved by an implicit second order accurate in time scheme, using Gauss-Seidel relaxation technique.
In pseudo-compressibility methodologies the additional term in continuity equation serves as the mechanism for an auxiliary coupling between pressure and velocity field. However, the addition of the pseudocompressibility term, leads the unsteady Navier-Stokes equations to lose their physical meaning, that is only recovered at steady state, where the divergence free condition of continuity equation is satisfied. This fact causes the need for special manipulations of pseudo-compressibility methods for the simulation of unsteady flow fields.
In our study, the introduction of the sense of pseudotime besides the physical time, gives the capability of solving unsteady incompressible flow fields by a pseudocompressibility methodology.
A new type of Flux Vector Splitting (FVS) time accurate methodology is developed and presented in details in the following sections, for the solution of unsteady flows inside domains with moving boundaries (Tsangaris and Pappou, 1999). This is essentially a flexible numerical method, which is capable of prediction of unsteady flows with self-excited unsteadiness. It is about a stable methodology that adequates for the solution of flows in domains with boundaries that are subject to arbitrary movements and strong deformations.

3.2.1 Evaluation of Time Derivatives in the Navier-Stokes Equations for Domains Discretized Using Moving Grids

If any of the flow domain boundaries are moving or are deformable, the arithmetic grid used for the spatial discretization of the flow domain must also be moving.

When a moving finite volume is used for the discretization of the flow domain in a Cartesian coordinate system (x, z; t), time derivatives of the variable Q are expressed in terms of the generalized system of coordinates $(\xi, \zeta; \tau)$ by using the relationship

$$\left(\frac{\partial Q}{\partial \tau}\right)_{\underline{\xi}} = \left(\frac{\partial Q}{\partial t}\right)_{\underline{x}} + \nabla Q \cdot \left(\frac{\partial \underline{x}}{\partial \tau}\right)_{\underline{\xi}} \qquad (3.15)$$

Here, the indices indicate the parameter that is maintained constant when performing the corresponding derivation. The left-hand side of equation (3.15) represents the time derivative of a flow quantity at the generalized coordinate system (ξ,ζ) i.e. at a grid node, since the Cartesian flow domain is transformed into a constant rectangular domain in the generalized coordinate system. The time derivative of the RHS in equation (3.15), with index $\underline{x}$, is evaluated at a fixed position in physical space; it hence corresponds to the time derivative of the unsteady term in the Navier-Stokes equations. The quantity $\left(\frac{\partial \underline{x}}{\partial \tau}\right)_{\underline{\xi}}$ or $\dot{\underline{x}}$ is obviously the grid velocity, or the velocity with which the point of calculation of the time derivative is transported from its initial to its new position. Thus, time derivatives in the Navier-Stokes equations can be evaluated in the following manner:

$$\left(\frac{\partial Q}{\partial t}\right)_{\underline{x}} = \left(\frac{\partial Q}{\partial \tau}\right)_{\underline{\xi}} - \dot{\underline{x}} \cdot \nabla Q$$

3.2.2 Unfactored implicit relaxation solution of the Navier-Stokes equations. The governing flow equations are the Navier-Stokes equations, which, expressed in terms of the generalized system of coordinates (ξ,ζ), and using the above relations for time derivatives in domains with moving boundaries, take the form

$$[\Gamma]\frac{\partial(JQ)}{\partial \tau} + \frac{\partial E}{\partial \xi} + \frac{\partial G}{\partial \zeta} = \frac{1}{Re}\left(\frac{\partial R}{\partial \xi} + \frac{\partial S}{\partial \zeta}\right) \tag{3.16}$$

where $[\Gamma]$ is the singular diagonal matrix, $[\Gamma] = \mathrm{diag}(0,1,1)^T$, and Q the unknown solution vector, $Q = (p \quad u \quad w)^T$, with p being the pressure, and (u,w) the velocity components in physical space. E, G and R, S are respectively the convective and diffusive flux vectors at the plane (ξ,ζ)

$$E = J\begin{pmatrix} u\xi_x + w\xi_z \\ \xi_t\, u + (u^2+p)\,\xi_x + u\, w\xi_z \\ \xi_t\, w + u\, w\xi_x + (w^2+p)\,\xi_z \end{pmatrix} \quad G = J\begin{pmatrix} u\zeta_x + w\zeta_z \\ \zeta_t\, u + (u^2+p)\,\zeta_x + u\, w\zeta_z \\ \zeta_t\, w + u\, w\zeta_x + (w^2+p)\,\zeta_z \end{pmatrix} \tag{3.17}$$

$$\begin{aligned} R &= J(0, \tau_{xx}\xi_x + \tau_{xz}\xi_z, \tau_{xz}\xi_x + \tau_{zz}\xi_z)^T \\ S &= J(0, \tau_{xx}\zeta_x + \tau_{xz}\zeta_z, \tau_{xz}\zeta_x + \tau_{zz}\zeta_z)^T \end{aligned} \quad \text{where} \quad \begin{aligned} \xi_t &= -(\xi_x x_\tau + \xi_z z_\tau) \\ \zeta_t &= -(\zeta_x x_\tau + \zeta_z z_\tau) \end{aligned} \tag{3.18}$$

are the metrics of the transformation, which are used to account for the movement of the grid. The grid velocity vector is

$$\vec{v}_{grid} = (x_\tau, z_\tau) \tag{3.19}$$

The symbol (t) is adopted to represent physical time, and must be distinguished from 'pseudo-time' (τ), the meaning of which will be explained later in this section.
Time integration of the physical unsteady flow equations, using a Crank-Nicholson scheme between time levels $(\bar{\cdot})$ and (n), yields

$$[\Gamma]\frac{\overline{JQ} - (JQ)^n}{Dt} + \theta\left[\left(\frac{\partial \bar{E}}{\partial \xi} + \frac{\partial \bar{G}}{\partial \zeta}\right) - \frac{1}{Re}\left(\frac{\partial \bar{R}}{\partial \xi} + \frac{\partial \bar{S}}{\partial \zeta}\right)\right] + (1-\theta)\left[\left(\frac{\partial E}{\partial \xi} + \frac{\partial G}{\partial \zeta}\right)^n - \frac{1}{Re}\left(\frac{\partial R}{\partial \xi} + \frac{\partial S}{\partial \zeta}\right)^n\right] = 0 \tag{3.20a}$$

where θ is a parameter controlling the time accuracy. $(\cdot)$ levels denote a sequence of intermediate time steps between the physical time level (n) and the level (n+1), at which the flow field must be calculated finally. An iterative procedure must be constructed for the final convergence of the problem at (n+1) state. For this purpose, the concept of pseudo-time is introduced besides that of physical time and additional derivatives in pseudo-time are evaluated for the pressure, as well as for the velocity components. By adding these auxiliary derivatives, the continuity and momentum equations take the form

$$\frac{\partial(\overline{JQ})}{\partial\tau}+[\Gamma]\frac{\overline{JQ}-(JQ)^n}{\Delta t}+\frac{1}{2}\left[\left(\frac{\partial\overline{E}}{\partial\xi}+\frac{\partial\overline{G}}{\partial\zeta}\right)-\frac{1}{Re}\left(\frac{\partial\overline{R}}{\partial\xi}+\frac{\partial\overline{S}}{\partial\zeta}\right)\right]$$
$$=-\frac{1}{2}\left[\left(\frac{\partial E}{\partial\xi}+\frac{\partial G}{\partial\zeta}\right)^n-\frac{1}{Re}\left(\frac{\partial R}{\partial\xi}+\frac{\partial S}{\partial\zeta}\right)^n\right] \tag{3.20b}$$

where τ is the pseudo-time. The additional quantity $\partial JQ/\partial\tau$ provides a path that is capable of yielding a time-accurate unsteady solution for incompressible flows. The physical unsteady Navier-Stokes equations are satisfied when the derivatives in pseudo-time vanish, i.e. when a 'steady state in pseudo-time' is reached.

In order to achieve such a steady state, equation (3.20b) is integrated in pseudo-time (τ), by using an Euler implicit scheme of first order accuracy in pseudo-time

$$\frac{\overline{JQ}^{m+1}-\overline{JQ}^{m}}{\Delta\tau}+[\Gamma]\frac{\overline{JQ}^{m+1}}{\Delta t}+\frac{1}{2}\left[\left(\frac{\partial\overline{E}}{\partial\xi}+\frac{\partial\overline{G}}{\partial\zeta}\right)^{m+1}-\frac{1}{Re}\left(\frac{\partial\overline{R}}{\partial\xi}+\frac{\partial\overline{S}}{\partial\zeta}\right)^{m+1}\right]=$$
$$[\Gamma]\frac{(JQ)^n}{\Delta t}-\frac{1}{2}\left[\left(\frac{\partial E}{\partial\xi}+\frac{\partial G}{\partial\zeta}\right)^n-\frac{1}{Re}\left(\frac{\partial R}{\partial\xi}+\frac{\partial S}{\partial\zeta}\right)^n\right]. \tag{3.21}$$

The superscript (m) denotes the time level in pseudo-time, which counts the number of internal iterations required to reach convergence at a steady state in pseudo-time. This steady state is the solution of the unsteady physical problem at the advanced time level (n+1).

A Newton iterative method is constructed for the prediction of Q^{m+1}, by linearizing the convective fluxes in pseudo-time about the known pseudo-time level (m)

$$\overline{E}^{m+1}=\overline{E}^{m}+\frac{\partial\overline{E}}{\partial Q}\Delta Q^{m+1},\quad \overline{G}^{m+1}=\overline{G}^{m}+\frac{\partial\overline{G}}{\partial Q}\Delta Q^{m+1}$$

and handling the diffusion as follows

$$\overline{R}^{m+1}=\overline{R}^{m}+\frac{\partial\overline{R}}{\partial Q}\Delta Q^{m+1}+\frac{\partial\overline{R}}{\partial Q_\xi}\Delta Q_\xi^{m+1}+\frac{\partial\overline{R}}{\partial Q_\zeta}\Delta Q_\zeta^{m+1}$$
$$\overline{S}^{m+1}=\overline{S}^{m}+\frac{\partial\overline{S}}{\partial Q}\Delta Q^{m+1}+\frac{\partial\overline{S}}{\partial Q_\xi}\Delta Q_\xi^{m+1}+\frac{\partial\overline{S}}{\partial Q_\zeta}\Delta Q_\zeta^{m+1}. \tag{3.22}$$

The above relations are substituted into equation (3.21). The unfactored equations are solved by a Newton method, i.e. by constructing a sequence of approximations $\left(q_l^{\,\mu}\right)$, such that

$\lim_{\mu>l} \overline{q}_l^{\mu} \rightarrow \overline{JQ}^{m+1}$, where μ is the sub-iteration state, with the correction formula $\overline{JQ}^{\mu+1} = \overline{q}_l^{\mu} + \Delta\overline{q}_l^{\mu+1}$. After a specific number of under-relaxation sub-iterations in pseudo-time, denoted by the superscript (μ), the vector $\overline{JQ}^{\mu+1}$ approaches the unknown solution vector of the problem, $(JQ)^{n+1}$, of the next physical time level (n+1) (see for details Tsangaris and Pappou, 1999). Using the above sequence of approximations and after the addition of the term $-\left([I]\frac{1}{\Delta\tau} + [\Gamma]\frac{1}{\Delta t}\right)\overline{q}_l^{\mu}$ to both sides of eqn (3.21), we obtain the final system for the solution of the unknown variable vector, in 'delta formulation'.

It must be noted that the unknown solution vector contains the Jacobian of the corresponding control volume during time marching. This results directly from the balance of flow quantities over moving and deformable control volumes.

3.2.3 The Flux Vector Splitting Method. The flux splitting methods are based on the splitting of the convective vectors. More specifically the Flux Vector Splitting method (FVS) consists in decomposing the convective flux vectors into two parts, according to the sign of the eigenvalues

$$E = E^{+} + E^{-} \quad , G = G^{+} + G^{-} \tag{3.23}$$

The upwind representation of the fluxes at the cell faces of the cell centered collocated grid, which is required for the discrete representation of the flux derivatives in the RHS of equation (3.21), is therefore achieved. This splitting gives the direction of the upwinding, when the discrete representation of the flux vectors E_ξ and G_ζ is obtained from the differences in the values of the fluxes E, G at the cell faces.

This approximation for the fluxes E_ξ and G_ζ for the FVS method with the use of upwind differencing, is accomplished in the case of compressible flows, where the fluxes are homogeneous functions of the primitive variables vector Q. The homogeneous property is valid because of the existence of an equation of state in the case of compressible fluid.

In the case of incompressible flow the homogeneous property for the flux vector is not valid. Thus, for the split representation of the flux vectors E_ξ and G_ζ, the following interpretation is used (Pappou & Tsangaris, 1997)

$$\frac{\partial E}{\partial \xi} = \frac{\partial E}{\partial Q}\frac{\partial Q}{\partial \xi} = A\frac{\partial Q}{\partial \xi} \; , \; \frac{\partial G}{\partial \zeta} = \frac{\partial G}{\partial Q}\frac{\partial Q}{\partial \zeta} = C\frac{\partial Q}{\partial \zeta} \tag{3.24}$$

The discretization of the flux vectors requires the upwind representation of the vector variables Q at the cell faces, according to the sign of the eigenvalues of the Jacobian matrices A and C. Flux splitting is achieved by the splitting of the Jacobian matrices A, C into a positive and a negative part, which are related to positive and negative eigenvalues respectively, as for the case of compressible flow. In particular, the splitting for the present method takes the form

$$\frac{\partial E}{\partial \xi} = A^{+}\frac{\partial Q^{+}}{\partial \xi} + A^{-}\frac{\partial Q^{-}}{\partial \xi} = \left(T\Lambda^{+}T^{-1}\right)\frac{\partial Q^{+}}{\partial \xi} + \left(T\Lambda^{-}T^{-1}\right)\frac{\partial Q^{-}}{\partial \xi} \tag{3.25a}$$

$$\frac{\partial G}{\partial \zeta} = C^{+}\frac{\partial Q^{+}}{\partial \zeta} + C^{-}\frac{\partial Q^{-}}{\partial \zeta} = \left(NL^{+}N^{-1}\right)\frac{\partial Q^{+}}{\partial \zeta} + \left(NL^{-}N^{-1}\right)\frac{\partial Q^{-}}{\partial \zeta} \tag{3.25b}$$

where Λ, L are the diagonal matrices of the eigenvalues of the Jacobian matrices A and C respectively. These matrices are decomposed into a positive (Λ^+, L^+) and a negative (Λ^-, L^-) part, which contain the positive and the negative eigenvalues matrix respectively.
The (+) sign in the above equations denotes the Jacobians A, C that correspond to the positive eigenvalues, and dictates a backward upwinding for the quantities $\partial Q^+/\partial\xi$ and $\partial Q^+/\partial\zeta$, while the (−) sign denotes the 'negative' matrices A, C and a forward upwinding for the quantities $\partial Q^-/\partial\xi$, $\partial Q^-/\partial\zeta$.
The upwind discrete form of the derivatives of the vector variable can be considered in the following two ways:

(a) by making the assumption (in the ξ-direction for example):

$$\left(\frac{\partial Q}{\partial\xi}\right)^+_{i,k} = Q^+_{i+\frac{1}{2},k} - Q^+_{i-\frac{1}{2},k}\ ,\quad \left(\frac{\partial Q}{\partial\xi}\right)^-_{i,k} = Q^-_{i+\frac{1}{2},k} - Q^-_{i-\frac{1}{2},k} \tag{3.26}$$

The use of this approach requires the upwind representation of the vector variable at the cell faces, which can be obtained from the sign of the eigenvalues at the center of the current finite control volume.

(b) by calculating the derivatives $\partial Q/\partial\xi$ and $\partial Q/\partial\zeta$ at the cell faces, with central differencing of the values of the vector variables at the cell centers:

$$(Q_\xi)_{i+1/2} = Q_{i+1,k} - Q_{i,k}\,,\quad (Q_\xi)_{i-1/2} = Q_{i,k} - Q_{i-1,k} \tag{3.27}$$

and by consequently expressing, using upwind representation, the values of the derivatives of the vector variables at the center of the control volume, in accordance with the splitting of the Jacobian matrix A. The order of accuracy of the arithmetic scheme that is produced consequently increases.
Both splittings are obtained in accordance with the values of the Jacobian matrix A, which are calculated at the centre of the cell. Calculations have been made with both above schemes for various test cases.
The values of the flux vectors at the cell faces are approached using upwind schemes that are accurate up to third order, such as the MUSCL (Monotone UpStream Centered for Conservation Law) scheme and hybrid upwind interpolation, as described hereafter:
Hybrid upwind extrapolation uses formulations that are up to third order accurate.
The monotone upstream centered for conservation law form (MUSCL) is frequently used for incompressible flows. MUSCL is the following three-point scheme

$$Q^+_{i+1/2,k} = Q_{i,k} + \frac{g_{i,k}}{4}[(1-kg_{i,k})\nabla + (1+kg_{i,k})\Delta]Q_{i,k}$$

$$Q^-_{i+1/2,k} = Q_{i+1,k} - \frac{g_{i+1,k}}{4}[(1+kg_{i+1,k})\nabla + (1-kg_{i+1,k})\Delta]Q_{i+1,k} \tag{3.28}$$

with $\Delta Q_{i,k} = Q_{i+1,k} - Q_{i,k}$ and $\nabla Q_{i,k} = Q_{i,k} - Q_{i-1,k}$.

The spatial accuracy depends on the parameter k. For instance the values $k = -1$, $k = 0$, $k = 1/3$ and $k = 1$ respectively produce fully upwinded, symmetric, third-order biased and centered schemes. The limiter g is the Van Albada sensor $g_{i,k} = \frac{1}{2} \frac{\nabla Q_{i,k} \Delta Q_{i,k}}{(\nabla Q_{i,k})^2 + (\Delta Q_{i,k})^2 + \varepsilon}$ ($\varepsilon \cong 10^{-5}$) where ε is a small number.

The role of the limiter is the reduction of the order of accuracy in regions where strong discontinuities of the flow are observed. Such regions can be found in the vicinity of shock waves for compressible flows and around recirculation zones for incompressible flows.

The upwind representation of the inviscid terms leads to the construction of a fully implicit scheme for the solution of the flow equations, since it loads up the diagonals of the implicit factors of the system of flow equations that all of them are solved simultaneously due to the existence of the pseudocompressibility quantity in the continuity equation. In comparison with other methods, this is one of the most significant advantages obtained by using the pseudo-compressibility for solving incompressible flows. The unfactored discretized Navier-Stokes equations are eventually solved by a very stable, fully implicit second order accurate (in time) scheme using the Gauss-Seidel relaxation technique. In this way, the errors that coexist with an implicit method of approximate factorization are avoided.

3.2.4 Definition of grid velocities. When the flow field inside a domain with moving boundaries is investigated, there is movement or deformation of the numerical grid, since the grid itself is a discrete representation of the flow domain. As previously described, the grid velocities participate in the numerical solutions of unsteady flow fields inside domains with moving walls. The definition of the grid velocity is not obvious since it is not a physical quantity. Its determination must be done in such a way as to avoid the introduction of non-physical terms in the flow equations.

For the numerical solution of fluid flow problems in moving co-ordinates, an additional conservation equation, namely the Space Conservation Law (SCL), has to be solved simultaneously with the mass, momentum and energy conservation equations. The SCL is a fundamental equation for moving coordinates, that only permits changes in the control volume if these are induced by the cell faces velocities. The SCL takes the form

$$\frac{d}{dt}\int_V dV - \int_S \vec{v}_b \, d\vec{s} = 0 \tag{3.29}$$

where V is the time varying control volume and $\vec{v}_b$ is the boundary velocity of the control volume. In order to demonstrate the necessity of satisfying the SCL we consider the mass conservation equation for an incompressible fluid. The boundary velocity and the time variation of the control volume (and consequently the time variation of the flow rate) are inserted in the following integral form of the continuity equation, by taking the movement of the control volume into account

$$\frac{d}{dt}\int_V \rho \, dV + \int_S \rho (\vec{v} - \vec{v}_b) \, d\vec{s} = 0 \tag{3.30}$$

After some rearranging, the above equation becomes

$$\underbrace{\frac{d}{dt}\int_V dV - \int_S \vec{v}_b d\vec{s}}_{\text{SCL}} + \int_S \vec{v} d\vec{s} = 0 \tag{3.31}$$

The marked terms constitute the SCL, and the continuity equation takes its physical form, i.e. $\mathrm{div}(\bar{v})=0$ (incompressible flow), only if this conservation law is satisfied. The non-compliance of the numerical solution procedure with the SCL generally implies that artificial sources of flow quantities are created, and in this case a very important error can be induced.

In order to satisfy the SCL, the grid velocities are defined in such a way that the rate of change of the cell volume $(\delta V)_{SCL}/\Delta t$ obtained from the SCL, is exactly equal to its corresponding actual (geometrical) rate $(\delta V)_G/\Delta t$

$$(\delta V)_G = (\delta V)_{SCL} \tag{3.32}$$

The SCL can be used for the calculation of the control volumes at each time level, in order to define the numerical grid. In using this approach, the grid velocities must be considered as unknown quantities during time marching. This approach is quite usual, because the grid velocities are not physical quantities and cannot be calculated by using an obvious technique during flow elevation. The grid movement is usually handled by constructing a numerical grid at each time level, which may be prescribed, and by consequently calculating the grid velocities in compliance with the SCL.

As an example, we shall give hereafter the expressions for the calculation of the grid velocities referring to the 'east' and 'north' cell faces in accordance with the geometrical quantities of Fig. 3.1 (Demirdzic and Peric 1998, 1999):

$$u_{g,e} = \frac{S^o_{x,e} + S^n_{x,e}}{2S^n_{x,e}} \frac{\delta x_e}{\Delta t}, \quad w_{g,e} = \frac{S^o_{z,e} + S^n_{z,e}}{2S^n_{z,e}} \frac{\delta z_n}{\Delta t} \tag{3.33a}$$

$$u_{g,n} = \frac{S^o_{x,n} + S^n_{x,n}}{2S^n_{x,n}} \frac{\delta x_n}{\Delta t}, \quad w_{g,n} = \frac{S^o_{z,n} + S^n_{z,n}}{2S^n_{z,n}} \frac{\delta z_n}{\Delta t} \tag{3.33b}$$

It is a matter of simple algebraic operations to prove that the above expressions satisfy the form of the SCL of equation (3.32).

Cell faces displacement between time levels '0' and 'n'

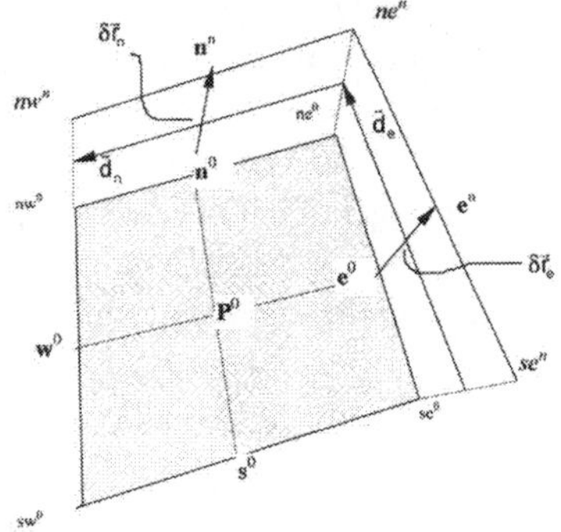

Cell face vectors:

$\vec{S}_e = (S_x, S_z)_e =$

$(z_{ne} - z_{se}, \ -x_{ne} + x_{se})$

$\vec{S}_n = (S_x, S_z)_n =$

$(-z_{ne} + z_{nw}, \ x_{ne} - x_{nw})$

Cell face displacement vectors:

$\delta\vec{r}_e = \vec{r}^{\,n}_e - \vec{r}^{\,o}_e =$

$(x^n_e - x^o_e, \ z^n_e - x^o_e)$

$\delta\bar{r}_n = \bar{r}^n_n - \bar{r}^o_n =$

$(x^n_n - x^o_n, \ z^n_n - x^o_n)$

Cell face velocities:

$\vec{u}_{ge} = (u_g, w_g)_e$

$\vec{u}_{gn} = (u_g, w_g)_n$

Figure 3.2. Control volume displacement at two successive time levels.

4 Applications for Blood Flow in Large Arteries

4.1 A Model Based on the Quasi 1D Flow Equations for the Prediction of Left Epicardial Coronary Blood Flow in Normal, Stenotic and Bypassed Coronary Arteries, by Single or Sequential Grafting

In the present study, a numerical model based on the quasi 1D flow equations was formulated and calibrated for the prediction of left epicardial coronary blood flow of a right dominant coronary circulation. It is used and verified for the prediction of normal , stenotic and bypassed coronary arteries, by single or sequential grafting. The model describes blood flow through a complicated circuit of deformable vessels, while simulates unexceptionably the pressure wave propagation inside tubes with moving-deformable walls and permits the prediction of collapse of the arterial wall under some specific flow condition (see for details Rammos et al., 1998).

The predicted results were compared with *in vivo* measurements in humans during cardiac surgery operations concerning the following main points:

1. The phasic characteristics of coronary pressure and flow in the normal coronary circuit and the effect of coronary artery disease.
2. The degree of efficacy of the revascularization procedure with single and sequential bypass grafts.
3. The haemodynamic comparison between these two bypass methods, with the belief that good early results predispose to favourable mechanical conditions that possibly offer adequate long-term graft patency rates.

For the fluid-dynamic study of left epicardial coronary circulation, a geometrical model based on anatomical and physiological data was used. It is the geometrical model of a right-dominant coronary circulation (RCA dominant) with data derived primarily from the data published in literature (Figure 4.1a). The used circuit of arteries consists of 16 vascular branches. The discrete geometrical parameters of each of the 16 branches, such as length and the local diameter at three positions (proximal, middle and distal) were obtained.

The mathematical model for the fluid-mechanic study of the left epicardial coronary circulation is based on the quasi 1D flow equations of paragraph 1.4 (equations 1.15 for mass conservation and 1.16 for momentum conservation). The primary variables of the problem are the velocity of the blood (v), the intracoronary blood pressure (p) and the area of the cross-section of the artery along the model (A). The blood kinematic viscosity ν was taken equal to 0.033 cm^2/s.

The system of governing equations is of hyperbolic type with characteristic curves on x-t plane, that can be solved with all the numerical techniques used in the case of compressible flows. The calculations in 1D problems are much less time consuming in comparison with corresponding calculation in two or three dimensions. This fact prepossesses for the use of explicit schemes for time marching such as Mac Cormack, Lax Wendroff.

To solve the above problem, the relationship between the interior pressure of each vessel and its cross-sectional area (elasticity) is required. This is a functional relationship between A and p, such as A=A(p), with the property $dA/dp > 0$. In this study we consider a linear relationship between A

and p that depends on the elastic properties of the vessel wall and its geometrical data in the following way:

$$A = A(p) = A_0\left[1 + \beta(p - p_d)\right]$$

The cross-sectional area $A_0 = A(p_d)$ of each artery, corresponds to the minimum pressure during the diastolic phase of cardiac cycle. All the above mentioned geometric properties of our model are assumed at this pressure. The elasticity coefficient β depends on the elastic properties of the material of the wall (Young's modulus, E), the, thickness of the wall (h) and the diameter (D_0) of the vessel in the following manner:

$$\beta = \frac{D_0}{Eh}$$

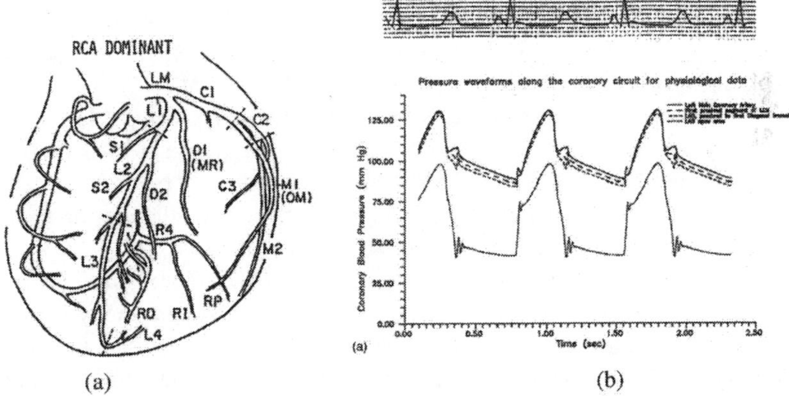

(a) (b)

Figure 4.1 Diagram of coronary artery circuit and branch nomenclature (a) and Pressure waveforms (b) in distinct locations along the normal left coronary system, combined with the corresponding ECG tracing .

The arithmetic scheme that was used in the present study is the time-dependent explicit, two step, finite difference Lax–Wendroff scheme of second order of accuracy in space and time for the solution of an hyperbolic system of differential equations.

The initial conditions for the coronary flow field were the diastolic pressure and zero flow rate for all the branches of the circuit.

The most important part of the arithmetic solution is the selection of suitable boundary conditions at the inlet of the circuitat the ramifications and at the distal ends of the network. More specifically:

-At the inlet of the field (left coronary ostium) the aortic pressure is imposed (Figure 4.1b).

- At the ramification of the network, no pressure losses, continuity of pressure or flow is assumed in order to satisfy the continuity equation.

-At the distal beds, pressure drops proportional to the peripheral resistance blocks are imposed. This terminal resistance is used to account for the cumulative effect of all distal vessels. Because of the topology of the terminal branches in the present circuit, the peripheral resistances are related directly to the state of the myocardial tension. This is achieved by expressing the peripheral resistance as being linearly and directly dependent upon left ventricular pressure, as is described in the literature (Rammos et al., 1998). Thus, during systole when the heart is contracted, the resistance is high, and during diastole when the heart is relaxed, the resistance is low. In this study, during the systolic phase, the left ventricular pressure waveform is imposed at the ends of the terminal branches that supply the left ventricular wall, and for the diastolic phase, the above model for the peripheral resistance is adopted. In this phase (diastolic), the resistance has the form suggested in literature. During the diastolic phase the resistance has the form:

$$R = R_B(1+U\overline{P})$$

Here R_B is the basal level of resistance, and $\overline{P}$ is the pressure function that is related to the myocardial tension:

$$\overline{P} = \frac{p_{LV} - p_{LVD}}{(p_{LV} - p_{LVD})_{max}}$$

Where p_{LV} is the instantaneous left ventricular pressure, p_{LVD} is the minimum diastolic left ventricular pressure and $(p_{LV} - p_{LVD})_{max}$ is the maximum increase in the left ventricular pressure. The quantity $\overline{P}$ ranges from 0 to 1. The quantity U represents the ratio of the maximum intramyocardial resistance to the basal level of resistance. It can also be viewed as being related to the strength of the myocardial contraction. The used values for the above parameters, in this study are $U = 15$ and $R_B = 10\times10^3(\text{dynes/cm}^2)/(\text{ml/s})$. These values of the parameters U and R_B are consistent with corresponding ones from the literature, causing the derived pressure and flow waveforms to be continuous at the transition from the systolic to the diastolic phase.

The two last boundary conditions are simplifying assumptions for the problem. The set of discrete equations is iteratively at each time level of the cardiac cycle and periodicity is achieved after almost 1.5 cycles. For the above procedure, each period is subdivided in 10000 time intervals.

Applying the present arithmetic model, pressure and flow waveforms are obtained at every point of each branch during the cardiac cycle for normal and diseased coronaries and after surgical correction.

Fluid mechanical information concerning the coronary circulation offers insight into pressure/flow data in vascular areas that are inaccessible by conventional invasive or non-invasive methods. It also provides information in evaluating the effects of cardiovascular drugs, andfurther illuminates the pathophysiology of many cardiovascular disorders such as valvular disorders, cardiomyopathies and various surgical treatment methods.

Observing the pressure-flow waveforms in normal vessels, the coronary flow pattern during systole correlates well with findings observed in various papers using intracoronary flowmeter or Doppler velocimetry techniques (see results and review of relevant literature in Rammos et al, 1998).

The same correlation can also be obtained by examining diastolic flow data. Therefore, the time of the beginning of the diastolic flow rise, the shape of the received flow and pressure waveforms, the average and peak values, and the timing of these events with the corresponding ECG tracings, showed a considerable similarity with *in vivo* data.

In coronary artery disease, the main contribution is the reconfirmation of a previously described finding of systolic flow augmentation in stenotic areas. It appears that an increase in blood flow during systole is an important compensatory mechanism in patients with severe perfusion deficits caused by coronary artery disease.

We emphasize the fact that this systolic flow augmentation component is only found in branches that have atherosclerotic changes and not elsewhere.

In these notes derived numerical results for the two different revascularization procedures are depicted in Figure 4.2. Flow/pressure waveforms correlate very well with relevant *in vivo* results obtained by pulsed Doppler and 'other flowmeter techniques' and by other clinical and angiographic studies.

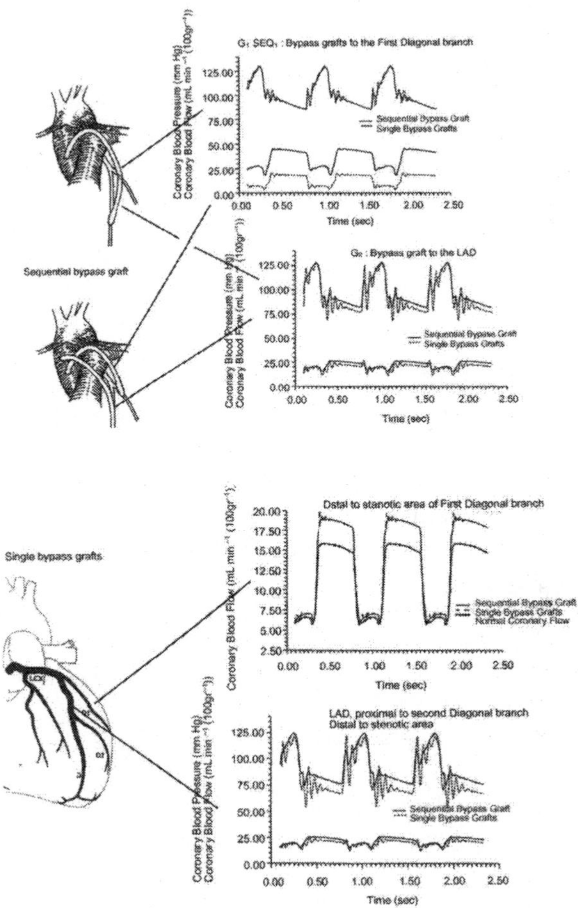

Figure 4.2 Comparative representation of pressure and blood flow waveforms for single and sequential bypass grafting at distinct positions along the anastomosed left coronary system. Specifically at the bypass grafts (a,b), post-stenotically at the first diagonal branch (c) and post-stenotically at the left anterior descending artery (d). Note that sequential provide considerably increased diastolic flow at the proximal anastomotic site (a) and slightly increased diastolic flow post-stenotically (d)

The introduction of the single grafts pre-stenotically does not alter in any way the status of the established coronary artery. Inside the stenotic vascular areas, after single grafting, blood pressure

is restored approximately to the aortic level, and the resultant flow is reduced to almost zero values with a minimal backflow component mainly during systole.
Post-stenotically, blood pressure almost acquires its normal pattern and the resultant flow changes to a diastolic predominant one, a fact that is observed repeatedly by other investigators.
After the positioning of a sequential saphenous vein graft, with the side-to-side (proximal) anastomosis at the first diagonal and the distal (end-to-side) anastomosis at the left anterior descending artery, the resultant flow is increased ~8–10% during diastole and 2–3% during systole, compared with single grafting with the vessel of the same type.
Comparing these two revascularization procedures, sequential grafts tend to provide higher diastolic pressure, and consequently greater diastolic flow especially in the proximal anastomotic site. The above results correlate well with the findings of clinical trials stating that in sequential grafts, side-to-side anastomoses have a decidedly better patency rate than end-to-side ones and better than single grafts in the proximal site. Furthermore, the cumulative early- and late-patency rate of the distal end-to- side anastomosis is not significantly better than that seen with conventional vein grafts.

4.2 Steady and Unsteady Flow Inside Stenosed Arteries

Arterial stenosis in which abnormal growths in the lumen of the arterial wall develop at various locations of the cardiovascular system represents one of the widespread diseases in humans. The deposit of cholesterol on the arterial wall and proliferation of the connective tissues in the wall, due to atherosclerosis, form plaques which grow inward and alter the blood flow. The appearance of large recirculating zones and the strong increase of the pressure drop are two main effects depending on the severity of the stenosis.
The two numerical methods described above, for the solution of steady and unsteady incompressible flows are used for the calculation of steady and unsteady flow inside a stenosed artery. The stenotic area is the 0.25% of the inlet area. The geometry of the artery and the used numerical grid (370x24) are depicted in Figure 4.3.
The present test case is selected as representative for the study of stenosed arteries in the arterial system.
The Re number for steady flow is set equal to 100 and the reference quantities for the non dimensionless numbers is the maximum inlet velocity, the inlet radius and the period of the unsteady inlet.
Concerning the unsteady case, an unsteady component is superimposed on the steady one. The peak Reynolds number of the unsteady profiles equals to 200 and the frequency leads to a Womersley number equal to 5, which is characteristic for biological flows.
Two velocity profiles for the axial velocity components along the flow field for steady flow case, are represented in Fig. 4.4 and the axial pressure distribution is depicted in Fig.4.5. The velocity and pressure results for the flow field calculated with each numerical method are compared and the flow fields seem to correlate significantly.
The pseudocompressibility methodology, Pappou and Tsangaris (1997) permits deeper convergence rate for pressure fields, and extremely faster convergence of the flow field. This is the major advantage of the implicit methods against explicit methods. SOLA method, as a pressure correction method ensures mass conservation at each time step even if the flow field is not convergent yet, Mathioulakis et al. (1997).

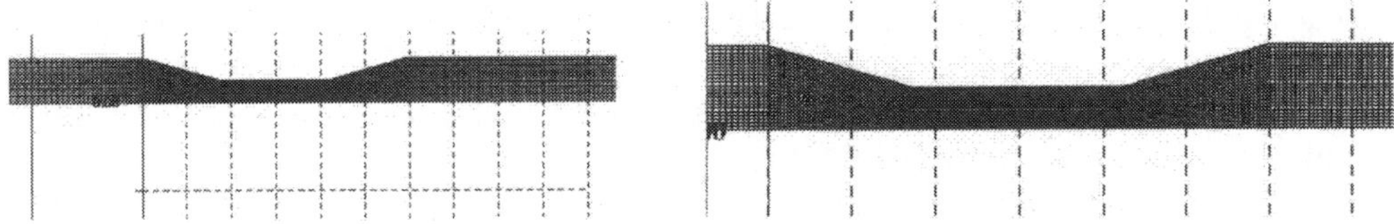

Figure 4.3: Geometry and numerical grid for the solution of steady and unsteady flow inside stenosed artery.

Steady state

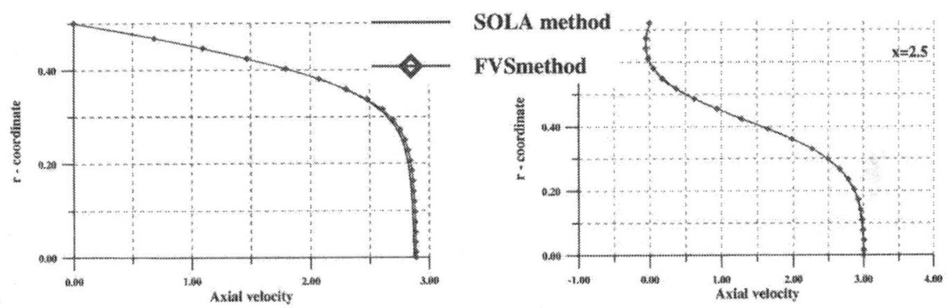

Figure 4.4: Radial and axial velocity profiles along stenosed tube for steady flow

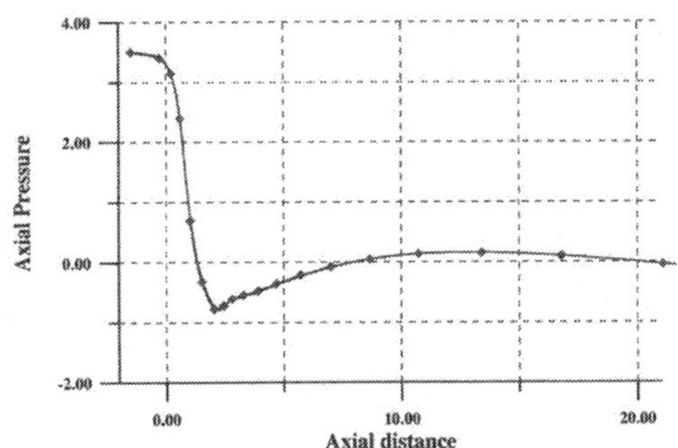

Figure 4.5: Axial Pressure distribution for steady flow

For the unsteady case of flow, velocity profiles at various time levels and axial positions are shown in Fig. 4.6.
When SOLA method is extended for the solution of unsteady flows, between two successive physical time levels a sequence of prediction-correction steps is intermediated, until a divergence free velocity field is obtained.
For the Womersley number of the examined test case 100000 time levels are required for the accurate time integration of unsteady flow equations. For the case of pseudocompressibility

methodology for the time discretization 10000 physical time levels per cycle were taken. In this case, as mentioned above, at each physical time level a pseudo-steady state in preudo-time is converged. At this situation only are satisfied the unsteady Navier-Stokes equations and the derivatives in preudo-time are vanished.

In Fig. 4.7 is depicted the axial pressure distribution along flow field for the unsteady flow case. The presented results are independent on the physical time step and arithmetic grid.

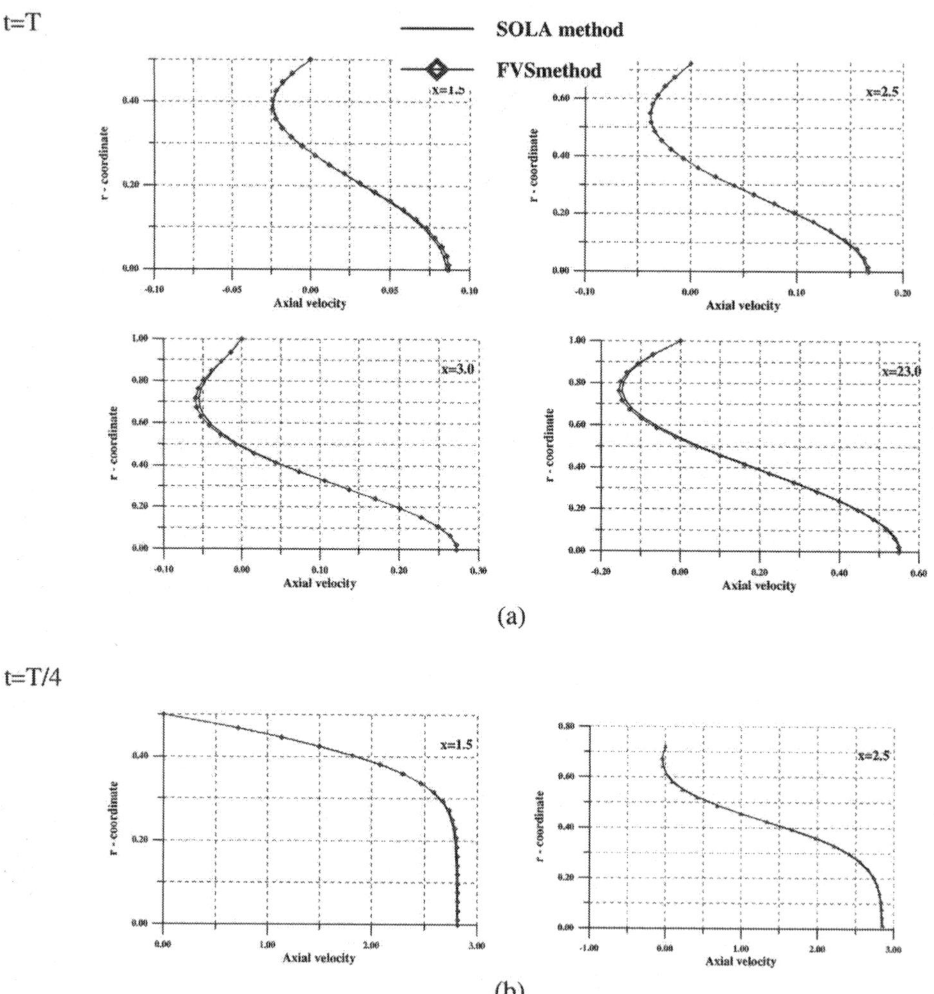

Figure 4.6: Axial velocity profiles along stenosed tube at various time instances during period

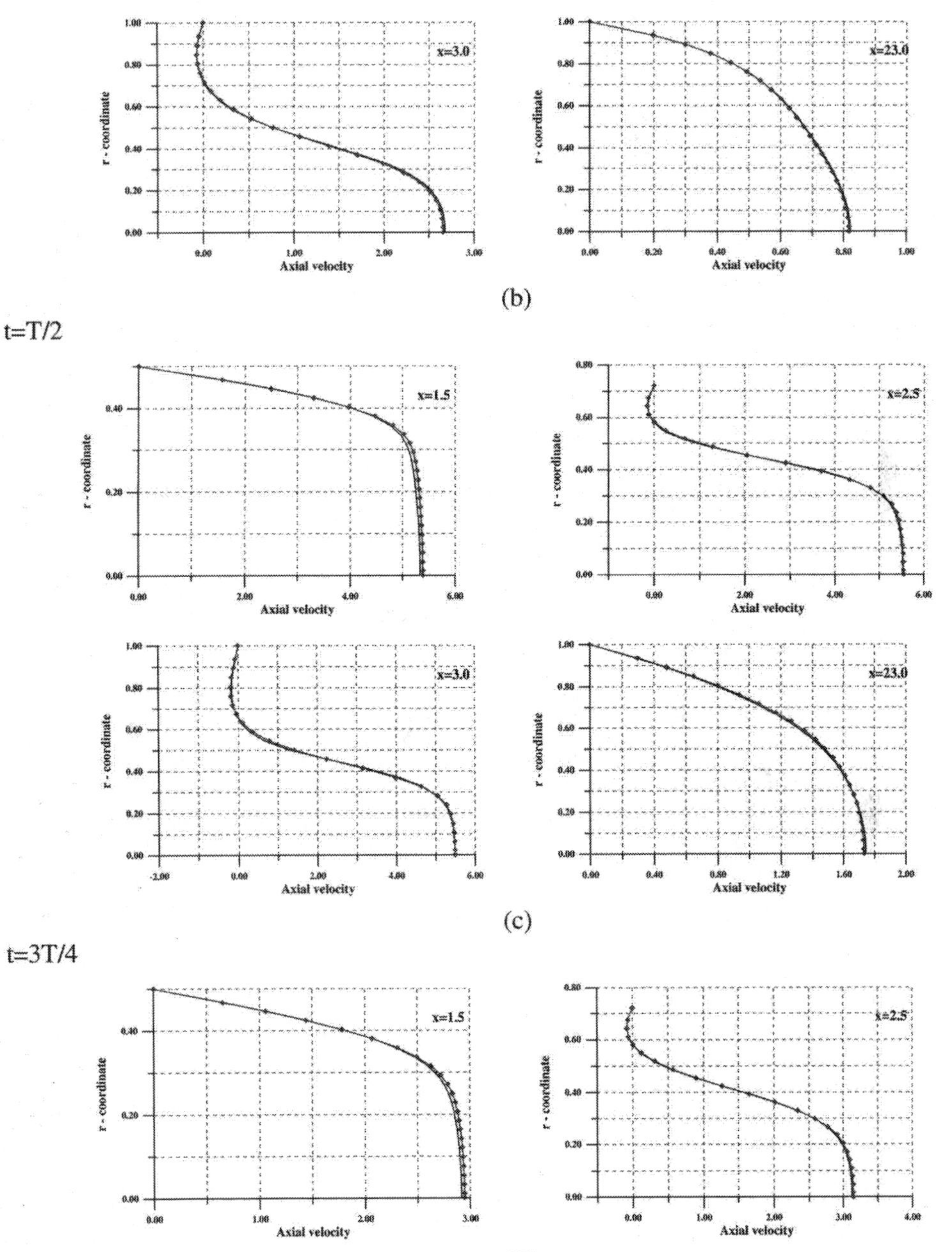

Figure 4.6 (continued): Axial velocity profiles along stenosed tube at various time instances during period

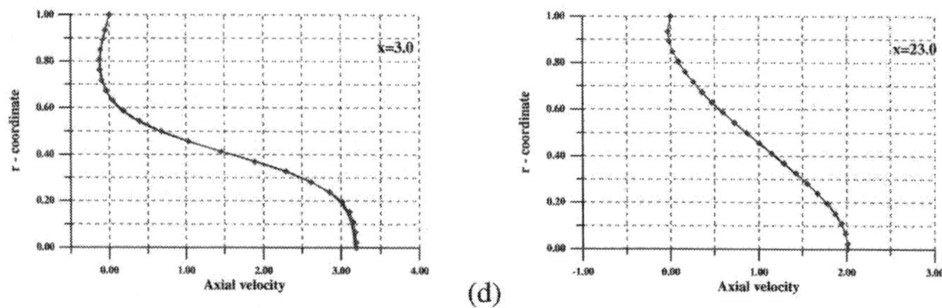

Figure 4.6: Axial velocity profiles along stenosed tube at various time instances during period

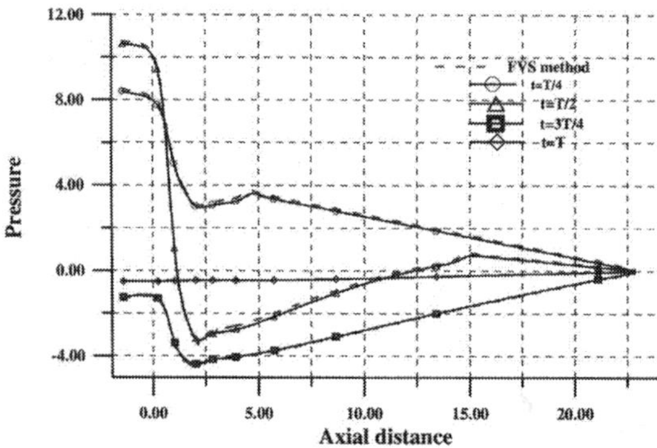

Figure 4.7 : Axial pressure distribution during period for the two numerical methods

4.3 Application to the problem of a deformable wall channel with steady and pulsating inflow conditions

4.3.1 Definition of the problem

A representative test case for the modeling and the solution of a two dimensional fluid-structure interaction problem, is the problem of a channel with a single partially deformable wall (Fig. 4.8). The deformable part of the wall is modeled using the membrane theory.

The geometrical model of this test case was preferred against an axisymmetric one for the simulation of unsteady flow inside vessels, where non-symmetric deformations occur.

This problem has been studied extensively in the relevant literature (Luo and Pedley 1995, 1996 1998, 1998), for steady inflow conditions. Different aspects of this case have been studied computationally, using the Reynolds number, the tension of the membrane, the external membrane pressure, and the inertia of the membrane, as control parameters for the resultant flow field.

The present study includes steady and unsteady flow conditions for the prediction of self excited and sustained unsteadiness. Two additional parameters are also investigated: (i) the ratio of the amplitude of the mean inflow velocity pulsation to the mean steady velocity λ and (ii) the Womersley parameter α_e or reduced frequency of the imposed inlet pulsating flow field.

This frequency is determined in relation to the frequency of the self-excited oscillations of the flow field for the steady inlet conditions.

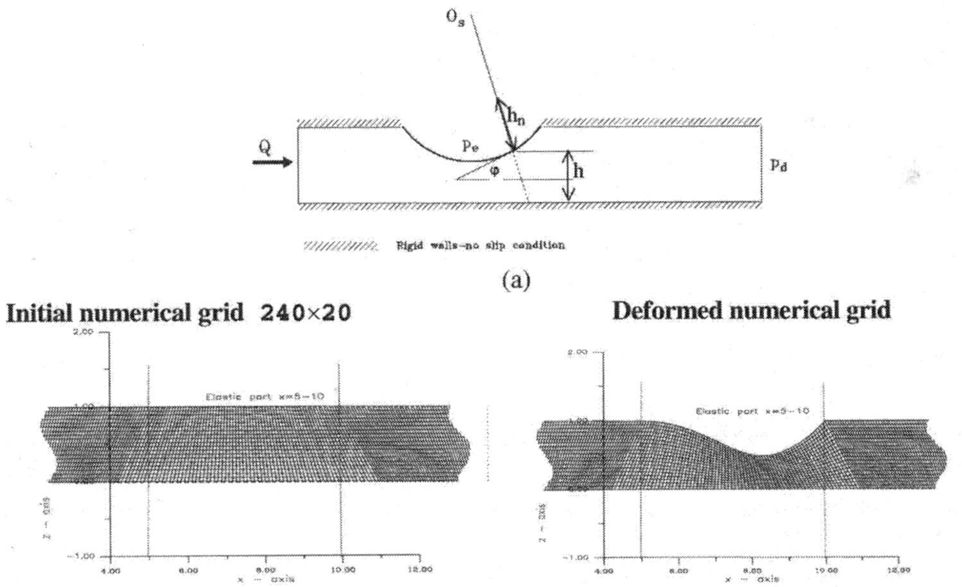

Figure 4.8. Geometrical characteristics of the deformable wall channel (a) and corresponding numerical grid representation (b).

For the movement of the membrane in its normal direction, we use the following simple equation in dimensionless form

$$m\frac{d^2h_n}{dt^2} = \kappa T + (\sigma_n - p_e) \tag{4.1}$$

Where h_n is the displacement of the membrane in its normal direction, T is the longitudinal tension of membrane per unit width, σ_n the normal fluid stress on the wall, p_e the transmural pressure at membrane and $\kappa = d\phi/ds$ (s counts the membrane length) the membrane curvature.

4.3.2 Numerical procedure for the solution of fluid-structure interaction problems

The major aim of a fluid-structure interactive scheme is to achieve full conjunction between fluid and structural analysis. In our numerical scheme for the solution of unsteady flows inside domains with moving boundaries, we use a dual time stepping procedure in 'pseudo-time', and at each physical time level an iterative procedure is constructed, as it was described previously. During this iterative procedure, the fluid problem and the structural problem must interact fully. For unsteady flow inside a channel, a segment of the boundaries of which is elastic, the internal unsteady flow, the external pressure over the elastic part and the moving elastic part must always be in equilibrium, satisfying both the Navier-Stokes equations and the membrane (elastic part) movement equation. In order to achieve this full conjunction between the fluid and the structure problems, we have constructed the following interactive procedure (Tsangaris and Pappou, 1999):

Step I: The solution of the flow field begins at physical time level (n). At this time level, an initial distribution $h(x)$ is assumed for the membrane height. In this way, at the initial state of the calculations, the membrane shape determines the flow domain and consequently the initial numerical grid. We note that the finite volumes are moving with velocity calculated between the present (n) and the previous $(n-1)$ states of the numerical grid. For the first cycle of the calculations (n=1), the grid velocities are supposed equal to zero.

Step II: The flow field is calculated by following the relaxation procedure that was described previously and by applying the following interaction procedure:

At each level of sub-iterations (m), we calculate the flow quantities $u^{n,m}$, $w^{n,m}$, $p^{n,m}$. During the first sequence of iterations in pseudo-time (m=1), the values of step I are used for the grid velocities.

Using the recently calculated pressure on the elastic segment of the boundary $p^{n,m}\big|_{wall}$, as well as the shear stresses on the elastic boundary, we solve the equation of equilibrium-movement of the membrane. During this step, we predict new values for the axial distribution of the membrane height $h^{n,m}(x)$.

Using the new values for the membrane height, we determine the flow domain, as well as the grid velocities, calculated between the levels $\left(x_{i,k}^{n,m}, z_{i,k}^{n,m}\right)$ and $\left(x_{i,k}^{n}, z_{i,k}^{n}\right)$.

We proceed to a control for the convergence of the flow quantities during time marching in 'pseudo-time' in accordance with the residual $\left|Q^{n,m} - Q^{n,m-1}\right| \le \varepsilon$. Another form of convergence control is also attempted for the wall shape at the elastic part via the residual $\left|h^{n,m} - h^{n,m-1}\right| \le \varepsilon$. If the prescribed convergence criteria are satisfied, the latest values of the problem variables are adopted for the new physical time level (n+1), and we set $Q^{n+1} = Q^{n,m}$ and $H^{n+1} = H^{n,m}$ (prediction of the flow field and the elastic wall shape at the physical time level (n+1)), before proceeding to the sequence time level (STEP I). On the other hand, if the problem does not converge, we continue the iterative procedure in pseudo-time until convergence is achieved.

4.3.3 Results for the inlet steady case

The case of steady flow being imposed on the channel inlet has been extensively studied in various papers by Luo and Pedley (1995, 1996 1998, 1998). We shall quote hereafter some of their results, and compare these with results obtained using our methodology. Table 4.1 gives the values of the dimensionless parameters for the inlet conditions in the steady case. If the value of the membrane tension T is gradually reduced, then for a certain combination of the Reynolds number and the external pressure, a critical value of T is reached, beyond which the flow field becomes unsteady. This means that the flow field starts oscillating and a vortex shedding procedure is formed, downstream the stenotic area that is formed during flow, so that a velocity and flow pulsation with a fundamental frequency f_s are initiated. In Figs. 4.5a,b the shape of the membrane and of the corresponding pressure variations across the upper boundary are shown for different time instants within the period of the induced oscillations. The results concern the dimensionless parameter values of Table 4.1 and with the value $\beta = 32.5$ (β is the ratio of a constant reference value of T_0 to the used value of T) for the membrane tension. This value is higher than the critical value, which ensures the onset of unsteadiness in the flow field.

Table 4.1: Testing data for the channel flow

L_e	L	L_o	Re	p_d	p_e	M	$\beta = T_0/T$
5	5	7	300	0	1.033	0	32.5

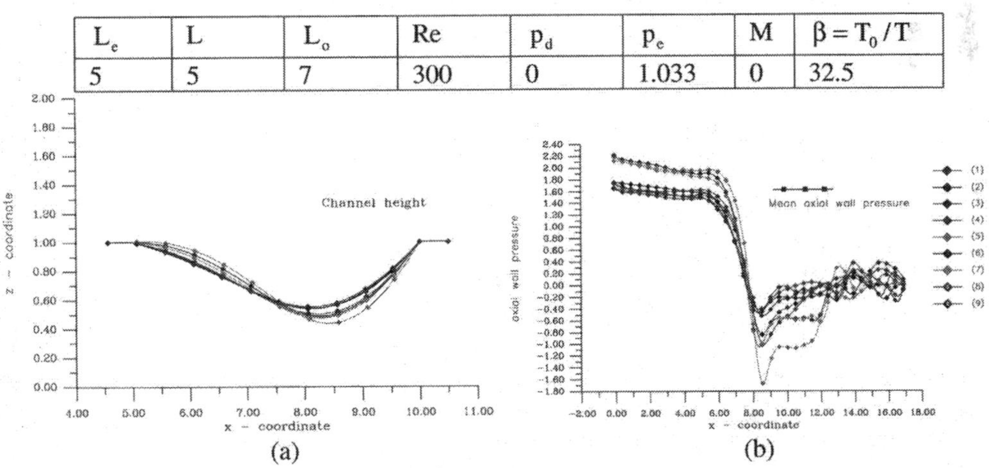

Figure 4.9. Membrane shape at different time instants for the case of steady inflow conditions (a) and pressure distribution across the upper boundary (solid wall and membrane) for different time instants (b).

In Table 4.2, a comparison between the results of literature (Luo and Pedley,1996) and those obtained using our methodology (FVS) is presented. This case concerns steady inlet conditions and inertialless membrane. Very good agreement between the two methodologies is generally observed.

Table 4.2: Comparison of the present FVS methodology with the results of Luo and Pedley (1996).

Re = 300	Maximum indentation		Dimensionless frequency f_s	
	FVS	Luo & Pedley	FVS	Luo & Pedley
$\beta = 30$	16%	20%	0.076	0.083
$\beta = 32.5$	29%	30%	0.043	0.048

4.3.4 Results for the pulsating inlet flow case and discussion

The problem of pulsating inflow volume rate is investigated. The new parameters that determine the pulsating test case are the dimensionless excitation frequencies of the pulsating inflow volume rate f_e, and the dimensionless amplitude of the volume rate flow at the inlet of the channel U_a or λ.

The flow rate at the inlet consists of a sinusoidal variation with excitation frequency f_e superimposed onto the steady flow rate. In dimensionless form is:

$$Q_i(t) = 1 + U_a \sin(2\pi f_e t) \tag{4.2}$$

The other parameters of the test case are identical to those used for the steady inflow case. Three different excitation frequencies and two different inlet flow amplitudes are examined, as shown in Table 4.3. Cases I to VI consist of various combinations of these parameters.

Table 4.3: Additional dimensionless parameters for the case of pulsating inflow conditions

Case	f_e	$\alpha_e = \sqrt{2\pi f_e \text{Re}}$	$U_a = \lambda$
I	0.0215 (=0.5 f_s)	6.366	0.33
II	0.043 (= f_s)	9.003	0.33
III	0.086 (=2 f_s)	12.732	0.33
IV	0.0215(=0.5 f_s)	6.366	0.66
V	0.043 (= f_s)	9.003	0.66
VI	0.086 (=2 f_s)	12.732	0.66

The following flow field quantities are investigated and presented in Figs 4.10 to 4.14. In these figures are depicted the following quantities mainly for the case V:

- Membrane shape during the period (Fig. 4.10, for two different flow cases).
- Time variation of membrane height at three positions along the membrane (Fig. 4.11a).
- Time variation of outlet flowrate versus corresponding pulsating inlet flowrate (Fig. 4.11b).
- Axial pressure distribution along the lower wall (Fig. 4.12a).
- Membrane vertical velocity in the period (Fig. 4.12b).

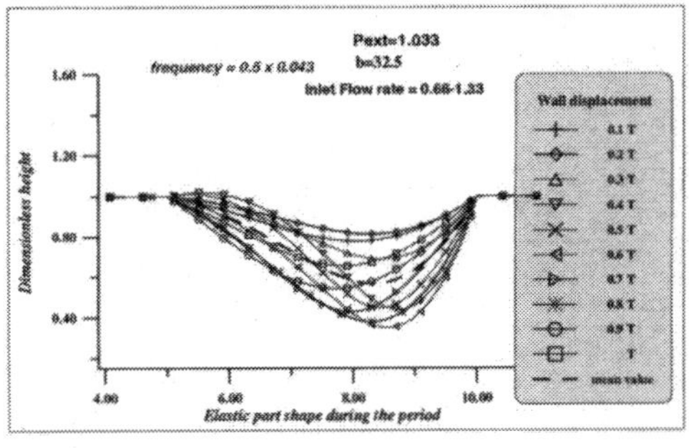

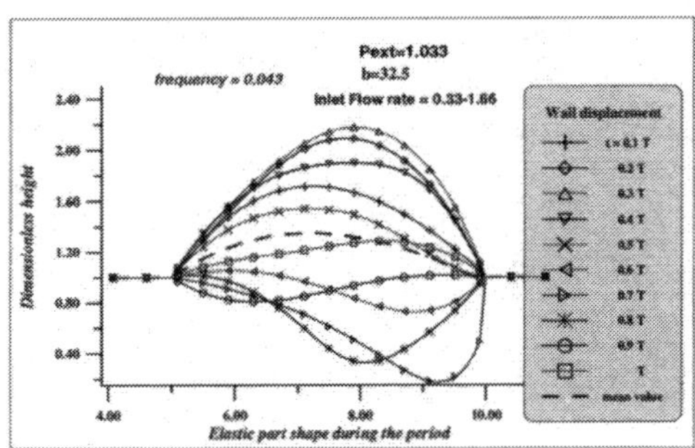

Figure 4.10. Membrane shape during the period for (a) $f_e = 0.0215$ and $U_a = 0.33$ and (b) $f_e = 0.043$ and $U_a = 0.66$.

- Time variations of time derivative of the domain volume (Fig. 4.13 for two values of the inlet flow amplitude).
- Visualization of the flowfield via isolines of pressure, stream and vorticity functions (Figs. 4.15 for streamlines).

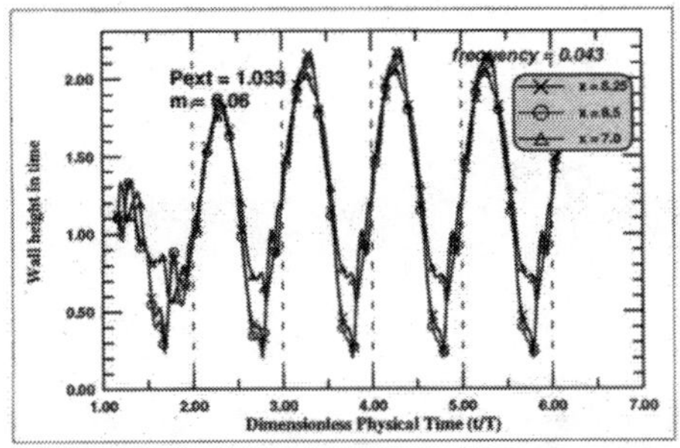

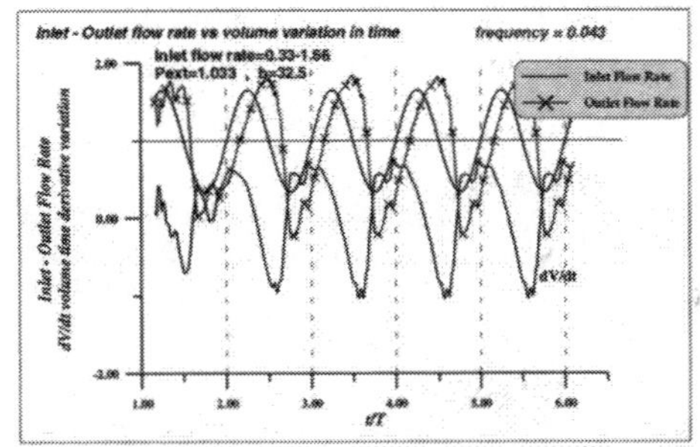

Figure 4.11. Time variation of membrane height for (a) and inlet-outlet flowrate for (b) $f_e = 0.043$ and $U_a = 0.66$.

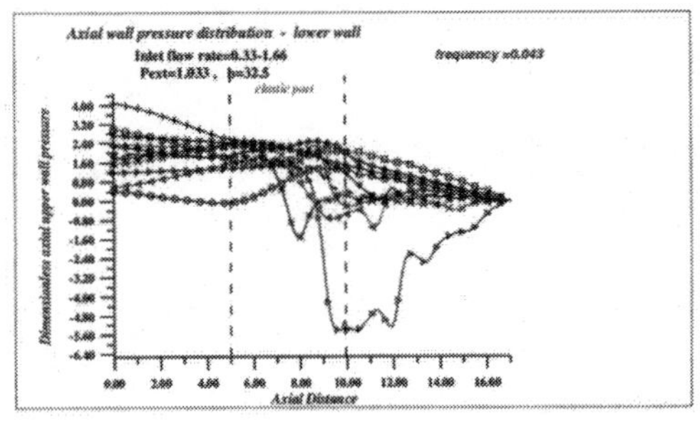

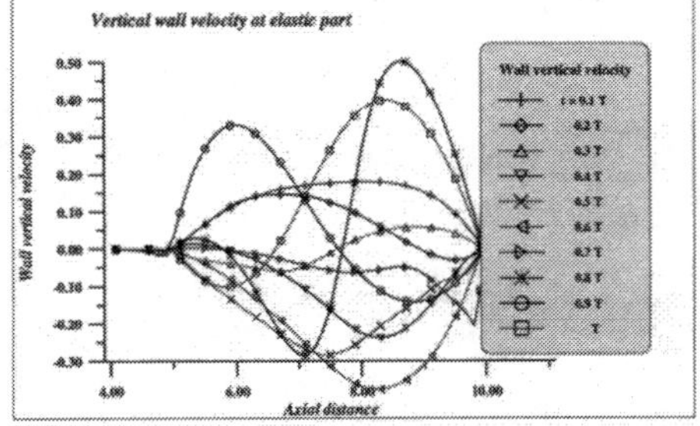

Figure 4.12. Axial variation of the pressure in the upper boundary (solid wall and membrane) (a) and membrane vertical velocity during the period (b) for $f_e = 0.043$ and $U_a = 0.66$.

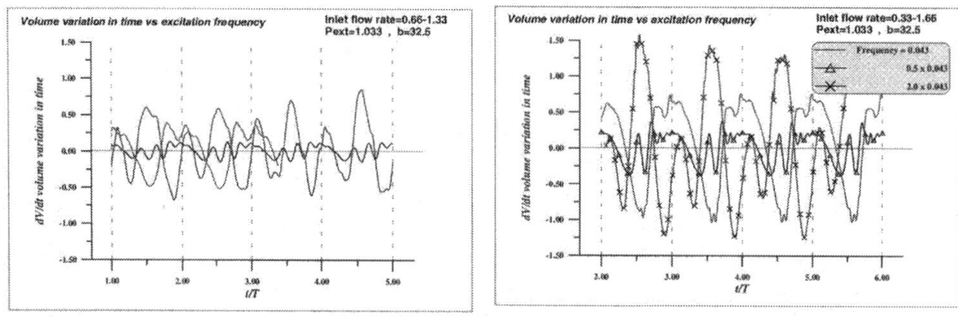

Figure 4.13. Variation of time derivatives of flow domain volume versus excitation frequency for (a) $U_a = 0.33$ and (b). $U_a = 0.66$.

A comparison of different characteristic quantities of the different steady case and cases of pulsating inflow conditions are summarized in Tsangaris and Pappou (1999) and briefly discussed in the following paragraphs.

For the observation of the flow results of an arithmetic methodology for unsteady flows through domains with moving walls, someone has to study carefully time evolution of the elastic wall shape (vertical wall velocity and acceleration) in relation to the ratio of inlet to outlet flow rate.

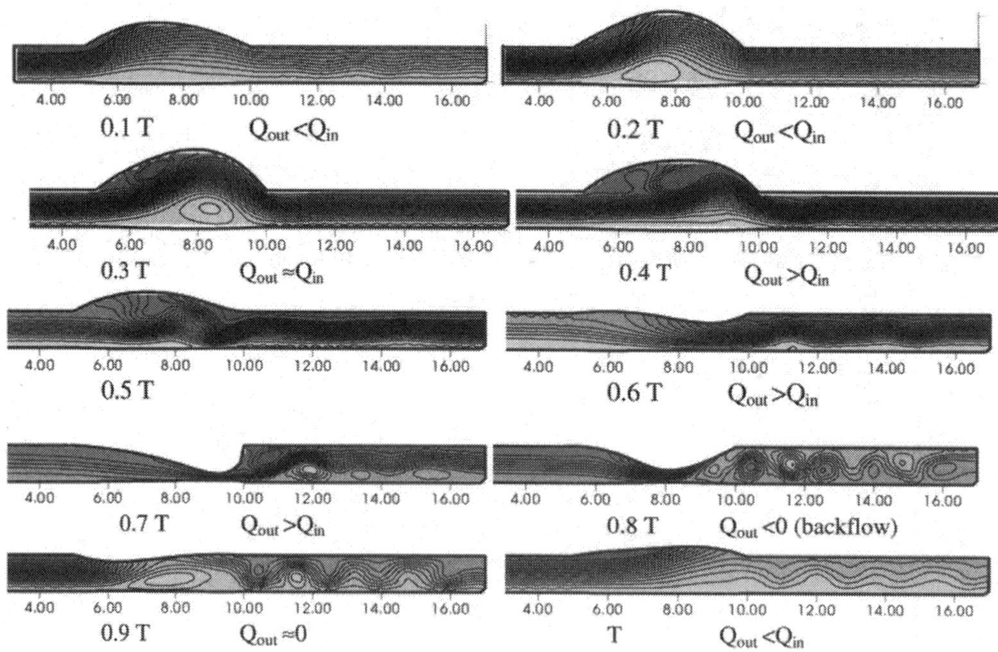

Figure 4.14. Streamlines for pulsating inflow conditions $f_e = 0.043$ and $U_a = 0.66$.

As we can see from the above representation of the numerical results, the imposed pulsating flow destroys the frequency of the self-excited vortices, obtained for the case of steady inflow conditions. For lower excitation frequencies (0.0215 and 0.043) a new fundamental frequency is imposed, which is equal to the excitation frequency. The fundamental frequency doubles for the high excitation frequency (0.086), for both examined values of the inlet flowrate amplitudes.
The minimum value of the membrane height slightly increases by increasing the excitation frequency, while it decreases with the amplitude of the inflow volume rate (Figure 4.10a,b). The maximum value of the membrane height increases both with the excitation frequency as well as with the amplitude of the inflow volume rate. As a result, the amplitude of the membrane displacement is amplified for higher values of the excitation frequency, as well as for higher amplitudes of the inflow volume rate. The site of maximum deformation, almost at x = 8.5, remains roughly constant for all values of the governing parameters in the unsteady case.

The pressure gradient is defined as the pressure difference between the inlet of the channel and the outlet (pressure there is kept constant and equal to zero). The amplitude of the pressure gradient increases with the excitation frequency, as well as with the amplitude of the inflow volume rate.
The amplitude of the volume flowrate at the outlet of the channel is amplified and the amplification becomes higher for higher frequencies and higher values of the amplitude of the inlet volume rate. This fact can be associated with analogous observations of the pressure gradient along the channel.
In addition, the lower value of the outlet flowrate takes negative values for higher values of the amplitude of the inlet flowrate, during the deceleration phase of the outflow.
The topology of the flow field varies with time. A general view of the flow field at different time instants within a period is depicted in Fig. 4.14. A recirculation zone appears at t = 0.2, during the period of inflation. The recirculation zone becomes stronger and moves downstream for t = 0.3T, when maximum dilatation of the membrane occurs. At this time instant the inflow rate is nearly equal to the outflow rate. During the contraction phase of the membrane (t = 0.3T-0.7T) when the flowrate at the outlet is higher than at the inlet, the flow downstream of the elastic part behaves quietly. At t = 0.7T, i.e. when the membrane reaches its deepest position, the outlet flowrate becomes equal to the inlet flowrate, and a strong recirculation zone appears attached to the upper wall, as well as downstream of the neck of the membrane. At the same time a series of vortices exist on both solid walls downstream of the stenosed area, and generation of a train of vorticity waves occurs. These vortices propagate with a finite velocity towards the exit. They exist only for a very short period (≈ 0.2T). At t = 0.8T the outflow is negative, which means that flow enters the system both from the inlet and the outlet: the membrane starts inflating, whilst the vortices still exist. During this phase, vortices seem to propagate upstream with a reduced velocity; their length decreases whilst their height increases, commandeering almost all the channel height. Also, during this phase, eddies detach themselves from the wall on which they were formed and tend to reattach to the opposite wall; they return to the original wall when the flow starts moving forward again. Vortex formation and propagation may cause fluttering in the waveforms of the flow quantities. At t = 0.9 the outlet flowrate is zero, while the inflow rate increases, hence inflating the membrane. This causes the generation of a vortex near the lower wall, just under the membrane. At the end of the period and at the beginning of the period (t = T, 0.1T), the membrane begins to inflate and all disturbances are swept away.

It should be noted that extensive recirculation zones are developed on the elastic wall during its inflation.
The flow field that we have just described is characterized by extremely complex phenomena highly non-linear.
In general, the methodology also converges for cases nearer to collapse, ($h_{min} = 0.15$) as well as for extensive inflation of the elastic part ($h_{min} = 2.17$).

5 Finite Difference methodologies for the blood flow in microcirculation

5.1 Blood flow in microcirculation

In the microcirculation the size of the vessels varies between the smallest, which are capillaries having typical diameters of 4-10μm to the largest which are the arterioles having diameters up to 150μm. The human red blood cells (RBC) have a biconcave discoid shape at rest, with a major diameter of about 8 μm, when suspended in isotonic medium (Caro el al., 1978). A deformable membrane of about 75 Å surrounds the RBC. The internal fluid is mainly the fluid hemoglobin having a much higher viscosity (6-7cP) than the plasma (1.2cP). It is obvious that in the microcirculation the blood cannot be considered as a homogeneous fluid, because the size of the RBC is comparable with vessel size. The RBCs interact with the vessel wall and they interact and deform each other and with flowing plasma, which surrounds them. As a result is the tendency of the RBCs to migrate away from the vessel wall, forming a plasma-skimming layer and effecting the apparent viscosity reduction, which is the well-known Fahraeus - Lindqvist effect.
In the following we modeled the plasma as a Newtonian fluid having a certain viscosity and the content hemoglobin of the RBC again as Newtonian fluid with a higher viscosity. The membrane of the RBC is simulated as a surface tension applied on the plasma- hemoglobin interface. The arteriole or capillary wall is simulated in the present as a straight vessel. As flow equation for both plasma and hemoglobin the Navier-Stokes equations are solved with the two different viscosities using the method of level-set.

5.2 Introduction to the level set method

Level-set is a method for implementation in numerical algorithms that feature moving interfaces between two or more phases. In general, the –strictly distinguished– phases can be of any kind, depending on the characteristics of the case being modeled. Thus, there are problems of two (or more) interacting (not mixing) fluids (Sussman et al., 1994), solidification of dendrites in matter microstructure (Hou el al.,1999), modeling of detonation shocks (Aslam et al., 1996), combustion and a series of problems not directly related with actual interface movement, such as image processing and handwriting recognition.
The method was introduced by Osher and Sethian (1998), and has been studied and improved ever since. It belongs to the so-called "interface capturing" methods, in contradiction to methodologies known as "interface tracking", which explicitly reconstruct the interface at each time step during execution. This repetitive reconstruction is one of the most 'troublesome' features of front tracking

methods, since it requires a considerable amount of program execution time. In addition, when it comes to interface merging or breaking up, some special programming techniques must be embedded (Agresar et al., 1998, Han and Tryggvason, 1999 and Cristini et al., 2001). Level-set methods have been proved to deal with those topological changes quite effortlessly, since the interface does not have to consist of a grid of points with changing topological relations between them, but instead, it is implicitly capture by means of a sign change of the level set function, during each time step.

The basic concept in level-set methods is the capturing of the interface through the use of a continuously smooth function φ defined over the whole computational domain. The initial definition of φ function has to be such a one so that the interface consists of the space points with zero φ values. In most cases it is convenient to define φ as the signed minimum absolute value distance from the interface. Other reasons for this kind of definition of φ function will be seen below.

Generally, function φ is convected in space either by its own curvature or by means of an external velocity field. The latter is the case of fluid interfaces, and the one of our own study. In fluid problems the external velocity field is exactly the one of fluid velocities, which is calculated by a suitable algorithm. In our work we have used a projection method for the numerical simulation (Zhu and Sethian, 1992).

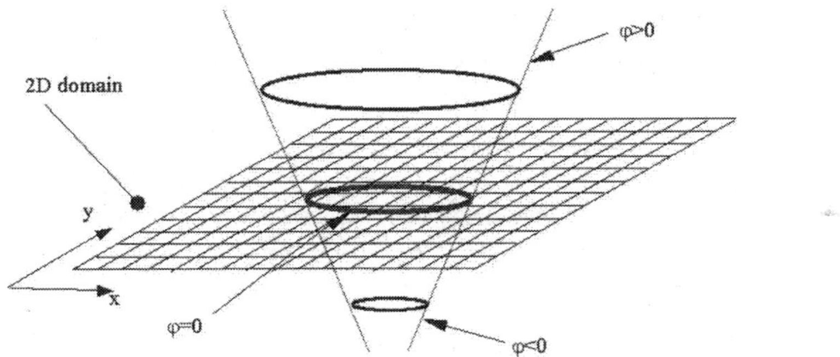

Figure 5.1: Concept of φ-function definition

During its convection, the interface continues actually to consist of zero level of φ (zero φ *level* is *set* onto the interface). The interface itself is not tracked explicitly (interface point calculations do *not* take place), but its position is captured implicitly by use of the φ values distribution. The points that the interface consists of are not 'calculated', simply because they are not needed; all of the information about density and viscosity allocation of the fluids is given by the corresponding φ value distribution over its definition domain. With reference to figure 5.1, a negative φ value indicates that the corresponding point lies inside the curve, or equally, it belongs to the interior fluid phase. Similarly, a positive φ value attributes features (density and viscosity) of the external phase to the corresponding point.

5.3 Projection

The projection method we have used for solving the flow equations can be found in Zhu and Sethian (1992) and Sussman et al. (1994). We have adapted it for the special case of equal liquid density, as our problem requires. That is because we aimed to model the interaction between two fluid phases (surrounding fluid and droplets) without considering buoyancy effects. The equations that govern the phenomenon are the Navier-Stokes equations, which describe viscous, newtonian fluids of steady density, with surface tension $\vec{F}$ on the interfaces, outside of any gravitational field, and the continuity equation for incompressible fluids.
Especially for two dimensions, the present projection method introduces a scalar stream function ψ:

$$u_t = \frac{\partial \psi}{\partial y}, \quad v_t = -\frac{\partial \psi}{\partial x} \tag{5.1}$$

Finally the method ends up to the next Poisson equation, which is solved to produce ψ:

$$-\nabla \psi^2 = \nabla \times L\vec{u} \tag{5.2}$$

From this point, with known velocity components (u, v) at time t, by solving equation (5.2) to ψ, and through (5.1) we obtain the corresponding values of the time derivatives (u_t, v_t). After that, the computation of new velocity values by means of their temporal derivatives takes place.

5.4 Interface smoothing

In order to avoid oscillations due to density and viscosity step variation through the interface, a certain width (namely $2{\cdot}\alpha$) is assigned to the interface. Along that width (in a perpendicular sense) the values of viscosity and density change smoothly between the two interface boundaries.

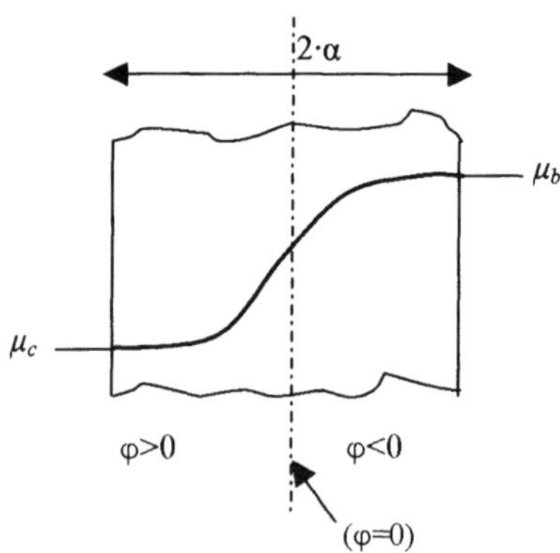

Figure 5.2: Smoothing of density and viscosity functions in interface

In figure 5.2 the variation of dynamic viscosity μ through the finite width of the interface is presented. We define:

$$\bar{\mu} = \frac{\mu_b + \mu_c}{2} \quad \Delta\mu = \frac{\mu_b - \mu_c}{2} \tag{5.3}$$

$$\mu(\phi) = \begin{cases} \mu_c & \phi > a \\ \mu_b & \phi < -a \\ \bar{\mu} - \Delta\mu \cdot \sin\left(\frac{\pi \cdot \phi}{2 \cdot a}\right) & -a \le \phi \le a \end{cases}$$

Density variation is utterly analogous to the above. Obviously it is of extreme importance that φ is kept a distance function, since it is used as a one by the interface smoothing equations.

In general, our fluid flow equations contain surface tension terms, which theoretically are strictly located on the interface. For the same reasons as above (avoidance of oscillations near the interface) the surface tension term $\vec{F}$ of flow equations is spread along width $2 \cdot \alpha$ as follows. Surface tension term of the Navier-Stokes equations is:

$$\vec{F} = \sigma \cdot \kappa \cdot \delta(d) \cdot \vec{n} \tag{5.4}$$

Where σ is the surface tension coefficient, κ is the interface curvature, d is distance from interface, $\delta(d)$ is the Dirac function:

$$\delta(d) = \begin{cases} 0, & d \neq 0 \\ \text{irregular point} & d = 0 \end{cases} \tag{5.5}$$

where:

$$\int_{\varepsilon}^{\varepsilon} \delta(x) dx = 1 \quad (\varepsilon > 0) \tag{5.6}$$

We define:

$$\delta(\phi) = \begin{cases} \frac{1}{2 \cdot a} \cdot \left(1 + \cos\left(\frac{\pi \cdot \phi}{a}\right)\right) & -a \le \phi \le a \\ 0 & |\phi| > a \end{cases} \tag{5.7}$$

Analytical geometry gives that, for a surface (3D case) or plane curve (2D case) which is expressed by the equation $\varphi=0$, we have for its curvature:

$$\kappa = \kappa(\phi) = \nabla \cdot \vec{n} \tag{5.8}$$

Where **n** is the outward normal unit vector onto the surface (or plane curve), defined as:

$$\vec{n} = \frac{\nabla\phi}{|\nabla\phi|} \tag{5.9}$$

The components of vector $\vec{F}$ for two dimensions are:

$$F_{(x)} = \sigma \cdot \kappa(\phi) \cdot \delta(\phi) \cdot \frac{\partial\phi}{\partial x}, \quad F_{(y)} = \sigma \cdot \kappa(\phi) \cdot \delta(\phi) \cdot \frac{\partial\phi}{\partial y} \tag{5.10}$$

In the projection algorithm we used, the surface tension term appears in the form: $\nabla \times \vec{F}$ which in two-dimensional cases gives:

$$\nabla \times \vec{F} = -\frac{\partial F_{(x)}}{\partial x} + \frac{\partial F_{(y)}}{\partial y} \tag{5.11}$$

And consequently (from equations 5.8) it yields that:

$$\nabla \times \vec{F} = -\sigma \cdot \left(\left(\kappa(\phi) \cdot \delta(\phi) \cdot \frac{\partial \phi}{\partial y} \right)_x - \left(\kappa(\phi) \cdot \delta(\phi) \cdot \frac{\partial \phi}{\partial x} \right)_y \right) \tag{5.12}$$

We introduce the Heavyside function H (φ): $\delta(\phi) = \frac{\partial H(\phi)}{\partial \phi}$ (5.13)

Thus, we obtain: $\nabla \times \mathbf{\vec{F}} = -\sigma \cdot \left(\kappa(\phi)_x \cdot H_y - \kappa(\phi)_y \cdot H_x \right)$ (5.14)

Where function H (φ) has been smoothed as:

$$H(\phi) = \begin{cases} 0.5 & \phi > a \\ -0.5 & \phi < -a \\ 0.5 \cdot \left(\frac{\phi}{a} + \frac{\sin}{\pi} \left(\frac{\pi \cdot \phi}{a} \right) \right) & -a \le \phi \le a \end{cases} \tag{5.15}$$

We point out here that surface tension term is computed by using the φ function values. As can be concluded from the above, it is essential that φ remains a distance function, or else all of the surface tension and "interface smoothing" formulas will result in invalid or irregular values.

5.5 Convection of the level set function

Level set method has the following main features: since the fluid interface consists of the same fluid elements, the material derivative of φ function of those elements should always be zero:

$$\left. \frac{D\phi}{Dt} \right|_{\phi=0} = 0 \tag{5.16}$$

The above equation in eulerian form is:

$$\left. \frac{\partial \phi}{\partial t} \right|_{\phi=0} + \vec{u} \cdot (\nabla \phi)|_{\phi=0} = 0 \tag{5.17}$$

Where $\vec{u}$ is the fluid velocity vector. Strictly mathematically, equation (5.17) is valid only on interface points (φ=0), but for consistency reasons it has to be applied on all definition points of φ. It is proved that even though the above equation maintains the zero φ value on the interface, it does not keep φ as a distance function at points that do not belong to the interface. Thus, although the interface is convected properly (theoretically), after a few iterations all of our 'interface smoothing' equations (5.3, 5.7, 5.19) begin to produce unwanted oscillations, which end in a general algorithm failure.

5.6 Reinitialization of level set function

In order to keep φ as a distance function, the following method of φ reinitialization is proposed in literature, Sussman et al. (1994). Supposing that $\varphi_o(\mathbf{x})$ is the solution that equation (5.17) gives. In accordance with the above, this function will actually have zero values on the interface, but not exactly the signed distance values away from it. The scope of the reinitialization procedure is to rebuild function φ values away from the interface in a corrective manner, so that they will again equal the proper-signed distance values. The following equation is introduced, and it is integrated in pseudo-time for every actual time step iteration:

$$\phi_t = S(\phi_o)\cdot\left(1-\sqrt{\phi_x^2+\phi_y^2}\right) \tag{5.18}$$

where $S(\varphi_o)$ is the sign function of φ_o:

$$S(\phi_o)=\begin{cases}-1 & \phi_o<0\\ 0 & \phi_o=0\\ +1 & \phi_o>0\end{cases} \tag{5.19}$$

By applying the above equation to φ_o we get the corrected (or 'reinitialized') values of the level set function φ, without disturbing the interface values of the level set function.

Except from the distance reinitialization problem, in many cases (Sussman et al., 1994, and Zhao at al., 1998) of multi-phase fluid motion, failure of mass conservation of one of the phases has been observed. In Chang et al. (1996) the following procedure, alternative to the aforementioned level set function reinitialization is proposed. One of the phases usually corresponds to a bubble, drop or cell, so its mass (or area in 2D cases) can be easily measured, and should be conserved as well. The following 'perturbed Hamilton-Jacobi' equation, in pseudo-time, is used instead of the reinitialization equation (5.18):

$$\frac{\partial\phi}{\partial t} + (A_o - A(t))\cdot(-P+\kappa(\phi))\cdot|\nabla\phi| = 0 \tag{5.20}$$

where A_o is the initial value of the drop (cell) mass (or area in 2D), A(t) is the calculated mass (area) during the pseudo-time loops, $\kappa(\varphi)$ is the curvature, and P is a real number introduced for convergence reasons only. In our work we found that P= -5 was the most suitable value.

During our experimental simulations we discovered that none of the above reinitialization techniques was enough as long as it was exclusively used. Specifically, as long as the 'distance reinitialization' procedure was implemented, relatively great mass loss or even gain was observed. On the other hand, when the 'mass conservation reinitialization' algorithm was applied, although the mass of the drop (cell) was conserved at a predefined tolerance, the finite width of the phase front did not remain as initially defined, thus causing excessive front diffusion after a number of iterations.

In order to deal with the aforementioned problems we used both of the above reinitialization techniques simultaneously, so each one eliminated the main drawback of the other. However, not both of the techniques are used evenly. The 'distance reinitialization' procedure is used as the basic reinitialization technique, and the 'mass conservation reinitialization' is implemented when the measured mass error exceeds a predefined tolerance level. Increase of execution time is inevitable of course.

5.7 Discretization

We used a central finite difference scheme with a staggered cartesian grid, and generally second order accuracy in space. This kind of grid was used also in Sussman et al. (1994). The form of the grid cells is the following (Figure 5.3)
The boundaries of the computational domain consist of grid cell sides, so there is no need for specially created pseudo-cells for applying the boundary conditions of the problem. The space step is denoted as S.

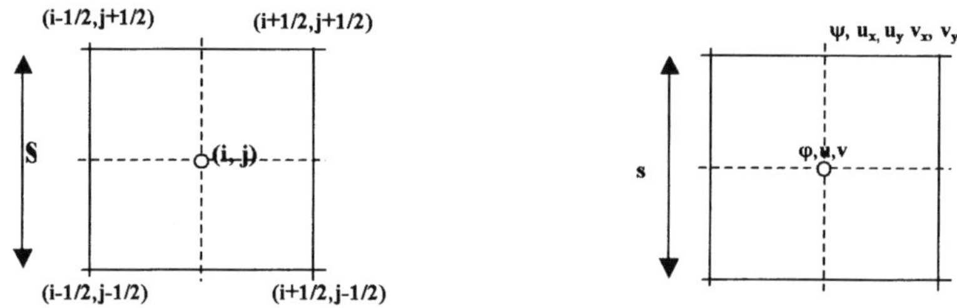

Figure 5.3: Staggered grid definition

In our implementation, fluid velocity components (u, v), and level set function values (φ) are defined at cell centers. The flux function (ψ) and spatial velocity derivatives (u_x, u_y, v_x, v_y) are defined at cell corners, as can be seen in figure 5.3.
In order to accelerate the algorithm execution we used a *moving grid*, which follows the drops (cells), and its dimensions are appropriately adjusted so that the drops (cells) are always contained in it. In that case the level set function is defined only in that moving grid, thus saving memory space and execution time.

5.8 Advantages of the Level Set method

Level set methods –as front capturing- have some certain advantages primarily over the front tracking methods (e.g. boundary integral method), which are:

- Automatic handling of cusps, splitting and merging of interfaces when that happens. This is quite important in fluid interface problems especially.
- Satisfying independence of the algorithm used for solving the velocity field equations, as only the values of velocities are needed for the convection of the level set function.
- Implementation to three dimensions is straightforward. The only part that possibly has to be changed is the fluid velocity algorithm.

Except from the above advantages, which generally characterize front capturing methods over front tracking ones, there is another major advantage of level set methods over the other commonly used front capturing method known as 'volume of fluid' (VOF). That has got to do with the definition of level set function as a distance function, from which local front curvature is easily computed, thus giving surface tension terms where needed.

5.9 Simulation and dimensionless numbers

We simulated the viscous motion and deformation of cells (one of two phases) contained in another fluid (second phase) flowing through straight closed channels. All of the cells begin with a cylindrical (cartesian 2-D) or spherical shape (axisymmetric). Our algorithm is restricted to two dimensional cartesian or axisymmetric coordinates, due to the specific algorithm for fluid velocity calculation being used.

The dimensionless numbers which characterize our case are the following (Kaliakatsos and Tsangaris, 2000):

- ***Size Ratio*** of cell:

$$\kappa = \frac{r}{h} \text{ or } \kappa = \frac{r}{R} \tag{5.21}$$

where r is the initial radius of the cell, h is the semi-distance between channel walls in two dimensional cartesian case, and R is the radius of the tube in the axisymmetric case.

- ***Dynamic Viscosity Ratio***:

$$\lambda = \frac{\mu_b}{\mu_c} \tag{5.22}$$

where μ_b is the dynamic viscosity of the phase corresponding to the cell, and μ_c is the one related to the external fluid.

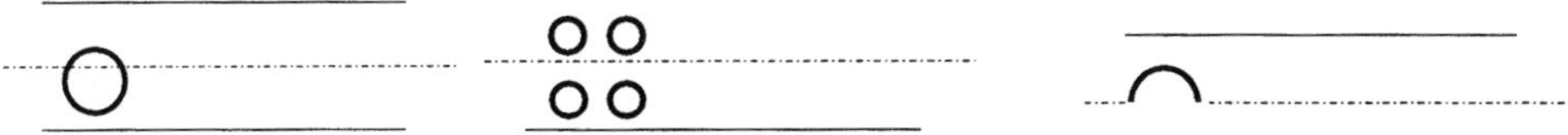

Figure 5.4: Initial position of cells for plane and axisymmetric flow

- ***Capillary number***:

$$Ca = \frac{\mu_c U}{\sigma} \tag{5.23}$$

where μ_c is the dynamic viscosity of the surrounding fluid, U is the average fluid velocity at the entrance of the channel, and σ is the surface tension coefficient. Capillary number quantifies the ratio of viscosity forces to surface tension forces acting on a fluid element.

- ***Reynolds number***:

$$Re = \frac{\rho_c U D_h}{\mu_c} \tag{5.24}$$

where ρ_c is the density of the external phase, and D_h is the hydraulic diameter which is defined for cartesian 2-D and axisymmetric cases respectively: $D_h = 4h$ or $D_h = 2R$ where h is the semi-width of the channel, and R is the radius of the cylindrical tube.

- ***Initial Eccentricity*** of the cell (in two dimensional cartesian case only):

$$\delta = \frac{e}{h} \tag{5.25}$$

where e is the distance between the center of the cell and the symmetry plane of the 2-D channel, and h is the semi-width of the channel.

Some of the aforementioned quantities are shown in figure 5.5:
The following dimensionless lengths, velocities and time have been used:

- *Length*: 'x' and 'y' for the horizontal and vertical axes, we used as a reference length either the semi-width of the 2-D channel, or the radius of the cylindrical tube in axisymmetric cases, respectively

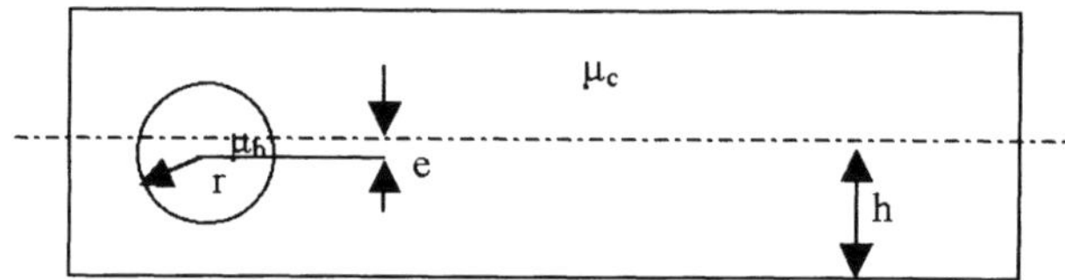

Figure 5.5: Sketch of the initial position and shape of cell

- *Velocity*: as a reference we used the average velocity of the fluid U.
- *Time*: we used the ratio of the above reference length to the reference velocity.

5.10 Results

We have run our program for cartesian two dimensional and axisymmetric coordinates. The former case is presented right below.

5.10.1 Cartesian Two Dimensional Cases

Firstly, we can see the effect of surface tension to the deformation of a large eccentrically positioned cell (channel walls correspond to lines y=0 and y=2). The *relative mass error* for the first case is depicted in figure 5.6; the corresponding error graph for the other case (larger surface tension) is almost identical.

In figure 5.7 the 'same' cell is depicted, but with larger surface tension acting onto the interface; as a primary result we get much less deformation of the cell for time values corresponding to those of figure 5.6. Apparently, the cell tends to maintain its original cylindrical shape as surface tension increases.

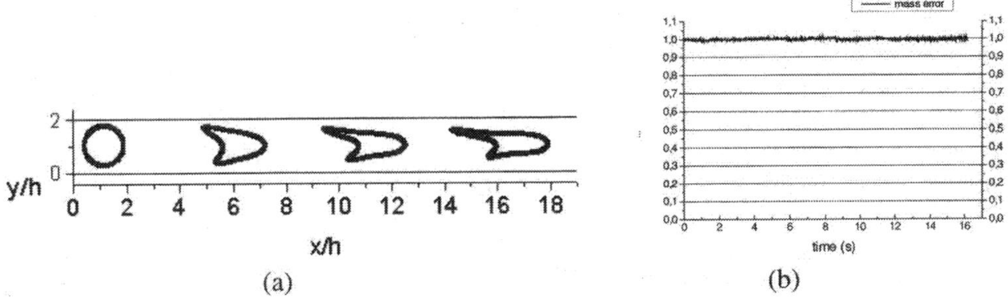

Figure 5.6: One large cell with small surface tension (σ=0.1)for: κ =0.73, λ=5.0, Ca=1.0, Re=40, δ=0.033,and dimensionless times t′=0.0, 4.0, 8.0, 12.0 (a), and relative cell mass error.(b)

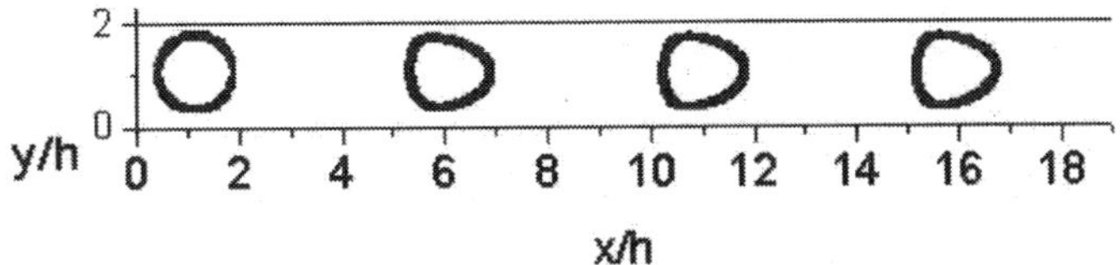

Figure 5.7. Large cell with large surface tension (σ=0.5) for κ=0.73, λ=5.0, Ca=0.2, Re=40, δ=0.033, and dimensionless times t′=0.0, 4.0, 8.0, 12.0

In the next example, figure 5.8, the movement and deformation of three identical cells of a medium size, which are initially positioned along the same line of eccentricity δ=0.067. It is obvious from the graph that the following cells are affected by the wake of the previous ones. More specifically, following cells are subjected to a more intense deformation. We will be able to watch that effect in following graphs as well.

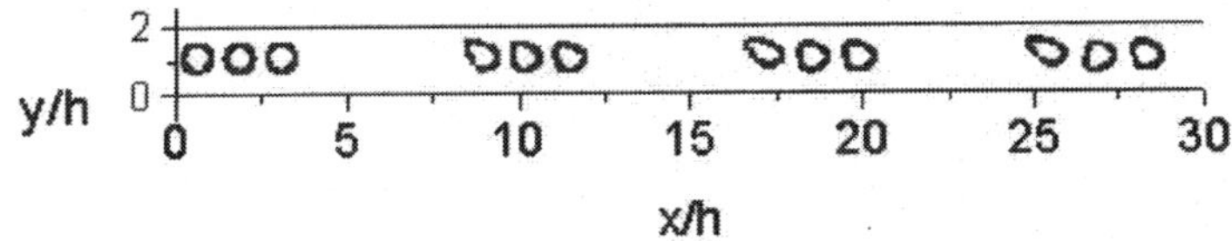

Figure 5.8. Group of three identical cells. κ=0.4, λ=3.0, Ca=1.0, Re=40, δ=0.067, t′=0.0, 6.0, 12.0, 18.0

As far as it concerns mass conservation, the relative error is kept less than 1%, which is the initially given tolerance. The corresponding graph is very much alike the previous one, so it is not depicted here.

In the next graph, figure 5.9, a similar group of cells in similar flow conditions but without any surface tension effects is presented. The dynamic viscosity ratio has been increased so as to counterbalance the lack of surface tension, thus restricting as possible very large deformations and cell breakup.

The graph of figure 5.9, presents the motion and deformation of cells belonging to a group of eight non-symmetrically positioned. It can be clearly seen that the cells tend to move to the symmetry plane of the channel. That behavior can be explained as follows: the cells due to their surface tension and larger density move to regions of smaller shear rates so that the corresponding 'deformation energy' is minimized. Regions of smaller shear rates are those of smaller velocity gradients, or else the regions closer to the symmetry plane of the channel.

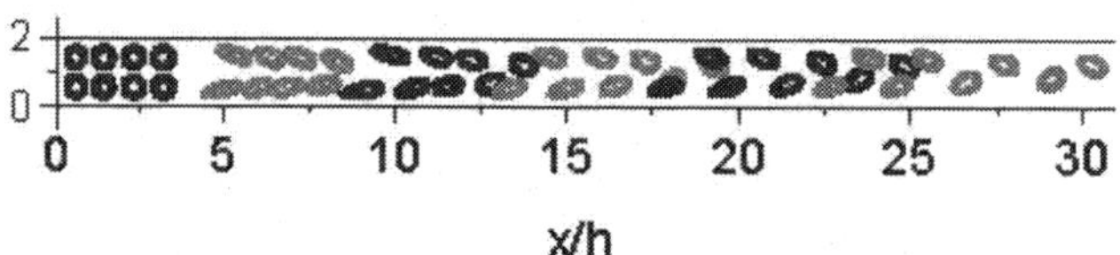

Figure 5.9. Group of eight cells (σ=0.09) for the following dimensionless numbers: κ=0.267, λ=3.0, Ca=1.11, Re=40, and dimensionless times t′=0.0, 4.0, 8.0, 12.0, 16.0, 20.0

The two different shades of cell contours are used in order to distinguish the positions corresponding to different time.

References

Agresar, G., Linderman, J.J., Tryggvason, G. and Powell K.G. (1998). An Adaptive, Cartesian, Front-Tracking Method for the Motion, Deformation and Adhesion of Circulating Cells. *Journal of Computational Physics.* 143:346-380.

Aslam, T.D., Bdzil, J.B. and Stewart, D.S. (1996). Level Set Methods Applied to Modeling Detonation Shock Dynamics. *Journal of Computational Physics.* 126:390-309.

Attinger, E.O. (1968). Analysis of pulsatile blood flow. In Levine, S.N., ed., *Advances in Biomedical Engineering and Medical Physics.* New York: Interscience . 1-59.

Beam, R. M., and Warming, R.F.(1976). An implicit finite-difference algorithm for hyperbolic system of conservation-law form. *International Journal of Computational Physics.* 22: 87-110.

Caro, C.G., Pedley, T.J., Schroter, R.C., and Seed, W.A. (1978). *The Mechanics of the Circulation.* New York: Oxford University Press.

Chang, J. L. C., and Kwak, D.(1984). On the method of pseudocompressibility for numerically solving incompressible flows. *AIAA Paper 84-252.*

Chang, Y.C., Hou, T.Y., Merriman, B. and Osher S. (1996). A Level Set Formulation of Eulerian Interface Capturing Methods for Incompressible Fluid Flows. *Journal of Computational Physics.* 124:449-464.

Choi, D., and Merkle, C. L. (1985). Application of time-iterative schemes to incompressible flow. *AIAA Journal.* 23:1519-1524.

Chorin, J. (1967). A numerical method for solving incompressible viscous problems. *Journal of Computational Physics.* 2:12-26.

Cristini, V., Blawzdziewicz, J. and Loewenberg M. (2001). An Adaptive Mesh Algorithm for Evolving Surfaces: Simulations of Drop Breakup and Coalescence. *Journal of Computational Physics.* 168:445-463.

Demirdzic, I., and Peric, M. (1988). Space conservation law in finite volume calculations of fluid flow, *International Journal for Numerical Methods in Fluids.* 8:1037-1050.

Demirdzic, I., and Peric, M. (1990). Finite volume method for prediction of fluid flow in arbitrary shaped domains with moving boundaries. *International Journal for Numerical Methods in Fluids.* 10:771-790.

Drikakis, D., and Tsangaris, S.(1991). An implicit characteristics flux averaging scheme for the Euler equations for real gases. *International Journal for Numerical Methods in Fluids*, 12:771-776.

Drikakis, D. and Tsangaris, S.(1993). On the solution of the compressible Navier-Stokes equations, using improved flux vector splitting methods. *Applied Mathematical Modeling,* 17:282-297.

Ferziger, J.H., and Peric, M. (1996). *Computational Methods for Fluid Dynamics.* Springer-Verlag, Heidelberg/Berlin, /new York.

Fortin, M., Peyert, R., and Teman, R. (1971). Resolution numerique des equations de Navier-Stokes pour un fluide incompressible. *Journal of Mechanics.* 10:357-390

Gow, B.S. (1973). The influence of vascular smooth muscle on the viscoelastic properties of blood vessels. In Bergel, D.H., ed., *Cardiovascular Fluid Dynamics*, Academic Press. 2:65-110.

Han, J. and Tryggvason G. (1999). Secondary breakup of axisymmetric liquid drops. I. Acceleration by a constant body force. *Physics of Fluids.* 11:3650-3667.

Harlow, F.H., and Welch, J. E. (1965). Numerical calculation of time dependent viscous incompressible flow with free surface. *Physics of fluids.* 8:2182-2189.

Hartwich, M., Hsu, C. H., and Liu, C. H. (1988). Vectorizable implicit algorithms for the flux-difference split, three-dimensional Navier-Stokes equations. *Journal of Fluids Engineering.* 110:297-305.

Hirt, C.W., and Cook, J.L. (1972). Calculating three-dimensional flows around structures and over rough terrain . *Journal of Computational Physics.* 10:324-340.

Hirt, C.W., Nichols, B.D., and Romero, N.C. (1975). SOLA – A numerical solution algorithm for transient fluid flows. In *Report No LA 5852.* Los Alamos Scientific Laboratory. 1-50.

Hou, T.Y., Rosakis, P. and LeFloch, P. (1999). A Level-Set Approach to the Computation of Twinning and Phase-Transition Dynamics. *Journal of Computational Physics.* 150:302-331.

Kaliakatsos, Ch., and Tsangaris, S. (2000). Motion of deformable drops in pipes and channels using Navier – Stokes equations. *Int. J. Num. Meth. Fluids.* 34:609-626.

Kwak, D., Chang, J. L. C., Shanks, S. P., and Chakravarthy, S. R. (1986). A three dimensional incompressible flow solving using primitive variables. *AIAA Journal.* 24:390-396.

Luo, X.Y., and Pedley, T.J. (1195). A numerical simulation of steady flow in a 2-D collapsible channel. *Journal of Fluids and Structures.* 9:149-174.

Luo, X.Y., and Pedley, T.J. (1996). A numerical simulation of unsteady flow in a two-dimensional collapsible channel. *Journal of Fluid Mechanics.* 314:191-225.

Luo, X.Y., and Pedley, T.J. (1998). The effects of wall inertia on flow in a two-dimensional collapsible channel. *Journal of Fluid Mechanics.* 363:253-280.

Pedley, T.J., and Luo, X.Y.(1998). Modeling flow and oscillations in collapsible tubes. *Theoretical and Computational Fluid Dynamics.* 10:277-294.

Mathioulakis, D.S., Pappou, Th., and Tsangaris, S. (1997). An experimental and numerical study of a 90° bifurcation. *Fluid Dynamics Research.* 19:1-26.

Merkle, C., and Athavale, M.A. (1987). Time accurate unsteady incompressible flow algorithms based on artificial compressibility. *AIAA Paper 87-1137.*

Osher, S., and Sethian, J. (1988). Fronts propagating with curvature-dependent speed: algorithms based on Hamilton-Jacobi formulations. *Journal of Computational Physics.* 79:12-49.

Pappou, Th., and Tsangaris, S.(1997). Development of an artificial compressibility methodology using Flux Vector Splitting. *International Journal for Numerical Methods in Fluids.* 25: 523-545.

Patel, D.J.(1973). The rheology of large blood vessels. In Bergel, D.H., ed., *Cardiovascular Fluid Dynamics*, Academic Press. 2:1-64.

Perktold, K., and Rappitsch, K. (1995). Computer Simulation of Arterial Blood Flows. Jaffrin, M.W., and Caro, C.G. NewYork:Plenum Press.

Peyret, R., and Taylor, T. D. (1983). Computational Methods for Fluid Flow. New York: Springer.

Rammos, K.St., Koullias, G.J., Pappou, Th.J., Bakas, A.J., Panagopoulos, P.G., and Tsangaris, S.G. (1998). A computer model for the prediction of left epicardial coronary blood flow in normal stetonic and bypassed coronary artery by single or sequential grafts. *Cardiovascular Surgery*. 6(6):635-648.

Rizzi, A., and Erikson, L.(1985). Computation of inviscid incompressible flow with rotation. *Journal of Fluid Mechanics*. 153:275-312.

Rogers, S. E., and Kwak, D. (1990). Upwind differencing scheme for the time accurate incompressible Navier-Stokes equations. *AIAA Journal*. 28: 253-262.

Soh, W. Y., and Goodrich, J. W. (1988). Unsteady solution of incompressible Navier-Stokes equations. *Journal of Computational Physics*. 79:113-134.

Steger, L., and Kutler, P. (1977). Implicit finite-difference procedures for the computation of vortex wakes. *AIAA Journal*. 15:581-590.

Steger, J. L., and Warming, R. F.(1981). Flux vector splitting of the inviscid gasdynamics equations with application to finite difference methods. *Journal of Computational Physics*.40:263-293.

Sussman, M., Smereka, P. and Osher, S. (1994). A Level Set Approach for Computing Solutions to Incompressible Two-Phase Flow. *Journal of Computational Physics*. 114:146-159.

Tsangaris, S., and Pappou, Th. (1999). Flow investigation in deformable arteries. In Verdonck, P., and Perktold, K., eds., *Intra and Extracorporeal Cardiovasvular Fluid Dynamics*.WIT Press 291-332.

Thomas, L., and Walters, R. W.(1985). Upwind relaxation algorithms for the Navier-Stokes equations. *AIAA Paper 85-1501-CP*.

Van Leer, B. (1982). Flux vector splitting for the Euler equations. In *Proceedings of the 8th International Conference on Numerical Methods in Fluids*. Aachen. Lecture Notes in Physics. 170:507-512.

Zhao, Hong-Kai, Merriman, B., Osher, S. and Wang L. (1998). Capturing the Behavior of Bubbles and Drops Using the Variational Level Set Approach. *Journal of Computational Physics*. 143:495-518.

Zhu, J. and Sethian, J.A. (1992). Projection Methods coupled to Level Set Interface Techniques. *Journal of Computational Physics*. 102:128-138.

Fluid Flow inside Deformable Vessels and in the Left Ventricle

Gianni Pedrizzetti[1] and Federico Domenichini[2]

[1] Dipartimento di Ingegneria Civile, Università di Trieste, Italy
[2] Dipartimento di Ingegneria Civile, Università di Firenze, Italy

Abstract. Topics regarding flow inside deformable domains are here considered either for artery and ventricle flows. The theory of finite elasticity is summarised in a perspective of application to fluid-tissue interaction problems. The formulation of a coupled fluid tissue problem is discussed. A linearized technique is then introduced as a model for wall elasticity in artery flow. This simplification transforms the coupled fluid-wall system into a cascade of uncoupled systems on a fixed domain. Computational examples are given in axisymmetric problems, for both finite and infinitesimal deformation cases, to show resonance and stability features of the interactive dynamics. The heart dynamics and the major ventricular diseases are briefly summarised in a mechanical perspective. The formulation of the flow in a model left ventricle is given, and numerical solutions for flow during the left ventricle filling (diastole) is studied and analysed in terms of vorticity dynamics. Results are also presented in relation to the physical phenomena observed in the clinical practice. In connection with this the mitral valve modelling is also introduced.

1. Flow inside Deformable Vessels

1.1 Introduction

The flow inside biological systems is commonly influenced at a relevant level by the mechanical behaviour of the surrounding tissues. The analysis required in biological fluid dynamics is usually that of a coupled fluid-wall problem. The heart walls contract during systole to pump out the blood into the circulatory system; the same walls have a mainly passive role during heart filling: their viscoelastic properties are correlated to the pressure required to comply them and fill the corresponding heart chamber. The arteries have elastic walls that move during a flow pulsation, they can store fluid mass at the systolic pressure and release it afterward.

Biological tissues that are in contact with fluid flows are in general soft tissues that can undergo relatively large deformations. For such a type of solids the infinitesimal theory of elasticity is often inadequate and the nonlinear theory of finite elasticity should be used (Green & Adkins, 1960). Although this is the actual theory for solid mechanics, it is much less diffused than its linear approximation, based on the assumption that deformations are infinitesimal, which is still used sometimes in media that exhibit finite deformations. This approach, that may be accepted to a first approximation, represents an oversimplification of the cardiovascular system mechanics and should not be pursed to advanced analysis.

A brief introduction to the theory of nonlinear elasticity is presented here by a fluid dynamics perspective. The main ingredients and concepts are given with the objective to support the reader for more rigorous readings on the subject. This theory is first introduced for three-dimensional soft tissues and then restricted to the case of membranes that represent adequate models for many cardiovascular districts. The approach is then simplified to axisymmetric membranes to give a more immediate feeling of equations and quantities involved.

The calculation of the simultaneous fluid-structure evolution reserves many technical difficulties and, despite the theoretical and practical relevance, very few results have been produced about fluid motion in elastic tubes (Luo & Pedley, 1996; Kumaran, 1996; Heil, 1997; Davies & Carpenter, 1997; Pedrizzetti, 1998). The coupled wall-fluid system is introduced, the formulation and some results are shown for the fluid flow in a circular vessel.

1.2 Nonlinear Elasticity in a Three-dimensional Tissue

The present discussion is mainly taken from the book by Green & Adkins (1960) and the review by Humphrey (1995); the reader should refer to these for a complete reference about the concepts that are reported below in a very concise form.

In the theory of (finite) elasticity it is of fundamental importance to identify the appropriate quantities that describe physical phenomena and that enter into the governing equations. Consider a tissue that is identified by the coordinates $\boldsymbol{x}$ of its material points (Lagrangian description), and call $\boldsymbol{X}$ the undeformed positions of these points. The deformation is in general described by the displacement gradient tensor

$$\mathbb{F} = \frac{\partial \boldsymbol{x}}{\partial \boldsymbol{X}}, \qquad \mathbb{F}_{ij} = \frac{\partial x_i}{\partial X_j}. \tag{1.1}$$

The tensor $\mathbb{F}$ contains information about either the rigid-body rotation and strain deformation. A measure of strain is given by the symmetric Green tensor

$$\mathbb{E} = \frac{1}{2}(\mathbb{F}^T \cdot \mathbb{F} - \mathbb{I}), \qquad \mathbb{E}_{ij} = \frac{1}{2}(\mathbb{F}_{ki}\mathbb{F}_{kj} - \delta_{ij}); \tag{1.2}$$

other equivalent definitions of strain are sometimes used in literature. The forces that act on tissues are usually described in terms of the well known Cauchy stress tensor $\mathbb{T}$. The projection of $\mathbb{T}$ along one direction $\boldsymbol{n}$ gives the stress vector $\boldsymbol{\tau}$ (force per unit area) acting, at that point, on a face with normal $\boldsymbol{n}$, i.e. $\boldsymbol{\tau} = \mathbb{T} \cdot \boldsymbol{n}$. It can be shown that, because of the conservation of angular momentum, the stress tensor is symmetric.

The mechanical equations involved for such quantities are the conservation of mass and of linear momentum. The conservation of mass relates the local variation of volume, given by the det$\mathbb{F}$, to the variation of material density ρ

$$\det\mathbb{F} = \frac{\rho(\boldsymbol{x})}{\rho(\boldsymbol{X})}; \tag{1.3}$$

in an incompressible material this reads

$$\det\mathbb{F} = 1, \quad \text{or} \quad \det\mathbb{E} = 0. \tag{1.4}$$

The balance of linear momentum

$$\nabla \cdot \mathbb{T} = \rho(\boldsymbol{a} - \boldsymbol{g}),$$

where $\boldsymbol{g}$ is a body force (e.g. acceleration due to the gravity) and $\boldsymbol{a}$ is the acceleration, is sometimes referred as the equilibrium equation. In solid mechanics it often used with a negligible right-hand-side term as

$$\nabla \cdot \mathbb{T} = 0. \tag{1.5}$$

This equation is then completed by the boundary conditions: these are the external loads, that define stress vector values or the flux of the stress tensor, at the volume edges. In order to close the mechanical problem (1.5), we must now specify a constitutive equation that gives the relation between stresses and strains. This is the place in the present continuum mechanics description where a solid tissue is differentiated from a fluid where stress depends on the rate of strain in place of the strain itself.

A constitutive equation defines the elastic properties of the tissue, that is the values of stress that corresponds to a given deformation. It is generally expressed in terms of the potential elastic energy W that is stored in correspondence of a strain state $\mathbb{E}$. An elastic constitutive equation is thus a function $W(\mathbb{E})$. From this, the Cauchy stress tensor is recovered by the relation

$$\mathbb{T} = \frac{1}{\det\mathbb{F}}\mathbb{F} \cdot \frac{\partial W}{\partial \mathbb{E}} \cdot \mathbb{F}^T, \qquad \mathbb{T}_{ij} = \frac{1}{\det\mathbb{F}}\mathbb{F}_{ik}\frac{\partial W}{\partial \mathbb{E}_{k\ell}}\mathbb{F}_{j\ell}. \tag{1.6}$$

If the material is isotropic the elastic energy does not depend separately on the different components of the stress tensor, but only on its invariants $W(I_1, I_2, I_3)$ where I_1 is the trace of $\mathbb{F}^T\mathbb{F}$, I_2 is the second invariant, $I_3 = \det(\mathbb{F}^T\mathbb{F})$; in an incompressible tissue $I_3 = 1$ because of (1.4), and W becomes a function of two variables $W(I_1, I_2)$. Several forms for the elastic potential are reported in literature. The simplest linear Mooney-Rivlin relationship

$$W(I_1, I_2) = C\{(I_1 - 3) + \alpha(I_2 - 3)\}$$

corresponds to a rubber material, which is called Neo-Hookian in the even simpler case when $\alpha = 0$. An exponential form of the dependencies is commonly found to be more adequate in biological tissues (Humphrey et al., 1992; Hsu et al., 1994; Kyriacou & Humphrey, 1996) where the stress grows very rapidly for large strain to avoid unacceptable deformation, a behaviour that is opposite to that of a Mooney-Rivlin material.

The equilibrium equation (1.5) eventually gives a nonlinear system in terms of the unknown displacements. The value of displacement give the unknown deformed geometry, therefore they are present in the metric coefficients of (1.5) as well as in the stress terms by a possibly nonlinear stress-strain relationship. Eventually equation (1.5) is a nonlinear system for the unknown deformation that can be directly solved in a few simple configurations, or is in general faced numerically.

1.3 Elastic Membranes

The solid mechanics problem is simplified in the case of a membranes. A membrane is a mathematical model for a solid shell of extremely small thickness such that it can be assumed as having no bending stiffness.

Let us consider that the membrane middle surface is instantaneously described by two curvilinear coordinates $\{x_1, x_2\}$ and that a third coordinate, x_3, normal to them, ranges from $-h/2$ to $+h/2$ along the membrane thickness of size h, so that the surface $x_3 = 0$ describes the middle surface. A membrane supports only stresses that belong to its surface (i.e. $\mathbb{T}_{i3} = \mathbb{T}_{3i} = 0$) and these are also assumed to be constant along the thickness. The equilibrium equation (1.5) simplifies considerably under these assumptions.

In order to give an immediate mathematical description, let us assume orthogonal coordinates $\{x_1, x_2\}$ over the membrane and a normal coordinate x_3, with corresponding metric coefficients h_i, $i = 1...3$, and let us assume, without loss of generality, that the coordinates are along the principal directions of stress (that coincide with those of strain because of isotropy). It is therefore natural to introduce the two membrane principal stresses (stresses per unit length) $\mathcal{T}_i$ $(i = 1, 2)$, defined by integrating the non-zero elements of $\mathbb{T}$ over the thickness

$$\mathcal{T}_i = \int_{-\frac{h}{2}}^{+\frac{h}{2}} \mathbb{T}_{ii}\, dx_3, \quad i = 1, 2 \text{ (no summation)}. \tag{1.7}$$

The integration of the membrane equilibrium equation (1.5) over the deformed membrane thickness, an operation that involves the external loads on the membrane surfaces as boundary conditions, gives the tangential equilibrium equations

$$\begin{aligned} \frac{1}{h_1}\frac{\partial \mathcal{T}_1}{\partial x_1} + \frac{1}{h_1 h_2}\frac{\partial h_2}{\partial x_1}(\mathcal{T}_1 - \mathcal{T}_2) = \tau_1, \\ \frac{1}{h_2}\frac{\partial \mathcal{T}_2}{\partial x_2} + \frac{1}{h_1 h_2}\frac{\partial h_1}{\partial x_2}(\mathcal{T}_2 - \mathcal{T}_1) = \tau_2; \end{aligned} \tag{1.8}$$

$\boldsymbol{\tau}$ is the total external wall shear stress vector, the sum of those applied on the upper and lower edges. The normal equilibrium can be written as

$$\kappa_1 \mathcal{T}_1 + \kappa_2 \mathcal{T}_2 = p, \tag{1.9}$$

where p is the transmural pressure and κ_i are the local curvatures of the membrane; in terms of metric coefficients they can be written as

$$\kappa_1 = \frac{1}{h_1 h_3}\frac{\partial h_1}{\partial x_3}, \qquad \kappa_2 = \frac{1}{h_2 h_3}\frac{\partial h_2}{\partial x_3}.$$

Equation (1.9) is known as the Laplace equation that in the case of a sphere or radius r $(\kappa_1 = \kappa_2 = 1/r,\ \mathcal{T}_1 = \mathcal{T}_2 = \mathcal{T})$ becomes the well known formula for a spherical bubble $p = 2\mathcal{T}/r$.

The membrane constitutive equation can be written in terms of the in-plane strains only. Calling λ_1, λ_2 the principal strains (first two eigenvalues of $\mathbb{F}$) for an isotropic membrane, the membrane elastic energy (potential energy integrated over the thickness) can be expressed directly as a function of these as $w(\lambda_1, \lambda_2)$ (Pipkin, 1968); the membrane stresses are then evaluated as

$$T_1 = \frac{1}{\lambda_2}\frac{\partial w}{\partial \lambda_1}, \quad T_2 = \frac{1}{\lambda_1}\frac{\partial w}{\partial \lambda_2}. \tag{1.10}$$

Once we specify a material, i.e. a form for $w(\lambda_1, \lambda_2)$, the membrane stresses become a function of the deformed geometry, and the problem of finding the deformations is that of solving the equilibrium equations (1.8-1.9) with the appropriate geometric boundary conditions.

1.4 Axisymmetric Membrane

Equations (1.8-1.9) become simpler when a simplified geometry is assumed. Let us consider axisymmetric membranes whose instantaneous deformed geometry is specified by the longitudinal and radial position of its material points $\{x(s), r(s)\}$, where s is a material coordinate along the membrane. The coordinates $\{x_1, x_2\} = \{s, \vartheta\}$, with metric coefficients

$$h_1 = \left(\left(\frac{\partial x}{\partial s}\right)^2 + \left(\frac{\partial r}{\partial s}\right)^2\right)^{\frac{1}{2}}, \quad h_2 = r(s),$$

are along the meridional and circumferential directions. The tangential equilibrium equation (1.8) is

$$\frac{\partial}{\partial s}(T_1 r) - \frac{\partial r}{\partial s} T_2 = h_1 r \tau_s, \tag{1.11}$$

and the normal equilibrium is (1.9) where the curvatures are

$$\kappa_1 = -\frac{\partial^2 r}{\partial s^2}\left(1 - \left(\frac{\partial r}{\partial s}\right)^2\right)^{-\frac{1}{2}}, \quad \kappa_2 = \frac{1}{r}\left(1 - \left(\frac{\partial r}{\partial s}\right)^2\right)^{\frac{1}{2}}. \tag{1.12}$$

Because of the axial symmetry of geometry and loads, the meridional and circumferential directions are eigendirections of the deformation gradient tensor, and the principal deformation ratios are

$$\lambda_1 = \left(\left(\frac{\partial x}{\partial s}\right)^2 + \left(\frac{\partial r}{\partial s}\right)^2\right)^{\frac{1}{2}}, \quad \lambda_2 = \frac{r}{R_0}; \tag{1.13}$$

R_0 is the value of the vessel radius in the unstressed state.

1.5 Fluid-solid systems

The previous equations can be used to describe the deformation of fluid boundaries, in this case they must be solved simultaneously with those governing the motion of the surrounding fluid, here assumed incompressible. The fluid-tissue coupling is given by the fluid stresses at the boundaries, pressure and shear stress, that act on the solid, and by the no-slip condition at the deformable boundaries. In general, the interaction is not a weak one and a method of solution must be designed in a way to solve both the fluid and solid equations at once. In addition it must be emphasised that the fluid dynamics in presence of elastic boundaries depends on the actual value of pressure rather than on pressure gradients only; therefore a further boundary condition for the fluid must be specified.

Example of different approaches, ranging from fluid-solid systems assembled as a whole nonlinear system to simpler iterative coupling, can be found in literature (Anayiotos et al., 1994; Perktold & Rapitsch, 1995; Luo & Pedley, 1996; Pedrizzetti, 1998). It is important to notice that some results include inertia or viscous effects on the solid equations in order to possibly make the system more realistic and also stabilise the numerical method.

One important point in the mathematical arrangement of a coupled system is the fact that the solid wall dynamics is commonly described in terms of material points (i.e. Lagrangian description). These points at the boundaries may not be the proper choice for the description of the fluid problem that is expressed on a fixed reference (i.e. Eulerian description) made adaptable to follow the moving domain, that does not necessarily coincide with a Lagrangian reference at the boundaries.

1.6 Axisymmetric Fluid System

Consider the flow of an incompressible viscous fluid, with density ρ and kinematic viscosity ν, moving in a rectilinear elastic vessel of circular cross section.

A x-axis is chosen to coincide with the tube axis, the boundary of the domain, assumed with an axial symmetry, is given at each instant by the membrane geometry fully specified by the radius $R(t,x)$ at each x-position.

Assume the duct axis as the x-axis of a cylindrical system of coordinates $\{x, r, \theta\}$, the axial symmetry makes the flow independent from θ. The governing equations for the fluid are the axisymmetric form of the Navier-Stokes equations

$$\begin{aligned}\frac{\partial v_x}{\partial t} + v_x\frac{\partial v_x}{\partial x} + v_r\frac{\partial v_x}{\partial r} &= -\frac{1}{\rho}\frac{\partial p}{\partial x} + \nu\nabla^2 v_x,\\ \frac{\partial v_r}{\partial t} + v_x\frac{\partial v_r}{\partial x} + v_r\frac{\partial v_r}{\partial r} &= -\frac{1}{\rho}\frac{\partial p}{\partial r} + \nu\Big(\nabla^2 v_r - \frac{v_r}{r^2}\Big),\end{aligned} \tag{1.14}$$

and the continuity equation (Batchelor 1967, Panton 1996)

$$\frac{\partial v_x}{\partial x} + \frac{1}{r}\frac{\partial r v_r}{\partial r} = 0. \tag{1.15}$$

The upstream and downstream boundary conditions are those of a uniform flow. On the axis of the duct, $r = 0$, symmetry conditions are given; at the wall $r = R(t,x)$ the fluid velocity vector must be equal to the membrane velocity. The accurate insertion of the latter conditions is crucial in every problem for the presence of the possibly separating boundary layer, the unique source of vorticity, and because the fluid and the structure interact through the value of variables evaluated at the wall.

We introduce a system of coordinates where the moving wall coincides with a constant coordinate curve; a shearing coordinate transformation is used (Eiseman, 1985) defining a new z-coordinate

$$z = \frac{r}{R(t,x)}; \tag{1.16}$$

in the new system of coordinate $\{t, x, z\}$ the z-direction is non-orthogonal both to the time direction and spatial x-direction. Equations (1.14-1.15) can be rewritten in this new system of coordinates by using composite derivatives rule, where an additional convective term given by the domain motion appears.

1.7 Axisymmetric Membrane Coupling

The membrane equations outlined in §1.4 are given in terms of the coordinates $\{x(t,s),\, r(t,s)\}$ of material points on the membrane, identified by the material coordinate s. On the contrary the wall geometry in the fluid problem is expressed as $R(t,x)$, using the Eulerian longitudinal coordinate x as independent variable.

This is a simple example where the description of the boundary in the fluid problem does not coincide with the description of the same boundary in the solid mechanics problem. To couple the fluid and structure mathematical formulations, the membrane problem can be rewritten in Eulerian terms: the function $S(t,x)$ is introduced to comply with two degrees of freedom, and the following Lagrangian-Eulerian transformation is used

$$R[t, x(t,s)] = r(t,s), \quad S[t, x(t,s)] = s\,. \tag{1.17}$$

Following (1.17), the principal deformation ratios (1.13), the tangential and normal equilibrium equations (1.11, 1.9, 1.12) can be expressed in terms of Eulerian variables, giving respectively

$$\lambda_1 = \left(\frac{1+R'^2}{S'^2}\right)^{\frac{1}{2}}, \qquad \lambda_2 = \frac{R}{R_0}, \tag{1.18}$$

$$RT_1' + R'(T_1 - T_2) = \left(1 + R'^2\right)^{1/2} R\,\tau_s, \tag{1.19}$$

$$\frac{-R''}{1+R'^2} T_1 + \frac{1}{R} T_2 = \left(1 + R'^2\right)^{1/2} p, \tag{1.20}$$

where the prime indicates a x-derivative.

The membrane velocities, i.e. the instantaneous material velocity of the points on the membrane $U_x = \partial x(s)/\partial t$ and $U_r = \partial r(s)/\partial t$, can be expressed in terms of Eulerian variables by time differentiation of relations (1.17)

$$\dot{R} = U_r - R'U_x\,, \quad \dot{S} + S'U_x = 0, \tag{1.21}$$

where the dot indicates a time derivative.

1.8 Discussion of one Illustrative Example

The previously outlined approach has been applied to the case of a circular rigid duct with an elastic membrane insertion (Pedrizzetti, 1998).

A fixed positive transmural pressure has been imposed downstream the membrane, thus the elastic wall is expanded outwards. A steady flow rate with an additional small sinusoidal oscillation is also imposed downstream of the elastic insertion

$$U(t) = 1 + \varepsilon \sin(2\pi S_t t). \tag{1.22}$$

The mathematical system is solved numerically by finite differences and a coupled time-marching method based on a vorticity-streamfunction axisymmetric formulation for the fluid. The vorticity equation is first advanced in time by the vorticity equation; the streamfunction and the wall vorticity are then updated at the same time of the whole membrane geometry. This requires the solution of a large nonlinear system of equations at each time-step. Details of the systems are reported elsewhere (Pedrizzetti, 1998).

The fluid-elastic system presents natural frequencies that appear because of elasticity and whose value depends from the interaction. When the system is forced close to a natural frequency a resonant behaviour may occur where each quantity reaches a steady oscillatory regime due to the continuous propagation of membrane waves. In Figure 1.1 the mean velocity upstream of the elastic insertion is reported as given by a pulsation (1.22) with $\varepsilon = 10^{-2}$ imposed downstream. This resonant behaviour of Figure 1.1 shows oscillation similar to (1.22) with an amplitude that is almost 40 times larger than the input one. This is actually an unstable solution that can be observed only in presence of an additional wall viscosity that bounds the exponential growth of the solution (Pedrizzetti 1998).

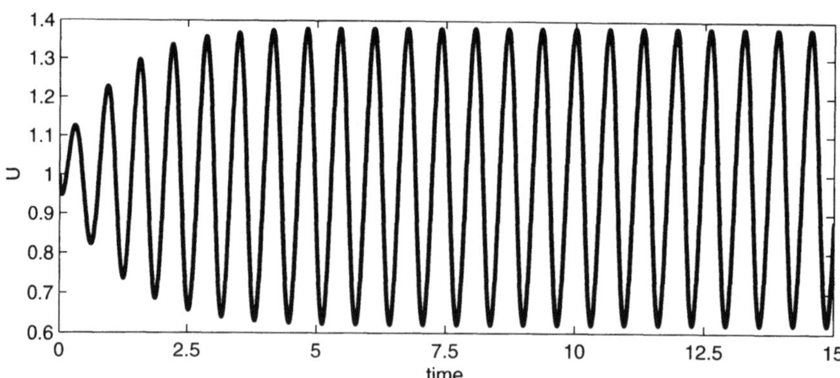

Figure 1.1. Evolution of the upstream mean velocity for downstream unsteady forcing (1.22) with $\varepsilon = 10^{-2}$, $S_t = 0.65^{-1}$, under resonant conditions.

One example of the interactive dynamics, in correspondence of the natural frequency of Figure 1.1, is shown in Figure 1.2 during one period of (1.22). The fluid-wall interaction takes the form of travelling waves along the membrane. The waves develop near the upstream end, where pressure fluctuations are larger, and propagates with positive celerity.

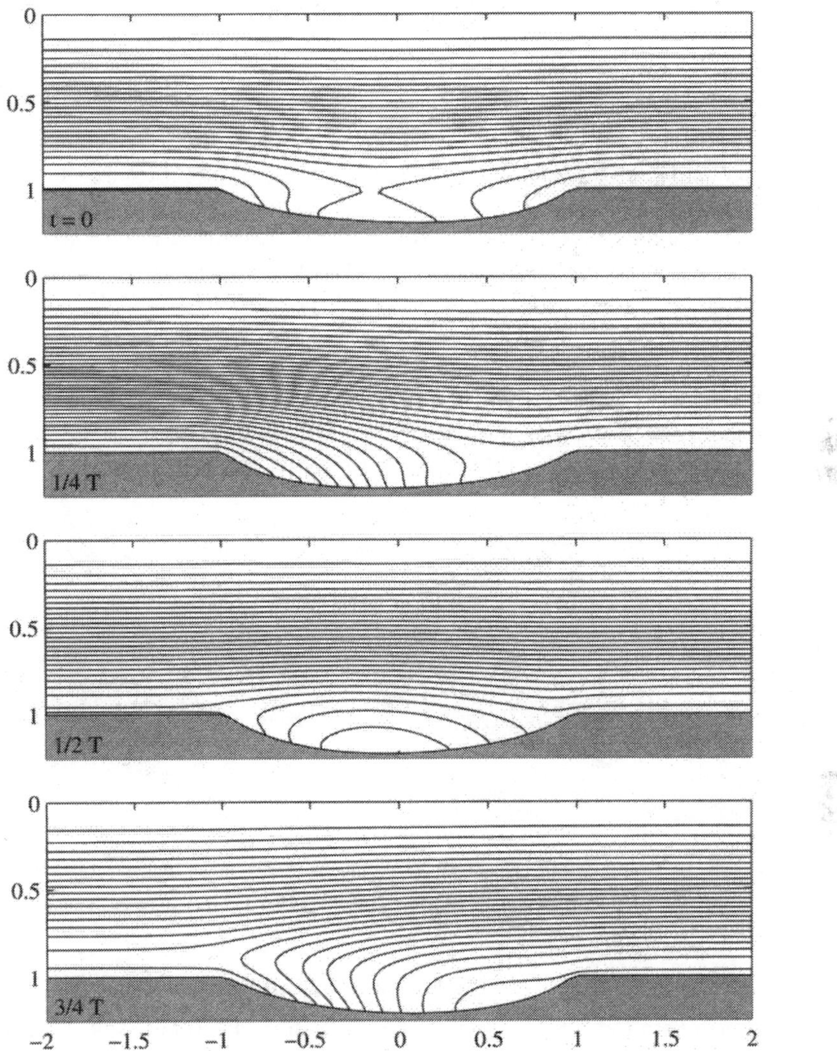

Figure 1.2 Axisymmetric oscillatory flow with downstream unsteady forcing (1.22) with $\varepsilon = 10^{-2}$, $S_t = 0.65^{-1}$, under resonant conditions, at four equispaced instants during a period.

The values of the natural frequencies, the form of the oscillations, and even the stability properties of the fluid-tissue interacting system, depend crucially on the way that fluid and structure are coupled. In this specific case, resonant progressive waves are found when either velocity and pressure are imposed downstream. The progressive waves decay rapidly from up to downstream so that they becomes unstable when the system is forced downstream and only the presence of wall viscosity avoids their undefined growth.

Viceversa, these waves decay rapidly when the system is forced upstream, and they are completely damped in absence of wall viscosity. Therefore, qualitatively different phenomena occur when the flow (1.22) is imposed upstream. In this case the system has shown natural frequencies associated to a synchronous membrane movement with fundamental wavelength equal to the membrane length. The membrane geometry at different instants is shown in Figure 1.3 in correspondence of travelling waves (left) and synchronous displacements (right).

The nonlinear coupled system may present several different solutions whose stability can depend on membrane properties like its viscosity (Lucey & Carpenter, 1992; Pedrizzetti, 1998) and inertia (Luo & Pedley, 1999). The resonant phenomena, as well as the self-excited oscillations observed under different conditions (Jensen, 1992; Luo & Pedley, 1996) correspond to strong fluid-structure interactions. These have been found to occur at frequencies that are higher ($St \sim 1$) than those found in the cardiovascular systems ($St \sim 10^{-2}$) where a weaker interaction is observed and expected. Therefore a simplified approach to the formulation of the fluid-structure coupled system may still be sought.

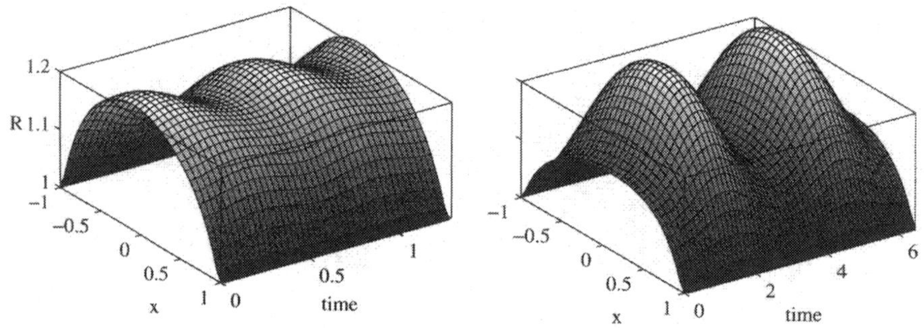

Figure 1.3. Space-time evolution of membrane the geometry. Travelling waves (left) and synchronous displacement (right) are two different resonant conditions of the same fluid-structure system.

1.9 Conclusive Remarks

Soft tissues may present relatively large deformations, under the action of normal biological loads, whose quantification must be accomplished within the framework of finite elasticity. This theory has been briefly outlined in its general three-dimensional form and in the reduced case of tissue represented by a membrane model. The main differences with the linear theory of elasticity, that is valid for infinitesimal deformations only, are the nonlinear stress-strain relationship and the fact that the equilibrium is imposed on the deformed geometry that is also the unknown of the problem. The basic issues of a fluid-tissue problem have been introduced and one example in an axisymmetric geometry has shown some peculiar behaviours that may arise in fluid-structure interactions.

2. Modelling the Flow inside Moderately Elastic Arteries

2.1 Introduction

The hydraulics of unsteady motion in elastic vessels has a high relevance in many applied fields, its primary applications are to the cardiovascular system to understand the evolution of pathology due to the interaction of the flow with the vessel deformation. The general solution of the coupled dynamics of a fluid flowing over elastic boundaries presents significant difficulties and every new result is often accompanied by the proposal of a different technique.

The presence of an elastic wall gives rise to various different phenomena depending on the specific value of the dimensionless parameters and on the boundary conditions. The presence of a compliant wall influences the stability of the boundary layer delaying the development of laminar-turbulent transition (Davies & Carpenter, 1997); the appearance of a negative transmural pressure, like in the venous system, leads to a reduction of the lumen in the vessel, which in turn reduces further the pressure possibly leading to a collapse of the vessel (Luo & Pedley, 1995; Heil, 1997). In addition, resonance phenomena can lead to natural self-induced oscillations (Luo & Pedley, 1996; Pedrizzetti, 1998); it should be said, however, that the spontaneous oscillations correspond to Strouhal numbers more than one order of magnitude larger than those typical of the cardiovascular system, therefore they do not represent the fundamental role of elasticity under physiological conditions. In most studies under realistic vascular conditions the numerical solution (Anayiotos et al., 1994; Perktold & Rapitsch, 1995) is sought by alternating the fluid and tissue equations within each time step. Such approach is very time-consuming, and often the differences found with the rigid case solution are relatively small when compared with the increase in the effort required by such a solution technique.

The theory of elastic wall deformations is divided into two main branches: the general nonlinear theory of finite deformations, introduced in the previous chapter, and the linear theory of small (strictly speaking infinitesimal) deformations. The latter allows a great simplification of the mathematical apparatus because of a linear strain-stress relationship and especially because the deformations are so small that the equilibrium is enforced on the undeformed configuration. A method for the numerical solution of the fluid flow in elastic vessels, which is generally well suited for large arteries, can be developed under the infinitesimal deformation approximation, like the linear theory for the wall deformation, and likewise it shares also the simplicity and easier applicability. The method is therefore an approximate one, it derives from *in vivo* measurements which reveal that the wall motion in large arteries is of relatively small entity (generally it is below ten percent of the *lumen* size) also in response to significant pressure variations. It is therefore natural to seek, at least at a preliminary stage, a solution of the fluid-wall system with a perturbative approach where the zeroth order solution is given by the flow in a rigid vessel, the first order correction gives the wall deformation and the induced flow modification without the need to solve the difficult coupled problem. Such an approach essentially assumes a locally infinite celerity, therefore it represents a good approximation for the fluid-wall interaction in sites of limited extent, these include typical situations associated with vascular diseases.

The linear theory (Lighthill, 1978) shows that the presence of an elastic wall leads to a phenomenon of propagation of elastic waves; it is however worth to notice that linear waves, of small amplitude and long wavelength, as well as other nonlinear phenomena based on perturbative analyses (Wang & Tarbell, 1992), are a result of the interaction between the wall deformation and the flow associated with it, leaving aside the deformation given by a bulk flow and its unsteady pressure field, which is of large entity with respect to its modulation associated with the wall motion itself, the former is in fact dominant in arterial districts of limited extent. Such an approach is here outlined in general terms, it is then applied to the case of pulsatile flow inside a circular tube with a smooth expansion by a numerical formulation in the axisymmetric approximation. The influence of wall elasticity on the flow and on the unsteady wall shear stress is studied in correspondence of parameters taken from the proximal region of the carotid artery bifurcation.

2.2 Generalities of the Linear Approach

Let us introduce here the method in its general aspects. We consider a three-dimensional flow inside a vessel with elastic walls which are described by the coordinates of material points. The elastic wall is deformed by the flow stresses acting on it and the deformed configuration is written as $\boldsymbol{X} = \overline{\boldsymbol{X}} + \delta\boldsymbol{X}$, where the overline indicate a reference fixed geometrical configuration. Let us assume that the wall inertia and viscosity are negligible so that deformation is evaluated by instantaneous static equilibrium law; this assumption is usually acceptable in large arteries where the Strouhal number is small. When the deformation is small (strictly speaking infinitesimal), it can be evaluated in terms of linear elasticity. Let us make the additional assumption, just for the sake of simplicity, that the shear stress contribution is negligible with respect to pressure and that the wall is regular enough that displacement occurs along the normal direction. Under these assumptions the deformation can be expressed in local terms as

$$\delta\boldsymbol{X} = (Eh)^{-1}\kappa(\overline{\boldsymbol{X}})\,\delta p\;\boldsymbol{n}\,; \tag{2.1}$$

where Eh is the elasticity constant which is given, in the case of a membrane, by the product of the Young modulus E and the wall thickness h; κ is a local geometric characteristics of the wall, $\boldsymbol{n}$ is the normal unit vector, and δp is the excess pressure, that is the difference between the actual wall pressure and the value that corresponds to the undeformed state $\overline{\boldsymbol{X}}$.

The law (2.1) is found under the hypothesis that the wall deformation is small, it is often true that the deformation is below 10% in realistic arteries. To better express the relative magnitudes, the law (2.1) is rewritten in terms of dimensionless quantities

$$\delta\boldsymbol{X} = \left(\frac{\rho U^2 L}{Eh}\right)\kappa(\overline{\boldsymbol{X}})\,\Delta p\;\boldsymbol{n}\,; \tag{2.2}$$

where $L, U, \rho L^3$, being ρ the fluid density, are the unit of length, velocity, and mass, respectively. Under the hypothesis that the deformation is small then, from equation (2.2), a dimensionless small parameter naturally appears

$$\varepsilon = \frac{\rho U^2 L}{Eh} \ll 1\,, \tag{2.3}$$

representing the proportionality between wall deformation and stress acting on the wall. The typical value of ε in major human arteries ranges from 10^{-4} to 10^{-2}.

Based on the presence of the small parameter ε we propose a perturbative approach to the interacting wall-fluid system. All variables are expressed in series of ε and equations are found for the various orders of approximation where, given equation (2.2), the undeformed configuration corresponds to the zeroth order. The zeroth and first orders are considered here, the former is the flow in a rigid vessel, the latter gives the wall deformation and the induced flow modification in a sequence that does not require the solution of a coupled system.

The fluid is governed by the mass and momentum balance equations, i.e. the Navier-Stokes equations, with boundary conditions that the fluid velocity at the walls is equal to the wall velocity.

The coupled mathematical problem, in dimensionless form, is formulated in a perturbative form as follows. Wall geometry, velocity, and pressure are expressed in series of ε as

$$\begin{aligned} \boldsymbol{X}(t) &= \overline{\boldsymbol{X}} + \varepsilon\, \delta \boldsymbol{X}^{(1)}(t) \; + O(\varepsilon^2)\,, \\ \boldsymbol{v}(\boldsymbol{x},t) &= \boldsymbol{v}^{(0)}(\boldsymbol{x},t) + \varepsilon\, \boldsymbol{v}^{(1)}(\boldsymbol{x},t) + O(\varepsilon^2)\,, \\ p(\boldsymbol{x},t) &= p^{(0)}(\boldsymbol{x},t) + \varepsilon\, p^{(1)}(\boldsymbol{x},t) + O(\varepsilon^2)\,, \end{aligned} \tag{2.4}$$

The $O(\varepsilon^0)$ terms represent the flow in a rigid vessel whereas the $O(\varepsilon^1)$ terms are the first correction due to the wall elasticity.

Introducing expressions (2.4) into the wall equation (2.2) the wall deformation is given at the first order with an explicit expression in terms of the order zero pressure

$$\delta \boldsymbol{X}^{(1)}(t) = \kappa(\overline{\boldsymbol{X}}) \left[p^{(0)}(x,t) - \overline{p}(x) \right]_{\boldsymbol{x}=\overline{\boldsymbol{X}}} \boldsymbol{n}\,. \tag{2.5}$$

where $\overline{p}(x)$ is the distribution of pressure that corresponds to the undeformed state $\overline{\boldsymbol{X}}$. Insertion of (2.4) into the fluid continuity equation ensures that continuity is enforced at any order, the Navier -Stokes equations is satisfied as it is for the zeroth order variables

$$\frac{\partial \boldsymbol{v}^{(0)}}{\partial t} + \left(\boldsymbol{v}^{(0)} \cdot \nabla\right) \boldsymbol{v}^{(0)} = -\nabla p^{(0)} + \frac{1}{R_e} \nabla^2 \boldsymbol{v}^{(0)}; \tag{2.6}$$

and couples zeroth and first order when $O(\varepsilon)$ terms are retained

$$\frac{\partial \boldsymbol{v}^{(1)}}{\partial t} + \left(\boldsymbol{v}^{(0)} \cdot \nabla\right) \boldsymbol{v}^{(1)} + \left(\boldsymbol{v}^{(1)} \cdot \nabla\right) \boldsymbol{v}^{(0)} = -\nabla p^{(1)} + \frac{1}{R_e} \nabla^2 \boldsymbol{v}^{(1)}. \tag{2.7}$$

The boundary conditions require that the fluid velocity at the wall is equal to the material wall velocity. Being $\frac{d\boldsymbol{X}}{dt}$ the velocity vector of material points on the wall, the boundary condition at the elastic wall reads

$$\varepsilon \frac{d}{dt} \delta \boldsymbol{X}^{(1)} = \boldsymbol{v}(t, \overline{\boldsymbol{X}} + \varepsilon\, \delta \boldsymbol{X}^{(1)}), \tag{2.8}$$

where $\boldsymbol{v}(t,x)$ is the fluid velocity, it is written in terms of values on the undeformed boundary by a formal Taylor expansion to give (Van Dyke, 1975)

$$\varepsilon\frac{d}{dt}\delta\boldsymbol{X}^{(1)}=\boldsymbol{v}^{(0)}(t,\overline{\boldsymbol{X}})+\varepsilon\left\{\left[\frac{\partial\boldsymbol{v}^{(0)}}{\partial r}\right]_{\overline{\boldsymbol{X}}}\delta\boldsymbol{X}^{(1)}+\boldsymbol{v}^{(1)}(t,\overline{\boldsymbol{X}})\right\}+O(\varepsilon^2). \tag{2.9}$$

Equating the same order terms in (2.9) gives the boundary conditions at the different orders

$$\boldsymbol{v}^{(0)}(t,\overline{\boldsymbol{X}})=0, \tag{2.10}$$

$$\boldsymbol{v}^{(1)}(t,\overline{\boldsymbol{X}})=\frac{d}{dt}\delta\boldsymbol{X}^{(1)}-\left[\frac{\partial\boldsymbol{v}^{(0)}}{\partial r}\right]_{\overline{\boldsymbol{X}}}\delta\boldsymbol{X}^{(1)}. \tag{2.11}$$

The problem at the zeroth order, specified by equation (2.6) with boundary condition (2.10), corresponds to the rigid wall problem that can be solved independently from the wall deformation. After the zeroth order solution is found the wall deformation is determined by the wall equilibrium (2.5). Afterward the first order correction can also be determined for the fluid on the basis of equation (2.7) with boundary condition (2.11).

The zeroth order solution is thus the flow in the undeformed rigid boundary and elasticity enters in the first order solution. The system can be solved independently of the actual value of elasticity, values for the perturbation parameter ε can be specified in post processing when the full solution of the flow in an elastic vessel is required.

2.3 Application to a Circular Vessel with an Expansion

We consider here the pulsatile flow inside a vessel of circular cross-section which presents an enlargement along its axis. The vessel expansion generates an adverse pressure gradient and a boundary layer separation should be expected (Pedrizzetti, 1996). The geometry is assumed as a model problem for the study of the influence of elasticity in presence of separated flow due to an expansion like in the proximal region of a carotid bifurcation as well as in several other situations.

Consider an elastic tube with rectilinear axis and circular cross section whose undeformed shape varies along the axis, and an incompressible viscous fluid, with density ρ and kinematic viscosity ν, moving inside it with a pulsatile flow volume rate of period T prescribed a specific position upstream of the region of interest

$$Q(t)=\frac{\pi}{2}(1+\sin 2\pi t); \tag{2.12}$$

here and in the following all variables are made dimensionless with the value of the radius of the tube infinitely upstream, R_0, and the corresponding maximum value of the section-averaged velocity, U_0. In addition the dimensionless time is normalised with the Strouhal number

$$S_t^{-1}=\frac{TU_0}{R_0}, \tag{2.13}$$

in order to have a unit period.

Assume the duct axis as the x-axis of a cylindrical system of coordinates $\{x, r, \theta\}$, the vessel undeformed geometry given by $\overline{R}(\boldsymbol{x}) = 1 + \frac{1}{4}(1 + \tanh x)$ and the dimensionless flow rate, prescribed at a position $x_0 = -10$. A pulsatile pressure is also imposed at the same position

$$p(x_0, t) = p_0 + \Delta p_0 \sin 2\pi(t - \phi); \tag{2.14}$$

which in general may not be in phase with the flow, the actual value of ϕ depending on the balance between inertial and viscous contributions to pressure losses all the way upstream the site under study.

The problem is solved under the approximation of axial symmetry, the governing equations are the axisymmetric form of the Navier-Stokes equations which are written in the vorticity-streamfunction formulation as

$$S_t \frac{\partial \omega}{\partial t_1} + J\left(\psi, \frac{\omega}{r}\right) = \frac{1}{R_e} \nabla^2 \omega; \tag{2.15}$$

where $\omega(x, r, t)$ is the azimuthal vorticity and $\psi(x, r, t)$ is the Stokes streamfunction; the Reynolds number is defined as $R_e = U_0 R_0 / \nu$, and vorticity and streamfunction are related by the Poisson equation (1.15).

The vessel deforms in the normal direction under the action of wall pressure $p(x, t)$. Accounting for the normal equilibrium only, equation 1.20, with $T_2(x, t) = 1 - R(x, t)/\overline{R}(x)$ and membrane incompressibility $T_1 = \frac{1}{2} T_2$, eventually gives

$$R(x, t) - \overline{R}(x) = \varepsilon \frac{2\overline{R}^2}{2 - \overline{R}\frac{d^2\overline{R}}{dx^2}} \left(p(x, t) - \overline{p}(x)\right); \tag{2.16}$$

where also $R'^2 \ll 1$ has been assumed. $\overline{p}(x)$ represents the pressure averaged during the period, $\overline{R}(x)$ is the undeformed geometry of the vessel, that is, from equation (2.16) the geometry corresponding to the average pressure $\overline{p}(x)$. The small parameter ε is defined by equation (2.3). It is evident from equation (2.16) that the system is independent from the static pressure constant p_0.

The system (2.15-16) is rearranged following the perturbative approach described in the previous section. The formulation is straightforward, details can be found in (Pedrizzetti et al., 2002). The system is solved by a second-order finite differences scheme employing a shearing coordinate transformation that replace the r-coordinate with $z = r/\overline{R}(x)$ to obtain boundary fitted coordinates. Time integration is performed with a third order explicit Runge-Kutta scheme (Pedrizzetti, 1996; Pedrizzetti et al., 2002; Roache, 1998).

2.4 Results: Elasticity and Phase Difference

Some results are summarised here, particular attention is posed to analyse the influence of elasticity. In what follows, the dimensionless parameter have been fixed to $R_e = 500$, $S_t = 0.02$, $\Delta p_0 = 10$, which correspond to typical values in major arteries.

The flow dynamics must be preliminarily analysed in the rigid vessel case; it constitutes the zeroth order term of the perturbative solution. The instantaneous systolic and diastolic flow fields are reported in Figure 2.1. The boundary-layer, developed during the acceleration, separates during systole and produces a small recirculating cell immediately downstream of the tube enlargement. The separation phenomenon corresponds to the roll-up of the boundary layer and the formation of a vortex structure eventually travelling downstream (Pedrizzetti, 1996). The positive separated vorticity induces a persistent negative wall shear stress in the portion of wall downstream of the enlargement. At diastole the boundary layer is thicker because of deceleration and the streamlines show about not net flow.

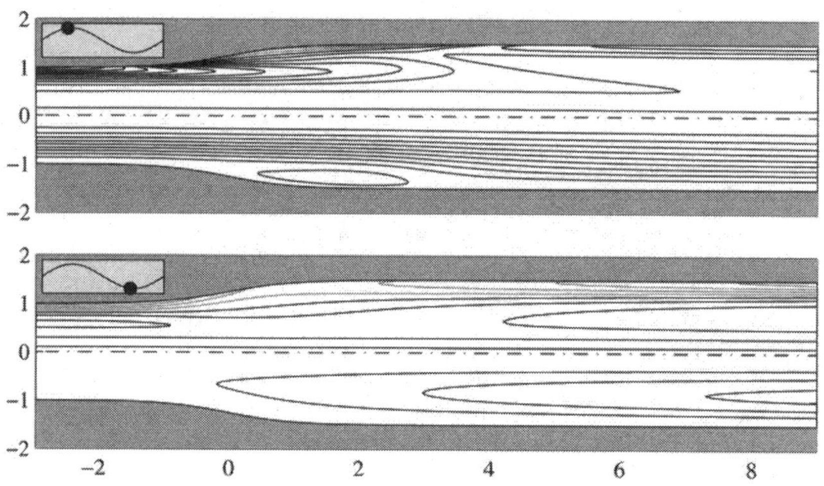

Figure 2.1. Instantaneous vorticity fields (upper half) and streamlines (lower half), rigid wall case, at systole and diastole. Vorticity levels from -5.25 to 5.25 step 0.5; streamfunction levels from 0 (symmetry axis) step 0.05; negative levels are grey.

The same flow fields in an elastic vessel are reported in Figure 2.2, at the same instants of the rigid case; a value $\varepsilon = 5 \times 10^{-3}$ has been selected to well visualise the wall deformation, flow and pressure wave are assumed to be in phase, $\phi = 0$. It can be observed that at systole the vessel dilates, therefore it tends to absorb the incoming flow; viceversa, the vessel shrinks when the flow decreases, thus increasing the actual discharge and its bulk velocity. During the acceleration, the synchronous deformation of the wall leads to a less extended separated region, while the contraction of the vessel during the deceleration phase of the imposed pulse tends to diminishing the value of the negative tangential stress.

The influence of elasticity on the wall shear stresses, computed as

$$\boldsymbol{\tau}(t, x, \overline{R} + \varepsilon\delta R^{(1)}(x,t)) = \tau^{(0)}(x,t,\overline{R}) + \varepsilon\left\{\left[\frac{\partial\boldsymbol{\tau}^{(0)}}{\partial r}\right]_{\overline{R}}\delta R^{(1)} + \boldsymbol{\tau}^{(1)}(x,t,\overline{R})\right\}, \quad (2.17)$$

is evidenced by the time-space development reported in Figure 2.3 for the rigid and the elastic cases, respectively. During the acceleration the flow separates at the enlargement and a negative wall shear stress appears on the wall. In the rigid case, $\varepsilon = 0$, the uniform flow upstream of the enlargement, and the appearance of separation at $t \simeq 0.1$ is clearly recognisable. The maxima negative stresses are induced by the separated vorticity translating downstream above the wall. The initial stage of the boundary-layer separation and the evolution of the positive stresses are not significantly affected by the wall elasticity. The negative shear at the wall is localised closer to the enlargement reflecting the different flow dynamics described above.

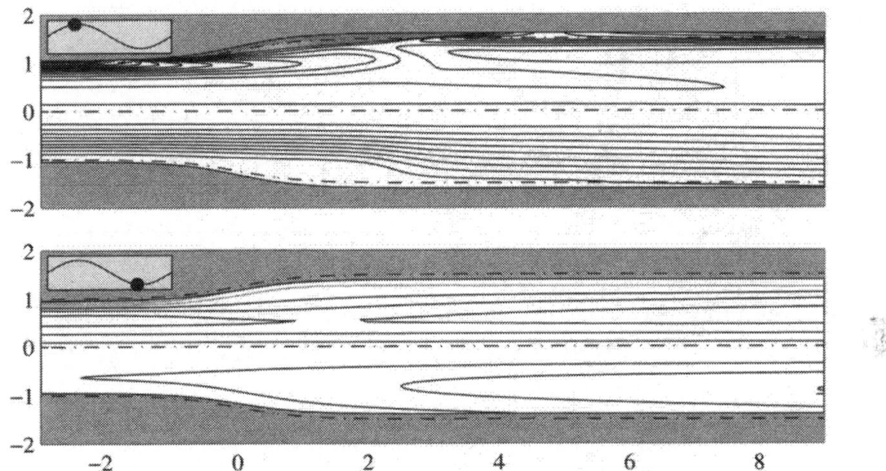

Figure 2.2. Vorticity fields (upper half) and streamlines (lower half), at systole and diastole, elastic wall case with $\phi = 0$. Vorticity levels from -5.25 to 5.25 step 0.5; streamfunction levels from 0 (symmetry axis) step 0.05; negative levels are grey. The dashed line represents the undeformed wall, $\varepsilon = 5 \times 10^{-3}$.

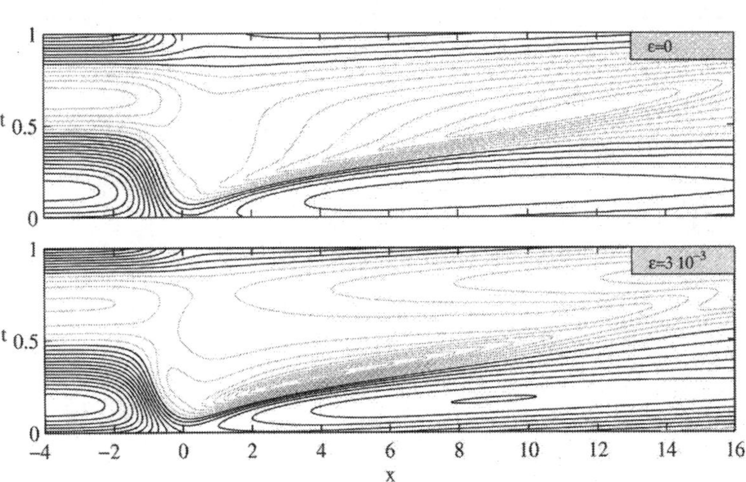

Figure 2.3. Space-time contour of the wall shear stress. Levels from -13.5×10^{-3} to 13.5×10^{-3} step 10^{-3}, negative values are grey. Rigid wall $\varepsilon = 0$, elastic wall with $\phi = 0$ and $\varepsilon = 3 \times 10^{-3}$.

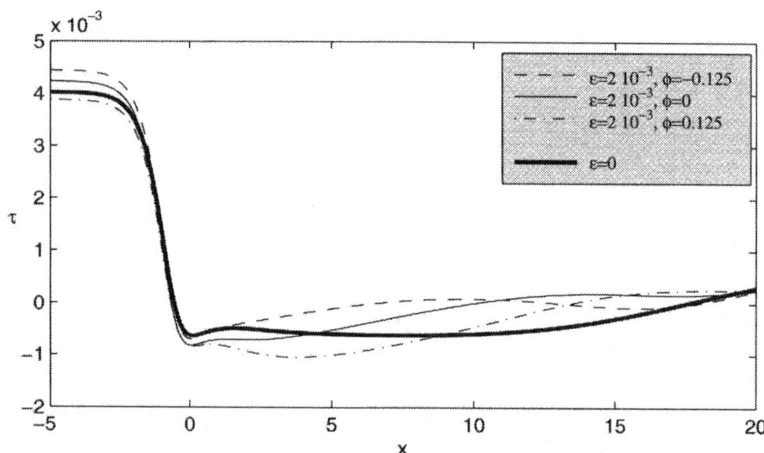

Figure 2.4. Distribution of the wall shear stress averaged over a period, the bold line represents the rigid wall $\varepsilon = 0$ case, others lines represent elastic wall cases with $\phi = [-0.125\ 0\ 0.125]$, $\varepsilon = 2 \times 10^{-3}$.

The distribution of the time-averaged wall shear stress is given in Figure 2.4; the bold line represents the rigid wall case $\varepsilon = 0$, the others lines represent the elastic results with ϕ ranging between -0.125 and 0.125. The zeroth order term shows a constant value of the mean stress in the upstream portion of the vessel, a steep decreasing at the enlargement followed by a gradually varying negative stress that becomes positive at large distance, approximately 35 times the variation of radius, from the expansion. The elastic case with $\phi = 0$, shows the effect of a shorter wake with a more pronounced negative stress just downstream the enlargement and a more rapid recover, after 25 times the variation of radius, to positive values. A significant dependence from the phase is observed, with a longer region occupied by negative shear when ϕ is positive and an apparently favourable behaviour for negative phases.

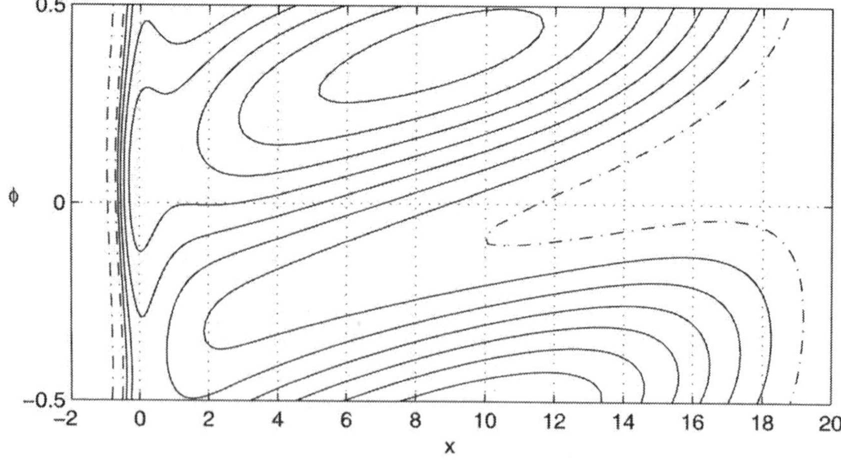

Figure 2.5. Space-phase contour of the wall shear stress. Levels from -15×10^{-4} to 15×10^{-4} step 2×10^{-4}, negative values are solid.

The global dependènce of the mean wall shear on the phase difference ϕ is reported in Figure 2.5, where the distribution of the mean wall shear stress along the wall and with phase is given. Solid lines represent negative values, the dashed one is the zero level. The trend derived from Figure 2.4 is shown to be systematic: negative phase differences around zero are characterised by a more localised and less intense boundary-layer separation. On the opposite, positive phase differences tend to elongate the recirculating cell and give rise to higher negative shear stress. Such a tendency shows its maximum for values of ϕ around 0.4, which are not of physiological interest.

2.5 Conclusive Remarks

A perturbative method for the simulation of the flow inside slightly deformed vessels is introduced. It is based on the assumption that vessel deformation are infinitesimal and thus it represents a possible fluid counterpart to the theory of infinitesimal elasticity. The equilibrium of the solid is evaluated on the undeformed configuration, similarly the fluid problem is solved on undeformed boundaries. The perturbative approach permits to find independently the zeroth order solution, corresponding to the rigid wall case, and the first one afterwards, related to the wall deformation, substantially uncoupling the fluid-wall problem. The method is well suited for elasticity and flow parameters often encountered in large artery flows.

The unsteady flow inside a rectilinear elastic vessel with a circular cross section varying along the tube axis has been studied with the numerical application of this method under the assumption of axisymmetric flow. In the present context, some results concerning the influence of wall elasticity on the boundary layer separation and wall shear stress have been discussed.

3. Fluid Dynamics Phenomena for the Left Ventricle

3.1 Introduction

The relevant increase in the number of symptomatic and asymptomatic patients in the cardiovascular area, has evidenced the need for an interaction between medical disciplines and technical-scientific ones in the development and improvement of diagnostic techniques and interpretative conceptual models. The objective of the interdisciplinary study should stand in the possible contribution to clinical practice: in the early detection of pathologies well before the symptomatic appearance of the disease, and in the systematic refinement of surgical techniques to reduce post-procedural complications.

The fluid mechanics has contributed significantly in the last years to the modelling of the flow in large arterial vessels (Botnar et al., 2000; Zhao et al., 2000; Bolzon et al., 2002, as recent examples, and references therein). A comparable accuracy of the numerical and experimental analysis has not yet been reached for the fluid dynamics of the heart. A cause for this delay may be found, among others, in the strong fluid-wall interaction which is fundamental in a cavity flow, like in the heart chambers, while it can be neglected, to a first approximation, in arterial flows.

The malfunction of the cardiac pump, in particular of the left ventricle, represents a primary cause of death in the modern society. Until a few years ago the clinical studies were focussed to the pumping phase: the left ventricle contraction known as systole, when the oxygenated blood is pumped in the circulatory system. Systolic pathologies represent a reduction in the heart's pump ability and in the available oxygen that may become insufficient to specific regions of the body. It is now known that a systolic pathology often represents only the terminal stage of a cardiac dysfunction, whose initial minor symptoms can be recognised during the diastolic phase (Mandinov et al., 2000), i.e. the ventricle filling with the blood that enters from the left atrium through the mitral valve. Therefore even in absence of evident advanced symptoms, the clinical practice pays particular attention to the features of the diastolic filling because they permit an early diagnosis of some diseases (Gaasch & Winter, 1994).

The role of the diastolic function in human health and disease remains nowadays enigmatic, depending on the difficulty in its assessment by physical examination or even by direct invasive measurement. Diastolic dysfunction can be defined as the inability of the hearth to accept adequate filling volume without an abnormal raise of the filling pressure; it is an early, and sometimes unique, manifestation of myocardial disease (Mandinov et al., 2000). Unfortunately, the non-invasive diagnostic tools currently available in the clinical setting are affected by the large number of haemodynamic dependent or independent variables which influence the measured parameters. For this reason, an unambiguous interpretation in terms of correspondence between definite pathological processes and typical semeiologic signs is difficult to achieve. An intriguing aspect of the left heart dynamics is the hypothesised occurrence of diastolic vortices within the ventricular chamber. Their presence and properties have been analyzed in theoretical (Reul et al., 1981; Wieting & Stripling, 1984), experimental (Steen & Steen, 1994), and numerical models (Vierendeels et al., 2000; Baccani et al., 2002*ab*), and introduced also in medical observations (Kim et al., 1994; Kim et al., 1995; Kilner et al., 2000). The presence of vorticity makes the fluid dynamics approach relevant for the understanding of the possible patterns and their relation with the mechanical properties of the surrounding tissues.

Some basic clinical points concerning in the left ventricle of the human heart are here introduced, particular attention is posed on the diastolic phenomena and to the mechanical interpretation of phenomena.

3.2 Healthy Heart Phases

The complete heart is made of two pumping devices. The right-heart pushes blood into the pulmonary circulation, the left heart pushes the oxygenated blood into the main circulatory system. They are potentially independent devices, however they share the properties of the same complete circulatory system, their mechanics is in phase and the geometry is intimately related because they belong to the same active organ. The dynamic of the heart is a complex phenomenon that combines cellular properties and activity, electrical excitation, soft tissue mechanics, and fluid mechanics, to cite a few matters involved. The interested reader is encouraged to go through the appropriate textbook, from solid mechanics to clinical physiology, for the different aspects. Here just an extreme synthesis is reported concerning those aspects that are pertinent for the following analyses.

The pumping phase of the heart starts with the electrical excitation of the sinoatrial node. This excitation propagates over the atrii and produces the atrial contraction (A). The electrical impulse, after a slow crossing of the atrioventricular node, propagates into the ventricles to excite the corresponding myocardial muscle fibers, that contract and produce the ventricular contraction, or systole (S). An unexcited partly passive phase follows when the myocardium relaxes and the heart expands its volume while filling under the pressure of the internal blood (early filling - E).

Let us revise the basic phenomena in the perspective of the fluid mechanics of the left heart. At systole (S) the ventricular volume shrinks, pressure in the ventricle rises and the blood is ejected through the aortic valve into the aorta. When the systolic phase is completed the myocardium start to relax, pressure decreases while both the aortic and the mitral valves are closed (isovolumetric relaxation). The pulmonary flow coming into the left atrium increases the atrial pressure until it opens the mitral valve, and starts the early filling phase (E) of diastole. Afterward the atrii excitation initiates and the following atrial contraction (A) gives a further ventricular filling jet. The ventricle excitation starts, pressure rapidly rises (isovolumetric contraction) with both valves closed until the new systole is initiated and the aortic valve opens (Fung, 1997; Opie, 2000). The time course of the left ventricular volume is shown in Figure 3.1 (above) for a healthy athlete. The corresponding systolic (aortic) and diastolic (transmitral) flow rates are also shown (Figure 3.1, below).

The relevance of fluid mechanics, and of fluid-structure interaction, during all the different phases of the left ventricle dynamics is evident and several different phenomena can be extracted. The fluid dynamics during the systolic phase is that of a cavity that shrinks and ejects the flow through a small opening, the aortic valve, the resistance of the flow reflects on the effort required by the muscle fibers during contraction. The actual results depending on several different factors like geometry, geometrical arrangements of muscle fibers, valve features, as well as on diastolic properties like filling pressure, favourable filling recirculation pattern to cite a few of them.

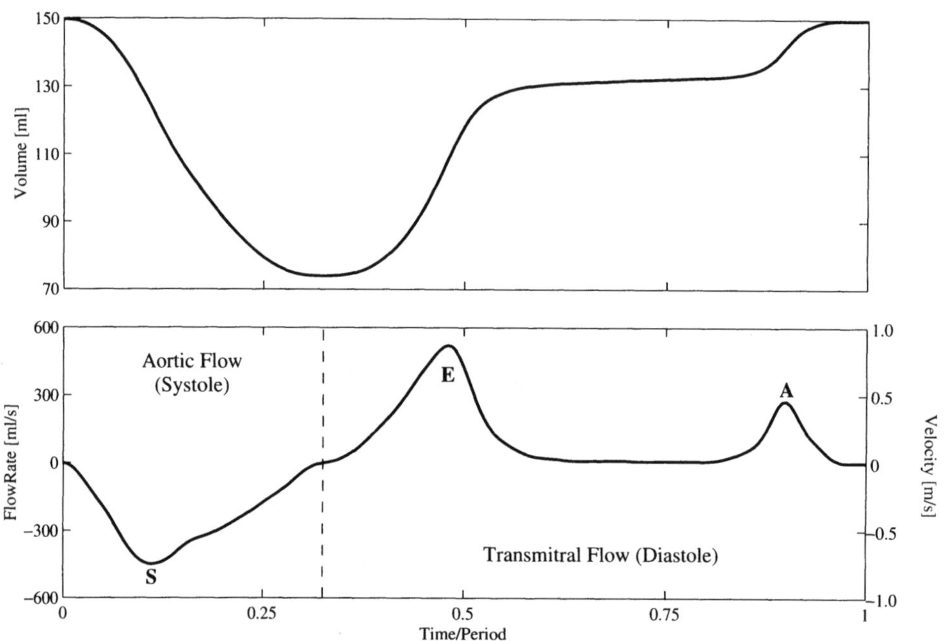

Figure 3.1. Time course of the left ventricular volume (above) in a healthy tall athlete. Its time derivative (below), indicative velocity values are given on the right.

During diastole the fundamental individual phenomena are the transvalvular jet, and the ventricular filling. The transmitral jet is that of a pulsatile flow, Figure 3.1, through an irregular, possibly mobile, orifice; a proper understanding of the fluid mechanics involved can support the interpretation of the clinically measured flow patterns and their relation with valvular function and dysfunction. The transmitral jet is a necessary ingredient for the fluid mechanics of ventricular filling, a complex unsteady fluid-structure problem that repeats every heartbeat (the ventricle is a cavity a few centimetres long that is filled, in a fraction of a second, by a transmitral jet that has a peak velocity about 1 m/s). The jet, that enters from the mitral valve, is surrounded by a shear layer that possibly rolls up into vorticity structures. The ventricular wall undergoes finite deformation under the action of the incoming flow, resulting in the volume increase. However the deformation is not only viscoelastic but presents also an active component, moreover the elastic properties change during filling while the myocardium empties from the blood. The diastolic flow can change substantially under different conditions and diseases, an improved understanding of the possible realistic patterns can support in the explanation of the phenomena underlying the clinical observation, eventually give an earlier diagnostics of dysfuctions and improve therapeutical choices.

3.3 Ventricular Diseases

The principal diseases in the heart correspond to reversible or irreversible changes of the wall properties (e.g. geometry, elasticity) commonly in a region of the ventricle, or of valve properties, that impair the proper interaction with the blood flow and eventually the pump ability. Revising the several possible diseases that can influence the functioning of the heart is out of the scope of these notes, for this the reader must refer to the appropriate literature (Little, 2000). The main, socially relevant, types of heart diseases are described here in relation to their influence on changes in the geometry and in the dynamics of the ventricle walls.

A first class of diseases that presents an immediate relation with the intraventricular fluid dynamics is that of *valvular diseases*. These, like stenosis, insufficiency, regurgitation, modify the geometry and the mobility of the valvular orifices. Therefore they influence the geometry of the transvalvular jet, and of the vorticity wake structure downstream.

Myocardial ischemia is the fundamental heart disease. It is the infarction of a portion of the myocardium, it is a consequence of the lack of oxygen to the myocardial fibers, and can be associated to an irreversible damage if extended in time. The infarcted region looses its muscular activity becoming essentially a passive tissue that does not contribute to systolic contraction. The lack of oxygen is commonly due to a total or partial obstruction somewhere in the coronary artery tree, the abrupt obstruction give rise to acute myocardial infarction (AMI), while a secondary ischemia can be caused by a progressive growth of a coronary stenosis. If the region interested to infarction is large this can completely impair the pumping ability.

Another important class of diseases is represented by the cardiopathy that lead to *ipertrophy* or *dilatation*. Ipertrophy commonly develops when a chamber is subjected to a higher pressure at the end of diastole (post-load) and an increased muscular work is then required during systole to empty the chamber. It can be a consequence of hypertension, aortic stenosis, mitral regurgitation, or it can develop primarily as a myocardial disease (cardiomyopathy). Ipertrophy produces some changes in the mechanical behaviour of the walls like, in primis, an increase of its stiffness without relevant geometrical changes at least in the first stage of the disease. Dilatation is primarily an increase of the chamber volume, commonly given by an increase of the filling pressure (pre-load). It can be a consequence of hypertension, aortic regurgitation and other causes, or it can develop primarily as a myocardial disease (dilated cardiomyopathy). The ability to receive and eject a same volume of blood with smaller deformation makes a dilated ventricle less kinetic and with reduced elasticity. The typical diastolic features found with either ipertrophy and dilatation are a delayed relaxation of the ventricle during its filling and eventually a restrictive (difficult to fill) behaviour. Ipertrophy and dilatation are not rigidly separated processes being consequences of similar causes and of the body adaptation, therefore they can appear individually and also develop in sequence or at the same time.

3.4 Clinical Doppler Measurements

Diagnoses are commonly performed by mean of non-invasive techniques (Magnetic Resonance, echo-Doppler ultrasound) which give an advanced but still incomplete picture of the fluid dynamics during the left ventricle filling. The typical diagnostic indicators have not a clear interpretation in terms of the corresponding mechanics and are thus difficult to be verified or quantified. An accurate modelling of the flow, and flow-wall interaction, inside the left ventricle is therefore necessary to improve the interpretative schemes of the clinical observations on the basis of the actual physical phenomena.

Echo-Doppler techniques can be used to produce different levels of results. In its simplest form the pulsed-wave Doppler evaluates one component of velocity at one point. It is commonly used at the transmitral position to produce a graph as shown in Figure 3.2 (left) of the type of that given below in Figure 3.1. The ratio of the peaks, E/A, is a first indicator of possible pathologies; in healthy subjects E/A $\gtrsim$ 1: the early filling is the principal component of the ventricular filling then completed by a not-negligible atrial contraction. The E/A ratio progressively decreases in presence of a delayed relaxation when the E-wave is unable to properly fill the ventricle and the a stronger atrial contraction develops to push the necessary blood volume. However this pattern often progresses to a generalised ventricle rigidity where the A-wave also is unable to produce adequate filling, higher pressure values develop to produce enough deformation then recovering the E-wave and a so called pseudonormal E/A profile. This is hardly recognisable from a normal one although it occurs in presence of a constrictive ventricle that often progresses to values E/A>1 and to an unnaturally strained ventricle that progressively looses its pumping ability.

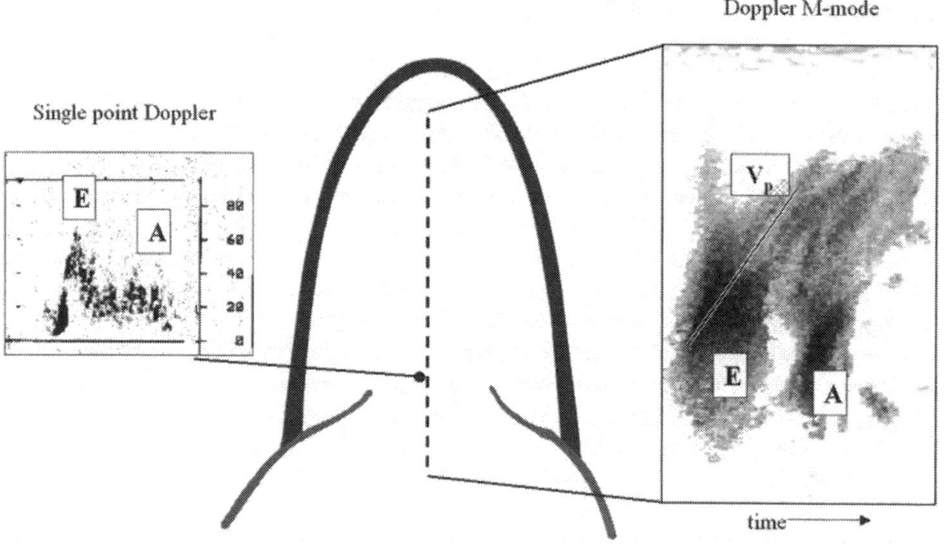

Figure 3.2. Example of pulsed Doppler (left) and Doppler M-mode (right) with indication of the corresponding measurement positions. The ventricle is shown upside-down to agree with the orientation of clinical measurements.

A more complete information is given by the two-dimensional Doppler when the velocity field is shown on a plane cutting the ventricle at several instant to produce a cineloop. The technology limitations imply, in addition to a moderate accuracy of the datum, that the growing spatial extension, from one point (zero-D) to 2D Doppler, suffers of a drastic decrease in the temporal resolution of the measured data; therefore the clinical observation gives in all cases an incomplete picture of the diastolic dynamics.

A recent Doppler analysis technique is the M-mode representation. The longitudinal component of velocity is measured along a transmitral line from the atrium to the ventricle apex, the time course of velocity distribution along such a line is represented in a two-dimensional colormap, Figure 3.2. This technique gives a space-time information that is summarised on a single image, and allows a relatively high time resolution. This representation, in addition to the E/A ratio at different spatial positions, allows to evaluate the extension of the transmitral jet (region of high velocity) and eventually to extract a number of clinical indicators (Garcia et al., 1998) that can be correlated to a possible pathology. Particular interest has been posed on the velocity of propagation, V_p, of the transmitral jet; this is the slope of the high velocity region (Figure 3.2) and appears to be a representative indicator for the ventricle mobility. The M-mode representation allows to discern more easily a healthy heart from one with delayed relaxation and a restrictive one (Takatsuji et al., 1997; Garcia et al., 1998; Garcia et al., 1999; Pai & Stoletniy, 1999; Møller et al., 2000; Møller et al., 2000). However many situations still remain ambiguous and require deeper examinations.

3.5 Improved Analysis of Measurements

A scientific support to clinical practice can be given in first instance by the manipulation of the clinical data to extract indicators that are not immediately readable from the image but that may be more closely related to the specific pathology. Several examples are becoming available (Bargiggia et al. 1991, Walker et al., 1995, Trambaiolo et al., 2000, Urheim et al., 2000, Masugata et al., 2001).

A simple one is shown here that helps in differentiating between a normal filling pattern and a pseudonormal that has an apparently similar M-mode. The pseudonormal pattern occurs in presence of a stiffer (restrictive) ventricle that require a higher effort to be filled by blood and is therefore accompanied by higher values of the intraventricular pressure gradient. Recent articles (Firstenberg et al., 2000; Tonti et al., 2001) give a clear assessment for the non-invasive estimation of transmitral pressure in the left ventricle, and comparisons between catheter measured pressure and values computed from the M-color Doppler are reported. It is essentially shown that transmitral pressure drop can be evaluated with good accuracy from non-invasive echocardiography by using the Euler equation projected along the transmitral line where Doppler M-mode is produced.

Let us indicate with x a coordinate along the measurement line, the M-mode datum is thus the velocity component $v_x(t, x)$. Projection of the Euler equation along x gives

$$\frac{\partial p}{\partial x} = \rho\left(\frac{\partial v_x}{\partial t} + v_x \frac{\partial v_x}{\partial x}\right) \tag{3.1}$$

where ρ is the fluid density and p is the pressure. Equation (3.1), that neglects viscous dissipation, and assumes a laminar flow along the chosen line, gives an estimate of the transmitral pressure drop accounting for the strong flow unsteadiness and for the spatial deceleration that occurs from the valve approaching the ventricle walls. The M-mode velocity and corresponding transmitral pressure are reported in Figure 3.3 for a normal ventricular flow and for a restrictive one. The differences in the velocity fields are not sufficient to evidence a relevant constriction in the second case. The intraventricular pressure shows a variation of $\pm$ 10 mmHg that is significantly higher than the normal values contained between $\pm$ 3 mmHg.

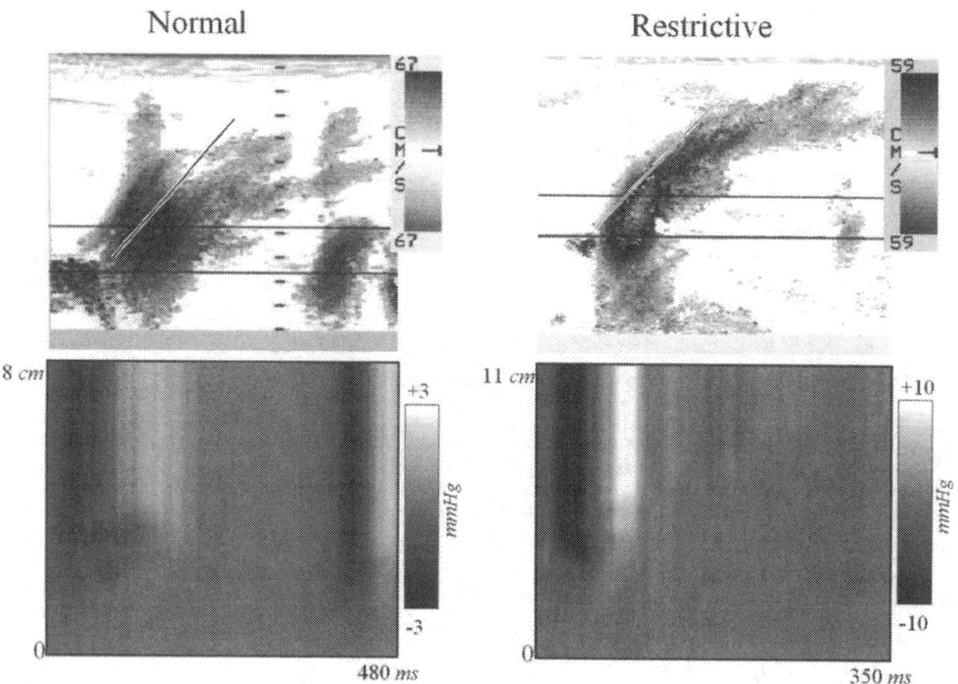

Figure 3.3. Doppler M-mode, above, and corresponding relative pressure distribution (below). The lower line indicates the mitral plane.

In conclusion an appropriate manipulation of the measured data, on the basis of conservation laws, allows to build additional indicators that may be more directly related to a specific pathology, improve the physical interpretation and discern situations. The relevance of this approach is destined to grow with the growing capabilities of diagnostic machinery, now ready to measure time-dependent 3D data, where the large amount of information acquired will necessarily require the development of synthetic representations.

3.6 Modelling

The objective of fluid dynamics is the creation of a mathematical modelling that may be used as a virtual laboratory to understand the physical phenomena involved, allowing the interpretation of clinically measured data.

The mathematical details about modelling the flow into the left ventricle are given in the next chapters, some results are reported here for illustrative purpose. The flow during two instants of the early filling phase are shown in Figure 3.4 for an ideal ventricle (axisymmetric and with a fixed mitral valve) with realistic parameters under healthy conditions, corresponding to those in Figure 3.1. During the accelerating stage, the jet is bounded by a circular vortex layer that rolls up into a vortex structure. In this phase the entering jet has a central, approximately irrotational, core surrounded by the vorticity layer. When the entering flow rate decreases the vortex detaches from the valvular trailing edge; then, during the diastasis between the E- and A-wave that has a negligible net inflow, the vortex travels downstream by self induced propulsion until it reaches the apex. The M-mode representation is shown on the right side of Figure 3.4. Despite the strong idealisation of the model the picture shows the correct major features. The irrotational bulk flow is recognised in the almost vertical patterns at the beginning of the E-filling during the accelerated flow. The propagating pattern is the trace of the translating vortex whose rotation gives a higher velocity at the centre, therefore the propagating velocity, V_p, is the vortex self-induced velocity.

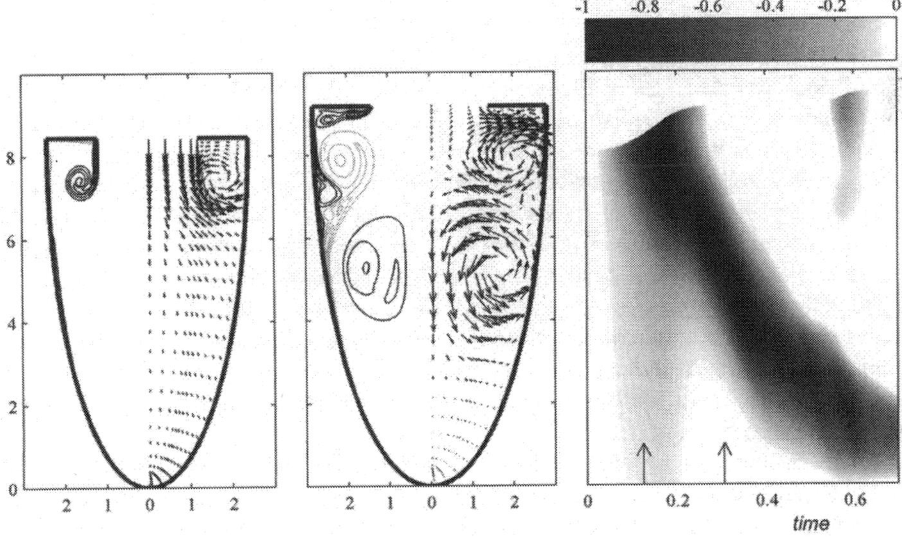

Figure 3.4. Flow fields in a model left ventricle during filling. Vorticity and velocity vectors are reported in the first two frames at $t = 0.125$ and $t = 0.3$; the corresponding virtual Doppler M-mode is shown in the right frame. Units are [cm], [s], and [m/s].

3.7 Conclusive Remarks

The phenomena involved in the left ventricle filling phase (diastole) are briefly revised. The major heart diseases and the corresponding clinical echo-Doppler measurements are introduced and discussed in relation with the corresponding fluid dynamics and fluid-structure interaction.

The fluid mechanics is contributing in the analysis of the measurements to extract information that are closely related to specific pathologies. Fluid dynamics modelling allows to build interpretative schemes of the clinical observations; these improve the understanding of the involved processes and of their possible relation with diseases.

4. Modelling the Fluid Dynamics of the Left Ventricular Filling

4.1 Introduction

The study of the heart dynamics has developed from lumped, zero-dimensional, models and slightly more complex one-dimensional schemes of inviscid flows to more realistic analyses based on the assumption of viscous motion; this is necessary to capture the vorticity-related dynamics, which characterises the impulsive inflow through the mitral valve, and its interaction with the boundary-layer developing at the moving ventricle walls. Numerical methods based on the finite elements or finite volumes schemes have been recently used to solve the fluid equations, in some cases coupled with models of the wall deformation (Taylor & Yamaguchi, 1995; Redaelli & Montevecchi 1996). Adaptive axisymmetric finite elements have been used to analyse the diastolic phase, with an initially ellipsoidal ventricle (Vierendeels et al., 2000). An alternative important method for the coupled solution of the fluid-wall problem has been developed by introducing the concept of immersed boundary elements (Peskin & McQueen, 1989*a*,*b*; Peskin & Printz, 1993; McQueen & Peskin, 1997). The fluid problem is solved on a regular domain (box), with a distribution of fictitious body forces; these replaces the presence of the wall, which in turn moves with the fluid within an iterative procedure. This method is especially well suited for the interaction with walls that are effectively immersed in the fluid (valvular leaflets, for example), it has also been applied to the dynamics of the left heart, reproducing the major features of the fluid-wall interaction (Lemmon & Yoganathan, 2000; McQueen & Peskin, 2000). These numerical methods can be applied to generic domains, having therefore characteristics of generality; on the contrary, the finite elements technique requires a significant computational effort to capture the local details of the flow field, while the immersed boundary elements method suffers of a reduced accuracy at the boundary, because the wall does not lie on coordinate curves and interpolation (a source of numerical diffusion) is necessary.

The numerical results (Vierendeels et al., 2000), as well as experimental ones (Steen & Steen, 1994), are in qualitative agreement between themselves and with clinical observations. A general picture of the diastolic fluid dynamics inside of the left ventricle can be given, at least for axisymmetric flows in normal (not pathologic) conditions. The mitral flow generates a vortex wake, a vortex ring in the axisymmetric approximation, that moves toward the ventricle apex and interacts with the boundary-layer vorticity at the ventricle walls. The problem appears primarily influenced by the geometry of the mitral valve and by the wall motion; this is strictly connected with the temporal law of the entering discharge, and it can be related to different pathologies of the valve or of the cardiac muscle. The ventricle geometry does not seem to play a fundamental role until the main geometrical proportions are retained, therefore a prolate spheroid has been suggested as sufficiently representative of the actual configuration on the average (Vierendeels at al., 2000; and references therein). Alternatively, one specific realistic geometry should be used.

A model left ventricle, a truncated prolate spheroid with the mitral valve corresponding to a thin circular orifice at the inlet, is here considered. The system of equations written in primitive variables is solved numerically by a mixed finite differences-spectral method in boundary-fitted moving coordinates. The mathematical formulation and the numerical method are briefly outlined; results obtained in the axisymmetric approximation are discussed.

4.2 Mathematical Formulation

The flow inside a model left ventricle for an incompressible fluid with density ρ and kinematic viscosity ν is here considered. The ventricle is assumed to be half of a prolate spheroid with major semiaxis $H^*(t^*)$ and principal diameter $D^*(t^*)$, being t^* the time, and where * denotes dimensional quantities. The mitral valve is modelled as an orifice of infinitesimal thickness in the equatorial plane, with a diameter $D_v^*(t^*)$; the ratio $m = D_v^*/D^*$ is kept constant in time. The system is forced by a given impulsive temporal law of the inlet flow-rate $Q^*(t^*)$, which gives the ventricle volume variation; the dynamics of the two degrees of freedom ventricle geometry, H^* and D^*, can be derived on the basis of a simple elastic wall modelling, as described in §4.3.

The problem is made dimensionless assuming the heartbeat period T as time scale, thus the early filling phase has a duration less than half unit time, and a reference inlet velocity U; this is given by the peak inlet discharge and the area of the orifice, with $m = 0.8$, at end systole ($t = 0$). Such choices allow an easy readability of the results and the immediate comparison with clinical data, being $T \simeq 1s$, $U \simeq 1m/s$. The reference unit of length is $L = UT$, the unit mass is ρL^3. In what follows dimensionless quantities are considered unless explicitly stated.

The governing equations for the fluid are the Navier-Stokes and continuity equations,

$$\frac{\partial \boldsymbol{v}}{\partial t} + (\boldsymbol{v} \cdot \nabla)\boldsymbol{v} = -\nabla p + \frac{1}{Re_T}\nabla^2 \boldsymbol{v}, \tag{4.1}$$

$$\nabla \cdot \boldsymbol{v} = 0, \tag{4.2}$$

$\boldsymbol{v}$ is the velocity vector, p is the pressure, $Re_T = U^2T/\nu$ is the Reynolds number corresponding to the chosen unit dimensions; a standard Reynolds number is $Re \sim Re_T D$.

The equations are written in a moving, boundary-fitted, prolate-spheroid system of coordinates $\{\mu, \eta, \theta\}$, related to the cylindrical system $\{r, z, \theta\}$ by (Morse & Feshbach, 1953)

$$r = \delta(t)\sinh(\alpha(t)\mu)\sin\eta; \;\; z = \delta(t)\cosh(\alpha(t)\mu)\cos\eta. \tag{4.3}$$

The functions $\delta(t)$ and $\alpha(t)$ specify the two degrees of freedom wall dynamics, being related to the geometric properties by $\delta = (H^2 - D^2/4)^{1/2}$, $\alpha = \tanh^{-1}(D/2H)$. The domain is $\mu \in [0, 1] \times \eta \in [0, \pi/2] \times \theta \in [0, 2\pi)$; the ventricle wall corresponds to the surface $\mu = 1$, the axis of symmetry to the coordinate curves $\mu = 0$ and $\eta = 0$, the equatorial plane is at $\eta = \pi/2$.

The vector form (4.1) is split in its scalar components, which can be formally written as:

$$\frac{\partial v_\mu}{\partial t} + \mathcal{C}_\mu(v_\mu, v_\eta, v_\theta, c_\mu, c_\eta, c_\theta, \delta, \alpha, \dot{\alpha}) + \mathcal{P}_\mu(p, \delta) - \frac{1}{Re_T}\mathcal{D}_\mu(v_\mu, v_\eta, v_\theta, \delta) = 0,$$

$$\frac{\partial v_\eta}{\partial t} + \mathcal{C}_\eta(v_\mu, v_\eta, v_\theta, c_\mu, c_\eta, c_\theta, \delta, \alpha, \dot{\alpha}) + \mathcal{P}_\eta(p, \delta) - \frac{1}{Re_T}\mathcal{D}_\eta(v_\mu, v_\eta, v_\theta, \delta) = 0,$$

$$\frac{\partial v_\theta}{\partial t} + \mathcal{C}_\theta(v_\mu, v_\eta, v_\theta, c_\mu, c_\eta, c_\theta, \delta, \alpha, \dot{\alpha}) + \mathcal{P}_\theta(p, \delta) - \frac{1}{Re_T}\mathcal{D}_\theta(v_\mu, v_\eta, v_\theta, \delta) = 0; \tag{4.4}$$

$\mathcal{C}$ represents the convective terms, $\mathcal{P}$ the pressure gradient, and $\mathcal{D}$ the diffusive terms. The vector field $\{c_\mu, c_\eta, c_\theta\}$ is the velocity of the coordinate system in the directions of the coordinate curves. Details about the complete mathematical formulation can be found in (Baccani et al., 2002*ab*). For three-dimensional solutions, a spectral representation of the variables in the θ-direction is employed to reduce the 3D problem into a series of 2D ones that are coupled by the nonlinear terms of the equation (4.4).

Equations (4.4) must be completed with the boundary conditions: the condition at the walls gives the fluid-wall coupling, the conditions at the inlet represent a model for the upstream atrial flow. The no-slip condition at the ventricle wall, $\mu = 1$, and on the closed part of the valvular plane, $\eta = \pi/2$ with $\mu \geq m$, reads

$$\boldsymbol{v} = \boldsymbol{c}. \tag{4.5}$$

At the axis of the ventricle, $\mu = 0$ and $\eta = 0$, physical boundary conditions are not available; symmetry consideration gives conditions to the terms of the Fourier expansion depending if they are odd or even ones (Canuto et al., 1988). At the inlet section, $\eta = \pi/2$ and $\mu < m$, the velocity field, which represents the properties of the incoming atrial flow, must be specified.

4.3 Wall Dynamics Model

An exhaustive analysis of the fluid-wall interaction could be obtained in principle by the coupled solution of the equations governing the fluid and wall dynamics (Luo & Pedley, 1996; Vierendeels et al., 1997; Pedrizzetti, 1998), thus requiring the definition of the structural properties of the walls; in the present case, the ventricle walls can be represented as a thick anisotropic shell with time-varying elastic properties (due to blood filling) in a non-linear finite deformation regime. Although the ventricle walls do not present a purely passive behaviour, different elastic wall models have been proposed for analysing the ventricular fluid-structure interaction (Peskin & Printz, 1993; Vierendeels et al., 1997; McQueen & Peskin, 2000). In addition, the unsteady transmural pressure profile in one reference position (commonly at the inlet) must also be specified. The tissue mechanical parameters and the pressure time-laws are not commonly estimated quantities, standard clinical measurements concern the inlet mitral velocity (Garcia et al., 1998) and the displacements of the ventricular walls measured in long-axis and short-axis views. Therefore we have chosen to consider the geometrical dynamics as an input data rather than the available mechanical properties; this in turn allows to avoid the solution of a coupled fluid-tissue system having also in mind that the fluid phenomena do not change when the same wall motion is imposed or is computed by a coupled solution.

The temporal law of the entering discharge can be estimated from the available data about the mitral velocity or geometry dynamics. A simplified model is then constructed to specify the two degrees of freedom of wall motion as follow.

The discharge law corresponds to the time variation of volume $V = \frac{\pi}{6} D^2 H$, and gives a relation between the diameter $D(t)$ and height $H(t)$ derivatives

$$Q = \frac{\pi}{6} D^2 H \Big(\frac{2}{D} \frac{dD}{dt} + \frac{1}{H} \frac{dH}{dt} \Big). \tag{4.6}$$

An additional relation must be specified to close this two degrees of freedom problem, it can be obtained from a simple relation that is extracted from a linear elastic membrane modelling. In the case of a prolate spheroid elastic membrane of vanishing thickness, the circumferential and meridional wall stresses at the equatorial plane are

$$S_1 = \frac{pD}{4}\frac{8H^2 - D^2}{4H^2}, \quad S_2 = \frac{pD}{4}; \tag{4.7}$$

where p is the isotropic transmural pressure. Assume as a first approximation the linear elasticity relation $S_1 = E\,dD/D$, $S_2 = E\,dH/H$, where E is an instantaneous Young modulus, inserting these in (4.7) and eliminating p and E we obtain

$$\frac{dD}{D} = \frac{(8H^2 - D^2)}{4H^2}\frac{dH}{H}. \tag{4.8}$$

Relation (4.8) has the correct asymptotic behaviours that it must reduce to a sphere $dD/D = dH/H$ when $D = 2H$, and to the behaviour of a cylinder $dD/D = 2\,dH/H$ when $H \gg D$. It has been tested with clinical data sets showing errors that could not be separated from the measurement uncertainty. Given the little available data and the possible large variation of real conditions, the approximation (4.8) represents a simple first approach to model the wall dynamics on the basis of the only knowledge of an inlet flow-rate and an initial condition for the ventricle geometry. Inserting (4.8) into (4.6) we obtain an ordinary differential equation for the time evolution of the diameter

$$\frac{dD}{dt} = \frac{6Q}{\pi}\frac{8H^2 - D^2}{20H^3D - 2HD^3}, \tag{4.9}$$

and, from (4.8), of the ventricle height. Based on these the temporal evolution of prolate spheroid parameter δ and α defined in §4.2 can then be obtained

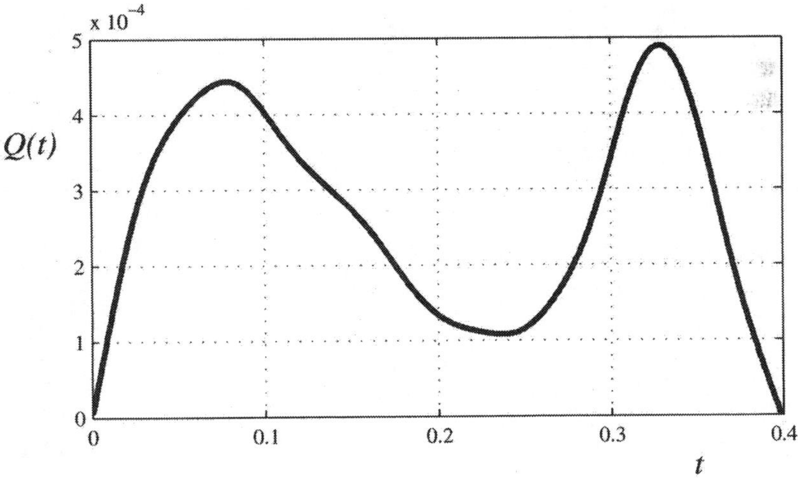

Figure 4.1. Time course of the left ventricular inflow in a young healthy subject.

4.4 Numerical Results

The mathematical model outlined in the previous section has been solved using a numerical method (Baccani et al., 2002*b*; and references therein). In what follows the results obtained in the axisymmetric approximation are reported; the equations (4.4) are simplified accordingly to this assumption. At the walls, the adherence condition is imposed; on the ventricle axis the boundary conditions can be written

$$v_\mu(0,\eta) = 0, \quad \left.\frac{\partial v_\eta}{\partial \mu}\right|_{0,\eta} = 0, \quad v_\eta(\mu,0) = 0, \quad \left.\frac{\partial v_\mu}{\partial \eta}\right|_{\mu,0} = 0. \tag{4.10}$$

At the mitral plane, the flow is assumed to enter the ventricle orthogonally, i.e. only the v_η component is different from zero; the v_η velocity profile is not imposed explicitly, rather the physically based condition that the entry flow is irrotational is imposed at the inlet. In fact, the flow inside the atrium accelerates to enter through the narrow mitral valve in the ventricle; the boundary layer is extremely thin because of the converging nature of the flow and of the extremely rapid unsteadiness parameters. In such a picture, given also the negligible mitral valve thickness, no finite boundary layer thickness can be estimated a priori.

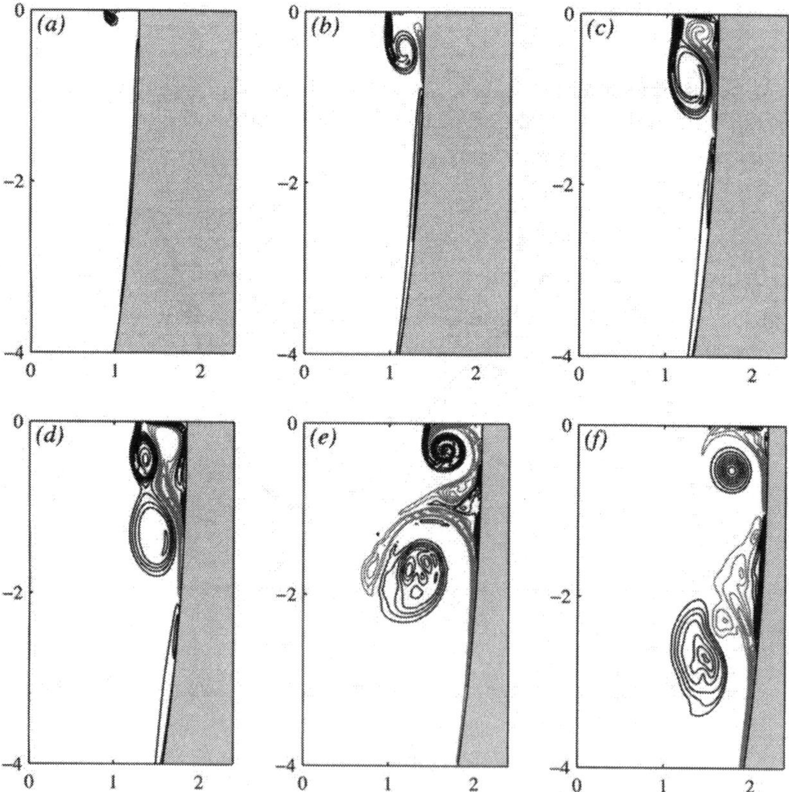

Figure 4.2. Vorticity fields in a model left ventricle during filling, healthy subject; positive (counterclockwise) values are black, negative grey.

The results reported in what follows have been obtained using a temporal law $Q(t)$ of Figure 4.1, derived from clinical data of a young healthy subject; the same $Q(t)$ has been used to analyse pathologic cases. The symptoms of the dilated cardiomyopathy are reproduced in a schematic manner, varying the dimensions of the ventricle at the beginning of the diastole: a more dilated ventricle corresponds to a more severe pathology. During the diastole, the ventricle walls are thus subjected to a decreasing motion, being the system forced by the same temporal variation of the volume (Baccani et al., 2002*a*). A typical Reynolds number is $Re_T = 2.5 \times 10^4$, the parameter m is assumed equal to 0.7.

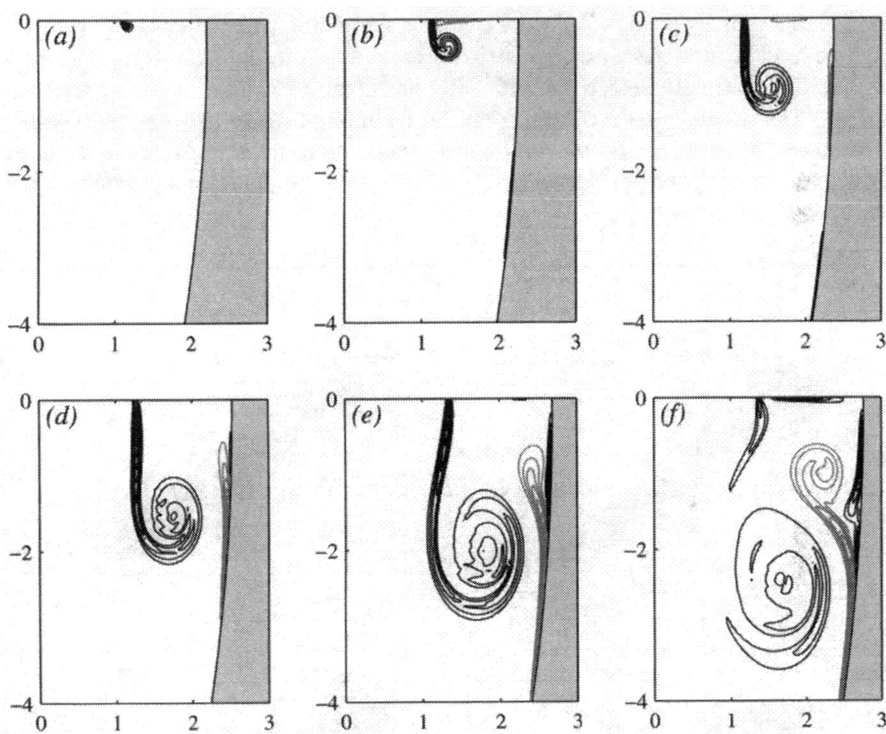

Figure 4.3. Vorticity fields in a model left ventricle during filling, pathological subject; positive (counterclockwise) values are black, negative grey.

The flow evolution in the reference healthy case is reported in Figure 4.2 in terms of vorticity distribution. During the initial stage of the early filling wave a positive (counterclockwise) vortex wake develops attached to the mitral edge, and begins to interact with the negative boundary-layer at the closed part of the mitral plane, Figure 4.2*a* at $t = 0.04$. The wake grows in size elongating towards the lateral wall of the ventricle, giving rise to the separation of the boundary-layer, Figure 4.2*b* at $t = 0.08$; during such a phase, the vortex wake begins to roll-up forming a well defined vortex. A stronger interaction with the wall viscous-layer and the tendency to detach of the vortex is observable in Figure 4.2*c* at $t = 0.12$, at the beginning of the early filling deceleration. The vortex induces a secondary separation at the lateral wall, with the eruption of the negative vorticity layer which completely detaches the primary positive vortex, Figures 4.2*d* and 2*e* at $t = 0.16$ and $t = 0.2$, respectively. The roll-up of a second positive vortex is thus induced, Figure 4.2*e*. Afterwards, the three-layer vorticity structure induced by the primary vortex translates toward the ventricle apex, and the second vortex detaches from the mitral edge, Figure 4.2*f* at $t = 0.24$ during the diastasis. A similar dynamics can be observed in correspondence of the second filling wave, giving a final complex vorticity distribution characterised by the presence of several detached vortices.

A different dynamics has been observed in pathologic cases; the instantaneous vorticity fields at the same times of Figure 4.2 are reported in Figure 4.3, corresponding to a low degree of pathology. The increased ventricle volume at the beginning of the diastole leads to a less intense interaction between the inlet vortex wake and the boundary-layer at the ventricle walls. Thus the wake elongates well inside of the chamber before to induce a significant separation, Figure 4.3*d*, and to finally detach from the valvular plane forming a more intense free vortex, Figure 4.3*f*. Such a behaviour has been found to be emphasised at growing pathology: a more dilated ventricle tends to inhibit the vortex-wall interaction, the wake remains attached to the mitral plane for longer periods, elongating further inside the ventricular chamber (Baccani et al., 2002*a*).

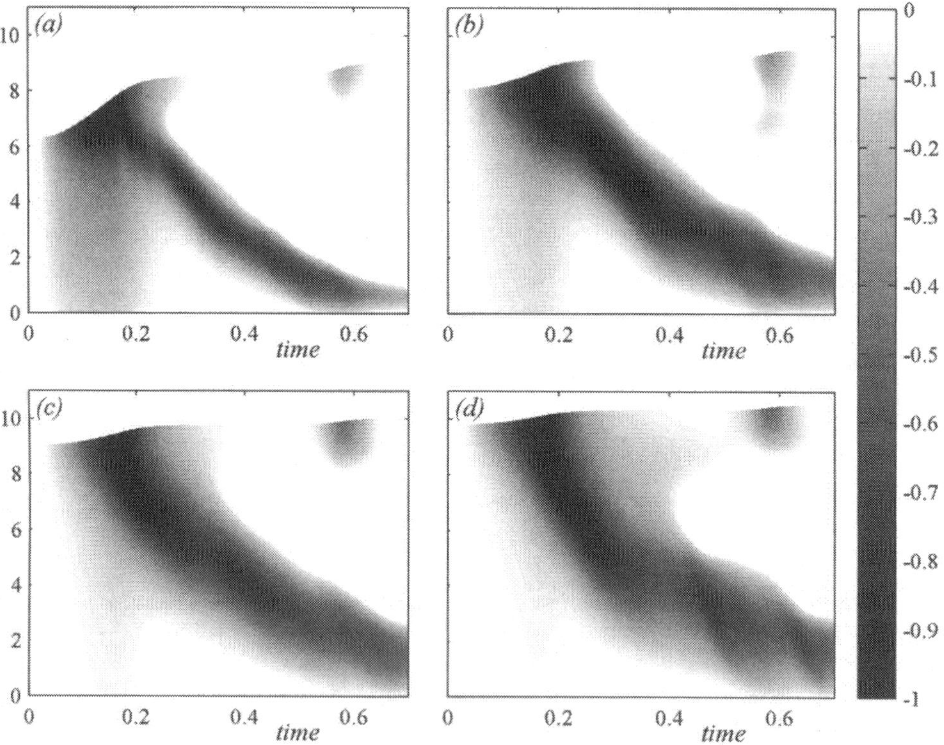

Figure 4.4. Virtual M-mode; healthy subject (a), increasing degree of pathology from (b) to (d). Units are [cm], [s], and [m/s].

A different view of the phenomenon can be given by the space-time maps of the axial velocity computed along the symmetry axis of the ventricle, which resembles the echo-Doppler M-mode images; these are reported in Figure 4.4 for the healthy case and for various degrees of the dilated cardiomyopathy. A common phenomenon is the well defined trace of the maxima, which travels from the mitral plane toward the apex; such a trace represents the footprint of the translating vortex on the axial velocity. More severe pathologies are related to more intense and slower vortices, so that the slope of the velocity trace decreases in agreement with what is clinically observed.

The clinical relevance of these results should be considered with caution. In fact the axisymmetric assumption corresponds to a peculiar unrealistically stable flow structure that may not be representative of the actual 3D flow. Important three-dimensional dynamics occurs, for example, because of the eccentricity of the mitral orifice. This has been shown to produce a more extended wake vorticity downstream the anterior leaflet (farther from the wall) and a smaller wake on the opposite side where the valve is contiguous to the ventricular wall (Bellhouse, 1972, Reul et al., 1981). Eventually this geometrical asymmetry (Kilner et al., 2000) gives rise to a large recirculating structure whose actual three-dimensional structure is still unclear. In Figure 4.5 the azimuthal vorticity on a centred transversal section of a ventricle with a slightly displaced mitral orifice is shown as computed numerically. The significant deviation of the vortex wake is a consequence of the small mitral displacement.

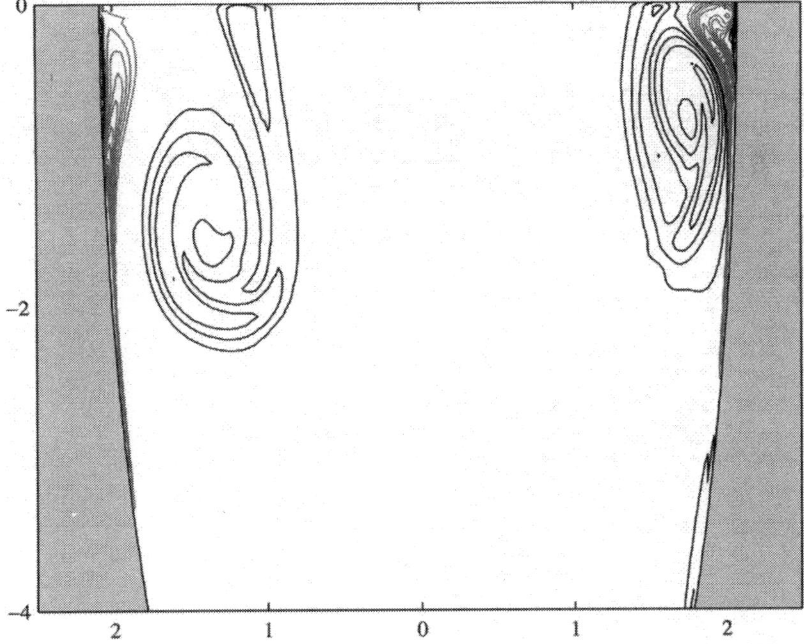

Figure 4.5. Azimuthal vorticity fields on a transversal plane in a model left ventricle during filling at the peak inlet flow.

4.5 Conclusive Remarks

The flow inside of an idealised left ventricle has been analysed in terms of vorticity dynamics, showing that the phenomenon as a whole is dominated by the presence of vortex structures. It has been shown that different vorticity patterns can be related to modifications of the wall ventricle motion; such differences can be detected from the analysis of the velocity space-temporal distribution along a transventricular segment, which resembles the typical output given by echo-Doppler measurements. Some final observations can be made about the mathematical approach to complex problems, such as the heart flow dynamics. It is clear that the reported results are not definitive, depending on the simplifying assumptions adopted, and that several further steps are necessary to achieve an exhaustive description of the left ventricular flow, such as realistic geometry, three-dimensional modelling, fluid-wall coupled dynamics, to cite few of them.

5. The Influence of the Mitral Valve on the Fluid Dynamics of the Left Ventricular Filling

5.1 Introduction

The principal fluid phenomena involved in the left ventricular flow during the diastolic phase are related to the presence of vortex structures, due to the separation of the thin boundary-layer which develops at the mitral valve. Such phenomena can be synthesised as the birth of vortex sheets attached to the valve during the acceleration of the flow, their detachment almost in correspondence of the peak entering discharge, their motion toward the ventricle apex, followed by the interaction with the boundary-layer at the side walls of the ventricle. This complex vorticity dynamics has been captured by numerical models, notwithstanding their limitation of axisymmetry, whose results furnish a first physically based description of the ventricular flow. The comparison between numerical results and clinical data has shown the capability of the mathematical modelling to clarify several phenomena observed during the clinical practice. In particular, the representation of the results in terms of space-time maps of the axial velocity, has clarified the physical meaning of the well known M-mode velocity pattern, which is strictly correlated with the vortex structures evolution.

The most evident difference is observed during the initial stage of the diastole, when the mitral valve motion plays an important role in the birth and development of the vortex sheets. The large part of the numerical models does not describe the valve motion, which is commonly assumed as a rigid orifice at the mitral plane. Furthermore, the velocity distribution at the inlet, in correspondence of the mitral valve, which depends on the characteristics of the atrial flow, is commonly arbitrarily specified. Therefore, two classes of problems arise from these considerations: the first one is the description of the valve motion in order to capture its influence on the ventricular flow, the second one is the study of the complete system atrium-valve-ventricle to analyse the characteristics of the jet flow profile entering into the ventricle.

5.2 Modelling the Valve Motion

In the preceding chapter, the mathematical model for an idealised ventricle and the results obtained by the numerical solution in the axisymmetric approximation have been presented. In that model, the valve is assumed to be a circular orifice of infinitesimal thickness. The velocity distribution has the component tangential to the mitral plane equal to zero, while the normal one is forced to give an irrotational entering profile. The model geometry does not allow the direct simulation of the valve opening, thus in the same scheme such a phenomenon can be modelled in first instance by the appropriate definition of a different inlet conditions. A simple model to reproduce the influence of the mitral valve motion on the left ventricular flow during diastole is here presented.

The time-law of the entering discharge $Q(t)$ is known from clinical data as in Figure 4.1, from the same set of data (anatomical M-mode), the time required during the early filing to complete valvular opening has been estimated, and the corresponding value Q_{lim} of the discharge at that instant is computed. It is then assumed, as a model, that the valve opening is proportional to the instantaneous discharge: the value $Q = 0$ corresponds to a closed valve, that is the angle β between the valve and the mitral plane (see Figure 5.1, left) is equal to zero; at

$Q = Q_{lim}$ a complete opening is reached, corresponding to $\beta = \pi/2$. During the intermediate stages $\beta(t)$ is linearly related to $Q(t)$ as shown in Figure 5.1 (right).

Once the angle β is instantaneously known it is still necessary to specify an inlet profile that properly depends from the valvular opening degree. At each instant of time, the projection on the mitral plane of the valve radius, say $D_v cos(\beta)/2$, gives the portion of the mitral orifice behind valvular leaflet; on this part, the profile of the entering velocity is assumed to be linear, simulating the adherence of the fluid to a rigid valve rotating with angular velocity $d\beta/dt$. An irrotational profile is imposed on the remaining fully open part of the mitral orifice, Figure 5.1.

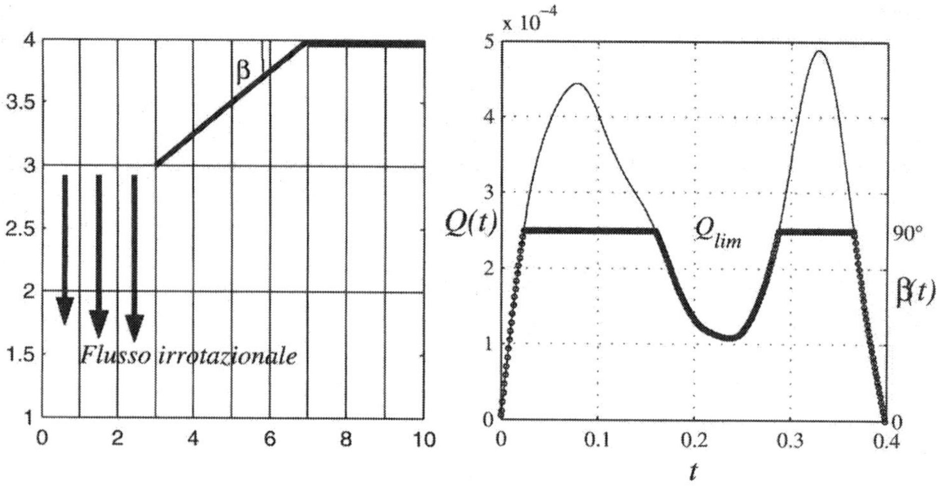

Figure 5.1. Inlet boundary conditions (left); time course of the left ventricular inflow (right).

The flow in absence of valvular modelling is shown in Figure 4.2. A different behaviour has been detected when Q_{lim} is different from zero. The initial dynamics is characterised by the presence of a vortex sheet, which reproduces in the present model the boundary-layer developing at the moving valve wall. During the opening period, such a sheet rotates from the initial position almost tangential to the mitral plane to the final orthogonal one, Figure 5.2*a* at $t = 0.04$. When the valve is completely open, the entering flow becomes everywhere irrotational, with the exception of the thin vortex wake attached to the edge on the mitral plane. During the following period, the wake dynamics resembles that reported in Figure 4.2, with its growth and interaction with the boundary-layer at the lateral wall (Figure 5.2*b* at $t = 0.08$). A vorticity field more complex than that reported in Figure 4.2 is eventually found (Figure 5.2*c*, $t = 0.2$). At this time, the valve is closing, the wake is attached to the mitral plane with a vortex sheet representing the trajectory of the valve edge.

A synthetic description of the flow dynamics is obtained from the spatial and temporal evolution of the velocity computed on the axis of the ventricle, which resembles the standard M-Mode output of the clinical measurements, Figure 5.3. A common feature is represented by the presence of two traces of the maxima of velocity, however a significant difference is observable in correspondence of the initial one. In absence of valvular modelling (Figure 5.3, left picture), the first branch is approximately vertical, the maximum velocity appears contemporarily in the entire ventricle depth, essentially because of incompressibility. The following phase shows the trace of the primary vortex motion, which detaches from the mitral plane and travels inside of the ventricle as discussed in §4.4 (see Figure 4.4). The initial filling phase is significantly influenced by the valve motion (see Figure 5.3, right picture). The initial trace is no more vertical and is characterised by the propagation of the velocity maximum from the mitral plane toward the apex. Such a propagation is the footprint on the axial velocity of the sheet evolution, which in the present model reproduces the valve dynamics. Two distinct inclined traces are thus recognisable, the first one is related to the opening dynamics, the second one to the motion of the primary vortex detached from the mitral plane. An eye-fitted linear approximation of such traces is reported in Figure 5.3, in order to estimate the propagation velocities. Such estimates give, for the second trace, the values 0.17 and 0.2, for the left and right pictures, respectively; this represent the vortex propagation velocity. The first trace is vertical in absence of valve and a value 0.55 has been obtained in the second case, this propagation velocity is connected with the leaflet velocity. These values can be assumed as a direct measure of the propagation velocity expressed in m/s, given the choice of the scaling quantities.

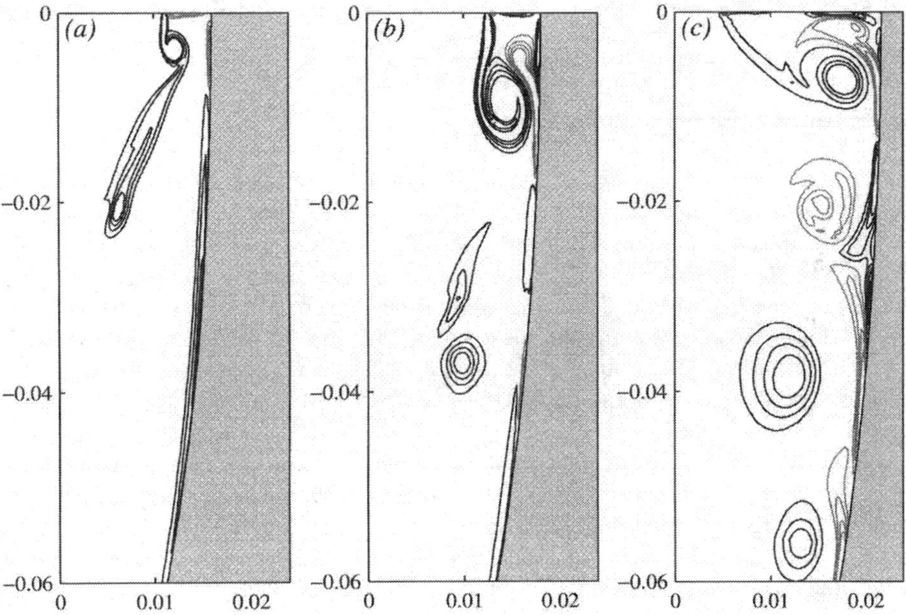

Figure 5.2. Vorticity fields in a model left ventricle with valve modelling; positive values are black, negative grey.

The influence of the valve motion on the initial dynamics is indirectly confirmed by experimental results, where a fixed valve does not induce any propagation (Steen & Steen,

1994), and almost contemporary maxima of the velocity on the symmetry axis are observed, showing a global behaviour similar to that reported in the left of Figure 5.3.

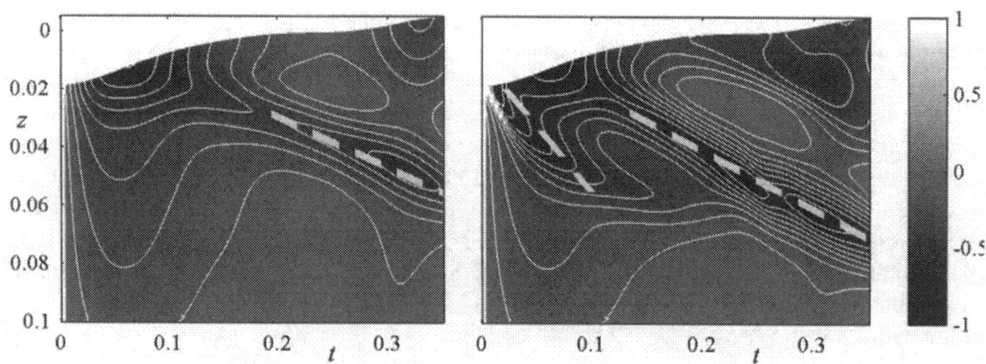

Figure 5.3. Virtual M-mode, without (left) and with valve modelling (right).Units are [cm], [s], and [m/s].

5.3 The Transmitral Flow Dynamics

As outlined in the previous sections, a crucial point in the modelling of the left ventricle fluid dynamics is the realistic modelling of the transmitral profile or, equivalently, the introduction of realistic inlet boundary conditions which strongly influence the whole vorticity dynamics: it has been shown that the simulation of the valve dynamics modifies the formation of the vortex structures, and consequently the intraventricular fluid dynamics. Therefore, a more realistic description of the diastolic ventricular flow should be pursed including three-dimensional effects in the transmitral jet. In this perspective, a first step is represented by the study of a three-dimensional flow through and downstream an orifice to understand the main ingredients for a three-dimensional modelling.

We assume the valve as a circular orifice of radius R_v^* inserted in a straight circular duct with radius R^*. The valve has a small, but finite, thickness; the valve position is defined by the eccentricity e^*, that is the radial distance between its centre and the duct axis. The flow motion is governed by the Navier-Stokes equations, completed by the no-slip condition at the walls, and by the inlet and outlet boundary-conditions. The problem can be made dimensionless assuming as time scale T the heart beat period, $T \simeq 1$ s, and the mitral radius R_v^* as unit length; the unit mass is therefore ρR_v^{*3}, being ρ the fluid density. Following these assumptions, a typical dimensionless radius of the duct is $R = 1.5$. The mean velocity upstream of the orifice has a maximum temporal value $U_0^* \simeq 0.2$ m/s, which in dimensionless unit gives $U_0 \simeq 20$, and a typical value 2000 of the Reynolds number $Re = 2U_0^* R^*/\nu$, where ν is the kinematic viscosity of the fluid.

Consider the duct axis as the z-axis of a three-dimensional system of coordinates. The inlet, $z = z_2$, is assumed 3 units upstream the orifice to form an upstream chamber (model atrium) of square transversal section. A constant velocity profile v_z is assumed there at each instant of time; a parametric temporal law

$$v_z(z_2, t) = U_0(tf)exp(1 - tf) \tag{5.1}$$

reproducing the characteristics of the early filling wave has been chosen, $f = 10$ is the deceleration frequency scale. The problem has been solved numerically solved using a finite volume method. The grid accuracy and the domain extension downstream have been found by grid refinement procedure. Typical numerical parameters, such as the number of cells and vertices, are 300160 and 310540, respectively; in order to describe the thin vorticity layers, the computational grid is made more fine close to the duct wall and in the vicinity of the mitral plane; the domain extends 15 unit lengths downstream the orifice. The time step Δt is chosen to guarantee the stability conditions, a typical value is $\Delta t = 0.0025$.

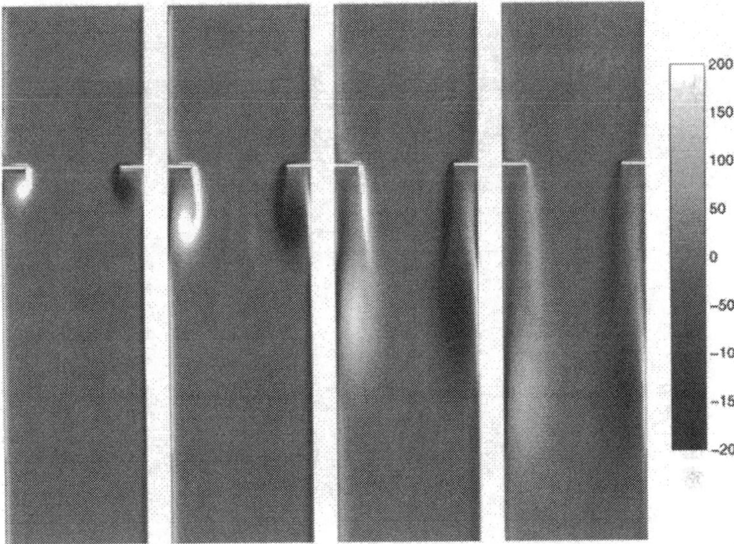

Figure 5.4. Centred transversal section of a three-dimensional flow through an orifice, vorticity component normal to the visualisation plane is shown; $t = [0.05\ 0.1\ 0.2\ 0.35]$ from left to right.

The instantaneous vorticity fields reported on a centred transversal section in Figure 5.4 correspond to the case $e = 0$, it is evident that the flow remains axisymmetric. The flow starts from rest, and immediately the thin viscous boundary-layer which develops at the mitral plane rolls-up into a vortex ring. The vortex wake is continuously forced by the incoming flow and smoothes out downstream for viscous effect; at the same time it induces a boundary-layer and a vortex-induced separation from the wall.

The appearance of three-dimensional phenomena can be evidenced assuming different values of the eccentricity e; therefore, the problem geometry is characterised by a plane of symmetry, the plane along which the valve is displaced. The results in the case $e = 0.1$ are reported below.

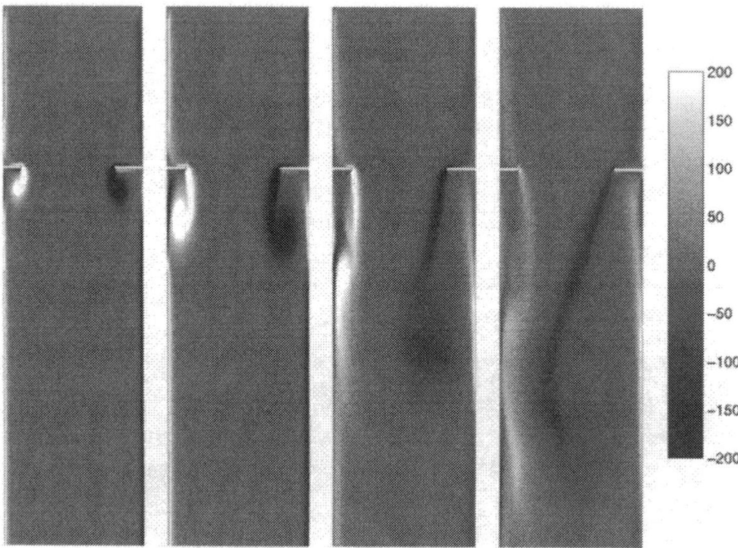

Figure 5.5. Centred transversal section of a three-dimensional flow through an orifice, vorticity component normal to the visualisation plane is shown; $t = [0.05\,0.1\,0.2\,0.35]$ from left to right.

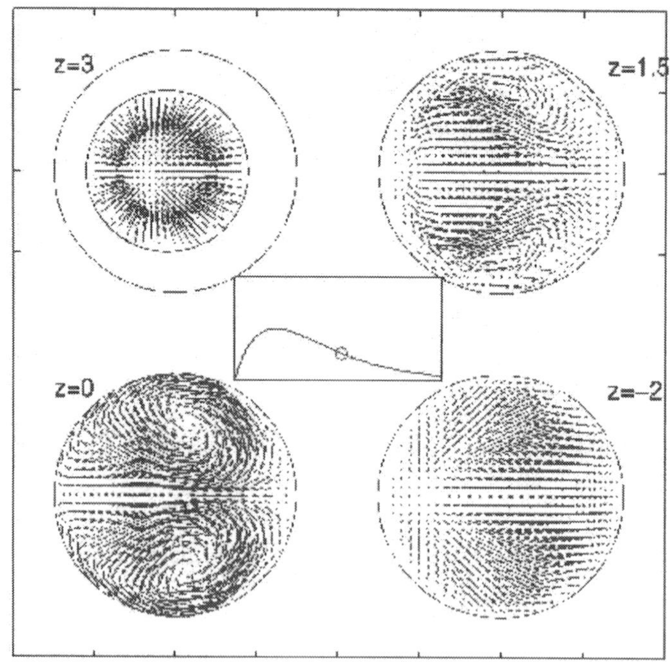

Figure 5.6. Flow through an orifice, cross-flow velocity fields at different sections, $t = 0.25$.

The vorticity component normal to the symmetry plane is plotted in Figure 5.5 at different instants of time. During the initial phase, the flow shows a behaviour similar to the axisymmetric one, with the instantaneous detachment of the boundary-layer from the orifice edge and the development of a vortex sheet. The eccentricity induces the deviation of the transmitral jet, the part of vorticity wake closer to the wall is splashes on the wall and is partly dissipated with the interaction with the induced boundary layer. The weaker right side of the vortex structure moves towards the centre of the duct; it does not have any significant interaction with the wall viscous layer and is eventually strongly deviated toward the opposite side.

The wake evolution is an example of three-dimensional vorticity dynamics; although in the axisymmetric case the vortex structures are stable circular vortex rings, a very different fully three-dimensional vorticity field is induced by the small eccentricity of the orifice. This dynamics can be read from the secondary circulation pattern shown in Figure 5.6. The jet at the orifice ($z = 3$) is an almost irrotational converging one. Downstream, after the main wake, the jet tends to enlarge, but its eccentricity gives a radial motion that splashes over the closer wall (on the left side in Figure 5.6) before than on the farther wall, as a consequence a pair of recirculating cells appear (clockwise above and counterclockwise below in Figure 5.6).

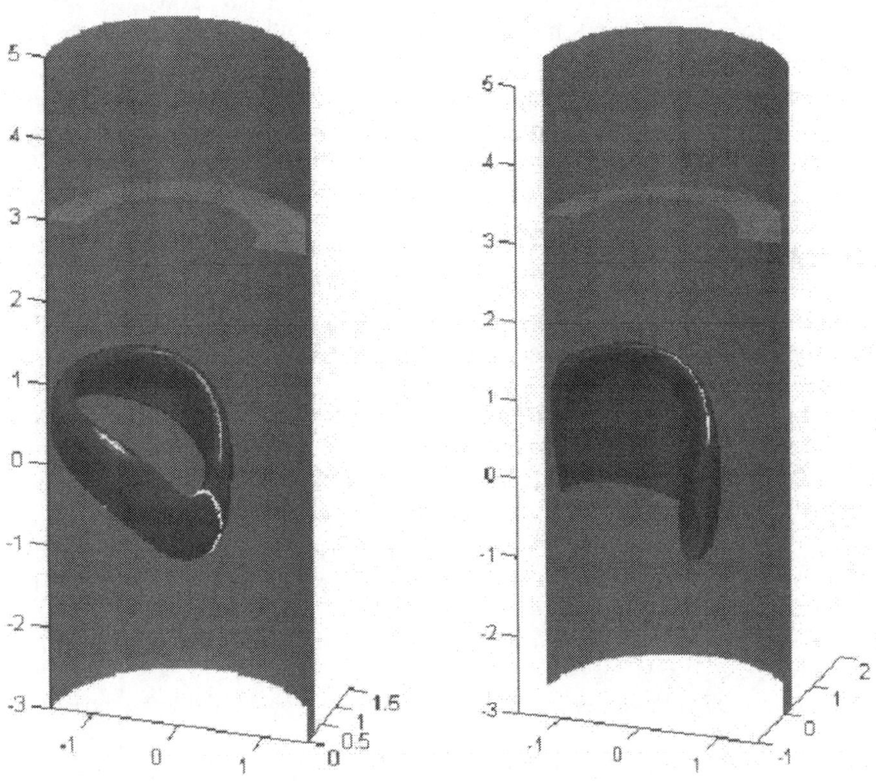

Figure 5.7. Three-dimensional flow through an orifice, one vortex tube at $t = 0.25$.

The secondary circulation's correspond to the birth of a vertical vorticity component, as new vorticity cannot be created within the flow, this vorticity must be part of the same vorticity wake shed from the orifice. In fact looking at the vortex tube (a tube whose boundary is made of vortex lines) in Figure 5.7, the three-dimensional shape of the vortex wake (or part of it) can be seen.

5.4 Conclusive Remarks

Simplified methods have been developed to analyse the ventricular flow during the diastolic phase. In an axisymmetric model left ventricle, the influence of the mitral valve motion has been reproduced with the introduction of special boundary-conditions. The mitral valve motion influences significantly the ventricular flow during the initial period of the filling E-wave. The transmitral M-mode representation shows two main branches that evidently represent the valvular opening and the vortex wake evolution, respectively.

The results evidence once more the necessity of an accurate definition of the inlet conditions for three-dimensional ventricular flow modelling. In the direction of understanding the main features of the transvalvular flow, a simple model for the three-dimensional flow through an orifice has been studied in a simple geometry. It has shown that in presence of a small eccentricity, a complete three-dimensional wake field develops thus affecting the evolution of the whole ventricular flow.

The present models representing basic studies that we consider necessary for the understanding of the main physical processes involved in the heart fluid dynamics and for the construction of synthetic interpretative schemes.

References

Anayiotos, A. S., Jones, S. A., Giddens, D. P., Glagov, S., and Zarins, C. K. (1994) Shear stress at a compliant model of the human carotid bifurcation. *ASME Journal of Biomechanical Engineering* 116: 98-106.

Baccani, B., Domenichini, F., Pedrizzetti, G., and Tonti G. (2002*a*) Fluid dynamics of the left ventricular filling in dilated cardiomyopathy. *Journal of Biomechanics* 35: 665-671.

Baccani, B., Domenichini, F., and Pedrizzetti, G. (2002*b*) Vortex dynamics in a model left ventricle during filling. Manuscript submitted for publication.

Bargiggia, G. S., Tronconi L., Sahn, D. J., et al. (1991) A new method for quantitation of mitral regurgitation based on color flow Doppler imaging of flow convergence proximal to regurgitant orifice. *Circulation* 84:1481-1489.

Batchelor, G. K. (1967) *An Introduction to Fluid Dynamics*. Cambridge, UK: Cambridge University Press.

Bellhouse, B. J., (1972) Fluid mechanics of a model mitral valve and left ventricle. *Cardiovas. Res.* 6:199-210.

Bolzon, G., Pedrizzetti, G., Zovatto, L., Grigioni, M., Daniele, C., and D'Avenio G. (2002) Flow on the symmetry plane of the total cavo-pulmonary connection. *Journal of Biomechanics* 35: 33-46.

Botnar, R., Rappitsch, G., Scheidegger, M. B., Liepsch, D., Perktold, K., and Boesiger, P. (2000) Hemodynamics in the carotid artery bifurcation: a comparison between numerical simulations and in vitro MRI measurements. *Journal of Biomechanics* 33: 137-144.

Canuto, C., Hussaini, M. Y., Quarteroni, A., and Zang, T. A. (1988) *Spectral Methods in Fluid Dynamics.* New York, NY: Springer-Verlag.

Davies, C., and Carpenter, P. W. (1997) Numerical simulation of the evolution of Tollmien-Schlichting waves over finite compliant panels. *Journal of Fluid Mechanics* 335: 361-392.

Eiseman, P. R. (1985) Grid generation for fluid mechanics computation. *Annual Review of Fluid Mechanics* 17: 487-522.

Fung, Y. C. (1997) *Biomechanics: Circulation.* New York, NY: Springer-Verlag, 2nd edition.

Gaasch, W. H., and Winter, M. M. (1994) *Left ventricular diastolic dysfunction and heart failure. Vol. 1.* Malvern, PA: Lea & Febiger.

Garcia, M. J., Thomas, J. D., and Klein, A. L. (1998) New Doppler echocardiographic applications for the study of diastolic function. *Journal of the American College of Cardiology* 32: 865-875.

Garcia, M. J., Smedira, N. G., Greenberg, N. L., Main, M., Firstenberg, M. S., Odabashian, J., and Thomas, J. D. (1999) Color M-Mode Doppler Flow Propagation Velocity is a Preload Insensitive Index of Left Ventricular Relaxation: Animal and Human Validation. *Journal of the American College of Cardiology* 35: 201–208.

Green, A.E., and Adkins, J. E. (1960) *Large Elastic Deformations.* Oxford, UK: Clarendon Press.

Grigioni, M., Pedrizzetti, G., Amodeo, A., Daniele, C., D'Avenio, C., Zovatto, L., and Di Donato, R.M. (2000) A comparison between numerical and PIV studies of the flow through a total cavopulmonary connection. *International Journal of Artificial Organs* 23: 579.

Heil, M., (1997) Stokes flow in collapsible tubes: computation and experiment. *Journal of Fluid Mechanics* 353: 285-312.

Hsu, F. P. K., Schwab, C., Rigamonti, D., and Humphrey, J. D. (1994) Identification of response functions from axisymmetric membrane inflation tests: implication for biomechanics. *International Journal of Solids and Structures* 31: 3375-3386.

Humphrey, J. D., (1995) Mechanics of the arterial wall: review and directions. *Critical Reviews in Biomedical Engineering* 23: 1-162.

Humphrey, J. D., Strumpf, R. K., and Yin, F. C. P. (1992) A constitutive theory for biomembranes: application to epicardial mechanics. *Journal of Biomechanical Engineering* 114: 461-466.

Jensen, O. E. (1992) Chaotic oscillations in a simple collapsible tube model. *Journal of Biomechanical Engineering* 114: 55-59.

Kilner, P. J., Yang, G. Z., Wilkes, A. J., Mohiaddin, R. H., Firmin, D. N., and Yacoub, M. H. (2000) Asymmetric redirection of flow through the heart. *Nature* 404: 759-761.

Kim, W. Y., Bisgaard, T., Nielsen, S. L., Poulsen, J. K., Pedersen, E. M., Hasenkam, J. M., and Yoganathan, A. P. (1994) Two-dimensional mitral flow velocity profiles in pig models using epicardial echo Doppler Cardiography. *Journal of the American College of Cardiology* 24: 532-545.

Kim, W. Y., Walker, P. G., Pedersen, E. M., Poulsen, J. K., Oyre, S., Houlind, K., and Yoganathan, A. P. (1995) Left ventricular blood flow patterns in normal subjects: a quantitative analysis by three-dimensional magnetic resonance velocity mapping. *Journal of the American College of Cardiology* 26: 224-238.

Kumaran, V., (1996) Stability of inviscid flow in a flexible tube. *Journal of Fluid Mechanics* 320: 1-17.

Kyriacou, S. K., and Humphrey, J. D. (1996) Influence of size, shape and properties on the mechanics of axisymmetric saccular aneurysms. *Journal of Biomechanics* 29: 1015-1022.

Lemmon, J. D., and Yoganathan, A. P. (2000) Three-dimensional computational model of left heart diastolic function with fluid-structure interaction. *Journal of Biomechanical Engineering* 122: 109-117.

Lighthill, J. (1978) *Waves in Fluids.* Cambridge, UK: Cambridge University Press.

Little, W. C. (2000) Assessment of Normal and Abnormal Cardiac Function. In Braunwald E., Zipes, D. P., and Libby, P., eds, *Heart Disease, A Textbook of Cardiovascular Medicine.* Philadelphia, PA: Saunders W. B., 6th edition, Chapter 15.

Lucey, A. D., and Carpenter, P. W. (1992) A numerical simulation of the interaction of a compliant wall and inviscid flow. *Journal of Fluid Mechanics* 234: 121-146.

Luo, X. Y., and Pedley, T. J. (1995), A numerical simulation of steady flow in a 2-D collapsible channel, *Journal of Fluids and Structures* 9: 149-174.

Luo, X. Y., and Pedley, T. J. (1996) A numerical simulation of unsteady flow in a two-dimensional collapsible channel. *Journal of Fluid Mechanics* 314: 191-225.

Luo, X. Y., and Pedley, T. J. (1998) The effects of wall inertia on flow in a two-dimensional collapsible channel. *Journal of Fluid Mechanics* 363: 253-280.

Mandinov, L., Eberli, F. R., Seiler, C., and Hess, O.M. (2000) Review: Diastolic heart failure. *Cardiovascular Research* 45: 813-825.

Masugata, H., Peters, B., Lafitte, S., Monet Strachan, G., Ohmori, K., DeMaria, A. N. (2001) Quantitative Assessment of Myocardial Perfusion During Graded Coronary Stenosis by Real-Time Myocardial Contrast Echo Refilling Curves. *J. Am. Coll. Cardiol.* 37:262–269.

McQueen, D. M., and Peskin, C. S. (1997) Shared-memory parallel vector implementation of the immersed boundary method for the computation of blood flow in the beating mammalian heart. *Journal of Supercomputing* 11: 213-236.

McQueen, D. M., and Peskin, C. S. (2000) A three-dimensional computer model of the human heart for studying cardiac fluid dynamics. *Computer Graphics* 34: 56-60.

Morse, P. M., and Feshbach, H. (1953) *Methods of theoretical physics.* New York, NY: McGraw-Hill.

Møller, J. E., Poulsen, S. H., Søndergaard, E., and Egstrup, K. (2000) Preload Dependence of Color M-Mode Doppler Flow Propagation Velocity in Controls and in Patients with Left Ventricular Dysfunction. *Journal of the American Society of Echocardiography* 13: 902-909.

Møller, J. E., Søndergaard, E., Poulsen, S. H., and Egstrup, K. (2000) Pseudonormal and Restrictive Filling Patterns Predict Left Ventricular Dilation and Cardiac Death After A First Myocardial Infarction: A Serial Color M-Mode Doppler Echocardiographic Study. *Journal of the American College of Cardiology* 36: 1841– 1846.

Opie, L. H. (2000) Mechanisms of cardiac contraction and relaxation. In Braunwald E., Zipes, D. P., and Libby, P., eds, *Heart Disease, A Textbook of Cardiovascular Medicine.* Philadelphia, PA: Saunders W. B., 6th edition, Chapter 14.

Pai, R. G., and Stoletniy, L. N. (1999) An Integrated Measure of Left Ventricular Diastolic Function Based on Relative Rates of Mitral E and A Wave Propagation. *Journal of the American Society of Echocardiography* 12: 811-816.

Panton, R. (1996) *Incompressible Flow.* New York, NY: John Wiley & Sons.

Pedrizzetti, G. (1996), Unsteady tube flow over an expansion, *Journal of Fluid Mechanics*, 310: 89-111.

Pedrizzetti, G. (1998) Fluid flow in a tube with an elastic membrane insertion. *Journal of Fluid Mechanics* 375: 39-64.

Pedrizzetti, G., Domenichini, F., Tortoriello, A., and Zovatto, L. (2002) Pulsatile flow inside moderately elastic arteries, its modelling and effects of elasticity. Computer Methods in Biomechanics and Biomedical Engineering (in press).

Perktold, K., and Rapitsch, G. (1995) Computer simulation of local blod flow and vessel mechanics in a compliant carotid artery bifurcation model. *Journal of Biomechanics* 28: 845-856.

Peskin, C. S., and McQueen, D.M. (1989) A three-dimensional computational method for blood flow in the heart: I Immersed elastic fibers in an incompressible fluid. *Journal of Computational Physics* 81: 372-405.

Peskin, C. S., and McQueen, D.M. (1989) A three-dimensional computational method for blood flow in the heart: II Contractile fibers. *Journal of Computational Physics* 82: 289-298.

Peskin, C. S., and Printz, B. F. (1993) Improved volume conservation for the three-dimensional Navier-Stokes equations involving an immersed moving membrane. *Journal of Computational Physics* 105: 33-46.

Pipkin, A. C. (1968) Integration of an equation in membrane theory. *Zeitschrift für Angewandte Mathematik und Physik* 19: 818-819.

Redaelli, A., Montevecchi, F. M. (1996) Computational evaluation of intraventricular pressure gradients based on a fluid-structure approach. *Journal of Biomechanical Engineering* 118: 529-537.

Reul, H., Talukder, N., and Muller, W. (1981) Fluid mechanics of the natural mitral valve. *Journal of Biomechanics* 14: 361-72.

Roache, P. J. (1998) *Fundamentals of Computational Fluid Dynamics.* Albuquerque, NM: Hermosa.

Steen, T., and Steen, S. (1994) Filling of a model left ventricle studied by colour M mode Doppler. *Cardiovascular Research* 28: 1821-1827.

Takatsuji, H., Mikami, T., Urasawa, K., Teranishi, J. I., Onozuka, H., Takagi, C., Makita, Y., Matsuo, H., Kusuoka, H., and Kitabatake, A. (1997) A new approach for evaluation of left ventricular diastolic function: spatial and temporal analysis of ventricular filling flow propagation by color M-Mode Doppler echocardiography. *Journal of the American College of Cardiology* 27: 365-371.

Taylor, T. W., and Yamaguchi T. (1995) Realistic three-dimensional left ventricular ejection determined from computational fluid dynamics. *Medical Engineering & Physics* 17: 602-608.

Trambaiolo, P., Tonti, G., Salustri, A., Fedele, F., and Sutherland G. (2000) New insights onto regional systolic and diastolic left ventricular function with tissue Doppler echocardiography: from qualitative analysis to a quantitative approach. *Journal of the American Society of Echocardiography* 14: 85-96.

Urheim, S., Edvardsen, T., Torp, H., Angelsen, B., Smiseth, O. A. (2000) Myocardial Strain by Doppler Echocardiography Validation of a New Method to Quantify Regional Myocardial Function. *Circulation* 102:1158-1164.

Van Dyke, M. (1975) *Perturbation methods in fluid mechanics.* Stanford, CA: The Parabolic Press.

Vierendeels, J. A, Verdonck, P., and Dick, E. (1997) Intraventricular pressure gradient and the role of pressure wave propagation. *Journal of Cardiovascular Diagnosis and Procedures* 14: 147-152.

Vierendeels, J. A., Riemslagh, K., Dick, E., and Verdonck, P. R. (2000) Computer simulation of intraventricular flow and pressure during diastole. Journal of Biomechanical Engineering 122: 667-674.

Walker, P. G., Oyre, S., Pedersen, et al. (1995) A new control volume method for calculating valvular regurgitation. *Circulation* 92:579-586.

Wang, D. M., and Tarbell, J. M. (1992), Nonlinear analysis of flow in an elastic tube (artery): steady streaming effects, *Journal of Fluid Mechanics* 239: 341-358.

Wieting, D. W., and Stripling, T. E. (1984) Dynamics and fluid dynamics of the mitral valve. In Duran, C., Angell, W. W., Johnson, A. D., and Oury, J. H., eds., *Recent Progress in Mitral Valve Disease.* London: Butterworths, 13-46.

Zhao, S. Z., Xu, X. Y., Hughes, A. D., Thom, S. A., Stanton, A. V., Ariff, B., and Long, Q. (2000) Blood flow and vessel mechanics in a physiologically realistic model of a human carotid bifurcation. *Journal of Biomechanics* 33: 975-984.

From Left Ventricular Dynamics to the Pathophysiology of the Failing Heart

Antonio Barsotti[1] and Frank Lloyd Dini[2]

[1] Dipartimento di Medicina Interna e Specialità Mediche. Cattedra di Cardiologia. Università degli
[2] Unità di Malattie Cardiovascolari II. Azienda Ospedaliera Pisana. Pisa. Italy.

Abstract. The mechanical function of the heart can be described by the pressure, volume and flow changes. The contracting left ventricle can be viewed as an impulse generator that rapidly accelerates blood to a high peak outflow velocity early in its ejection period. The driving force for the flow of blood is the total energy imparted to it that corresponds to the mechanical, external work performed by the ventricle. The shape and structure of the ventricle play a major role in the transfer of mechanical energy from myocytes, to myocardial wall, to the ventricular cavity, where it is used for cardiac pump function.

1 Introduction

The purpose of the heart as a pump is to restore the total fluid energy in order to maintain the forward movement of blood to supply substrates to and remove catabolites from the body organs. The heart pump is made up of myocardial cells (myocytes) supported within an extensive myocardial collagen network (Weber and Brilla, 1991). Within each myocyte, repeating units of sarcomeres arranged in series forms myofibrils. Myocyte cells provide the contractile force needed to generate pressure and to eject blood during systole through a valved outflow tract into the circulation. Diastole is the portion of the cardiac cycle that begins with isovolumic relaxation and ends with the termination of ventricular filling at the time of mitral valve closure. Ventricular filling is determined by the active relaxation of the myocardium, the ventricular compliance, the atrial contraction and the return of blood from the great veins. Diastole occupies two-third of the cycle at rest, but is reduced to only one third of the cycle at high heart rates.

For the heart to function as a muscular pump it requires energy. Its mechanical behavior depends from a constant supply of oxygen and nutrients through the coronary circulation. The energy requirements of the heart are best represented by myocardial oxygen consumption. Oxygen is utilized in the mitochondria by the process of oxidative phosphorilation to generate high-energy phosphates, such as adenosine triphosphate (ATP) and phosphocreatine. The most important substrates for the heart are fatty acids, glucose and lactate. The major determinants of myocardial oxygen consumption are heart rate, systolic wall stress and velocity of fiber contraction (Table I) (Opie, 1998).

Table I. Determinants of myocardial oxygen consumption.

❖ Heart rate
❖ Systolic wall stress
❖ Myocardial contractility
❖ Ventricular muscular mass

2 Relationship Between Wall Stress and Loading Conditions

The term cardiodynamics refers to the mechanical events that are associated with the contraction of the heart. The cardiac cycle includes both electrical (e.g., myocardial cellular depolarization) and mechanical events (e.g., myocardial cell shortening leading to pressure generation and, thus, volume changes). The electrical events precede and initiate the mechanical events. The excitation-contraction coupling is the term used to define the events that connect the depolarization of the cell membrane to the contraction of the muscle fibers.

The contracting left ventricle can be viewed as an impulse generator that rapidly accelerates blood to a high peak outflow velocity early in its ejection period (Levick, 1995). The amount of blood ejected from each ventricle with each beat (the stroke volume) is ultimately a function of the extent of myocardial fiber shortening. Left ventricular ejection is caused by a reduction in circumference and a decrease in the longitudinal axis. In the normal resting adult, each ventricle discharges approximately 70 to 90 ml of blood per beat. The ventricle does not empty completely, and the volume remaining at the end of systole, after the stroke volume is expelled, is called end-systolic volume. The ejection fraction indicates the amount of the initial volume (diastolic) that is expelled in systole. Ejection fraction (EF) is estimated from end-diastolic volume (EDV) and end-systolic volume (ESV) as follows:

$$EF = \frac{EDV - ESV}{EDV} \times 100$$

Ejection fraction is the most widely utilized index for evaluating left ventricular systolic performance. The normal values in adult at rest are between 50% and 70% (Figure 1).

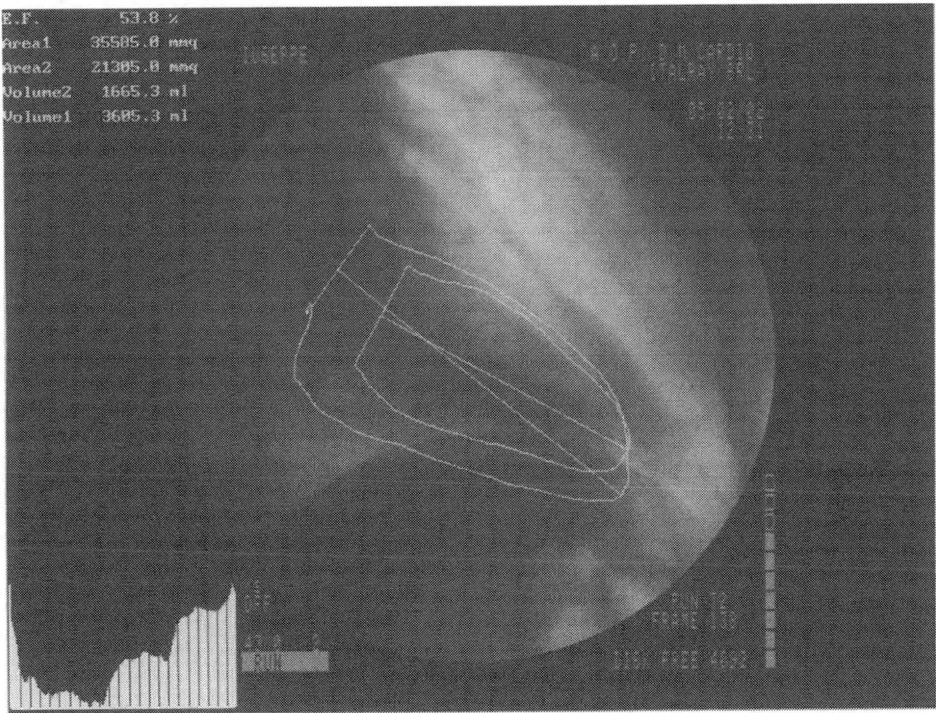

Figure 1. End-diastolic and end-systolic left ventricular volume measurements from angiographic images for the evaluation of ejection fraction.

Cardiac output is determined by the stroke volume and the number of beats per minute (heart rate):

$$Cardiac\ Output = Stroke\ Volume \times Heart\ Rate$$

Cardiac output is an extremely important cardiovascular variable that is continuously adjusted so that the cardiovascular system operates to meet the body's moment-to-moment transport needs. The heart of a normal 70-kg subject contracts at a rate of 70-75 beats/min at rest and pumps 5-7 liters.

Beyond heart rate, contractility and cardiac load (preload and afterload) regulate stroke volume and cardiac output (Ross et al., 1966, Ross and Braunwald, 1974, Shepherd and Vanhoutte, 1979, Katz, 1992).

The intrinsic ability of myocardial cells to develop force is termed contractility or inotropic state. In essence, contractility is the energy with which the myocytes contract and determines ventricular performance independent to loading conditions. This is reflected in the extent and velocity of shortening of myofibrils after the formation of cross-bridges between contractile proteins. Since the functioning of the intact heart is greatly affected by loading conditions, most of the understanding about contractility has been derived from studies carried out on isolated fragments of the heart, shortening longitudinally while carrying a fixed load.

Preload is related to the volume of blood in the ventricle just before contraction. It is relates to the degree of stretch of the myocyte cells prior to their activation and is proportional to the amount of energy delivered to generate force during contraction (Frank-Starling mechanism). Afterload is primarily determined by systemic vascular resistance that, by analogy of Ohm's law, is identified by the expression:

Systemic Vascular Resistance=(Mean Aortic Pressure–Mean Right Atrial Pressure)/Cardiac Output

Other factors that affect afterload include: aortic impedance, the arterial wall stiffness, the mass of the column of the blood in the aorta and the viscosity of the blood. Rhythmic contraction of the ventricular myocardium creates the force needed to overcome systemic vascular resistance.

The force present in the wall per unit cross-sectional area is defined as wall stress. The use of wall stress as a measure of myocardial function is based on the principle that, for equilibrium to exist at any point of the cardiac cycle, the forces acting within the ventricular wall must exactly balance the forces acting on the wall (Mirsky, 1969). The ventricular myocardium must develop a force equal to the end-diastolic wall stress (preload) before it can generate pressure and shorten against the systemic vascular resistance that is related to the force against which the heart contracts (afterload).

The shape and structure of the left ventricle play a major role in the transfer of mechanical energy from myocytes, to myocardial wall, to the ventricular cavity, where it is used for cardiac pump function. Relationship between ventricular configuration and heart dynamics during cardiac cycle is based on fiber orientation and their mechanical behavior within the myocardial wall (Hutchins et al., 1978). In the normal heart, fibers are arranged in a spiral fashion around the central cavity of the left ventricle. In the midwall, circumferentially oriented fibers are predominant. The subendocardial and subepicardial fibers run functionally parallel to the long axis of the left ventricle. During isovolumetric contraction, the left ventricle becomes more spherical, because longitudinal fibers shorten more effectively. During ventricular ejection, although the descent of the base continues, constriction of the circumferential muscle bundles in the middle layers prevails (Streeter et al., 1969).

Stress differs from pressure, which is also a ratio between force and area. Any given value of stress – that is a vector quantity - deforms a material along a single direction and in general varies across the material being deformed. Any given value of pressure – that is a scalar quantity – is a property applied to the contained fluid or gas (e.g., within the left ventricular chamber) that is pushing outward uniformly in all directions.

On the assumption that the left ventricle is a sphere with uniform wall thickness, the relationship between ventricular cavity size, wall stress, and intraventricular pressures can be predicted from the law of Laplace. This principle states that wall stress (WS) is directly proportional to intraventricular pressure (P) and the radius (r), and inversely proportional to wall thickness (h).

$$WS = \frac{P \times r}{2h}$$

The importance of Laplace's law lies in the fact that wall stress is a major determinant of both cardiac energy expenditure and efficiency. Since both pressure and radius decline gradually during ejection, the left ventricle unloads itself during ejection as its volume decreases. This unloading is mainly conditioned by the forces (afterload) that the ventricular myocardium must overcome in order to shorten. Systolic blood pressure also is used as an indicator of afterload. Since pressure is equal to force divided by area, systolic force is incorporated in the expression of pressure and thus systolic pressure does relate to afterload.

To date, estimates of left ventricular wall stress are possible from either angiographic or echocardiographic images using a geometric model of the ventricle assumed as a thicked-walled rolated ellipse (Dodge and Baxeley, 1969, Chapman et al.,1958). For a thick-walled ellipsoid model, three components of the wall stress – which are mutually perpendicular – can be identified: circumferential, radial and longitudinal (Fifer and Grossman, 2000). The circumferential stress may be defined as the force acting within the wall on the equatorial direction; which results from the shortening of internal minor axis due to the contraction of circumferential fibers. Radial stress is the force within the wall that depends from the radius of the left ventricle, in equilibrium with the one applied at the inner surface of the ventricular wall. Longitudinal stress is directed from apex to base and is counterbalanced by the forces generated by the contraction of the fibers that run parallel to the internal major axis. The significance of radial stress is readily understandable by considering the ventricular wall's curvature. As the radius of the cavity increases, ventricular curvature will decrease and – for any given distending pressure – myocardial tension has to rise, since a higher component of it must be angled toward the cavity to equalize the pressure acting on the wall. Hence, left ventricular pressure is mainly generated by circumferential stress in the wall, but the force acting on the endocardial surface area will be balanced by the radial component of stress within the ventricular wall (Gunther and Grossman, 1979).

3 The Concepts of External and Internal Ventricular Work

Two types of work must be considered in assessing the energetic of left ventricular function (Table II): the external work and the internal work. The amount of work performed correlates with the oxygen requirements of the myocardium (Suga et al., 1982).

The total external work (15-30% of total work) of the ventricle depends on the amount of pressure (pressure-volume-work) and kinetic energy it has to generate. The ejection of a volume under pressure represents a form of mechanical work: the pressure-volume work; this is because, in a fluid system, work (force × distance) is equal to pressure × volume (Carabello, 1993). The kinetic work – that accounts for about 5% of the external work – is spent to generate the kinetic energy that is needed to move the column of blood through the ventricular outflow tract into the circulation.

Stroke work, i.e., the pressure-volume work exerted by the left ventricle during each beat, can be calculated as the stroke volume times the pressure at which the blood is ejected:

$$Stroke\ work = mean\ arterial\ pressure \times stroke\ volume$$

Strictly speaking, the calculation of stroke work should account for the rates of change of pressure and volume during each systole. Stroke volume is equal to end-diastolic volume (EDV) minus end-systolic volume (ESV):

$$Stroke\ work = mean\ arterial\ pressure \times (EDV - ESV)$$

Because arterial pressure changes throughout ejection, stroke work is more accurately expressed by the integral PdV, where P is the pressure at which each increment (dV) of the stroke volume is ejected. Stroke work represents approximately 95% of total external work, whereas kinetic work accounts for 5%.

Table II. Sources of energy utilization of the myocyte cells.

❖ Basic processes responsible for the simple survival of the tissue, which include ionic homeostasis and protein synthesis
❖ Excitation-contraction coupling and cross-bridges interaction during systole
❖ Active relaxation during diastole
❖ Mithocondrial resynthesis of high energy phosphates

A useful way to examine the interactions between the left ventricle and the circulation is to construct a pressure-volume loop, which plots the changes in ventricular pressure and volume that occur during a single cardiac cycle (Suga and Sagawa, 1974, Sagawa et al., 1988, Kass and Maughan, 1988).

Inscription of the pressure-volume loop begins at end-diastole and is inscribed in a counter-clockwise direction passing through four phases. The first, isovolumic contraction, begins at the end-diastole and ends at the opening of the aortic valve when the ventricle encounters its afterload. The second phase – during which pressure is converted into flow - begins at the opening of the aortic valve and ends at the closure of the aortic valve. The diastolic portion of the cardiac cycle – that includes isovolumic relaxation and the filling phase, ensues. Isovolumic relaxation begins when the ejection of blood from the ventricle has ceased; the ventricular volume remains constant until the mitral valve opens. The next phase is the filling period during which the blood flows into the ventricle from the central veins and the left atrium (Figure 2).

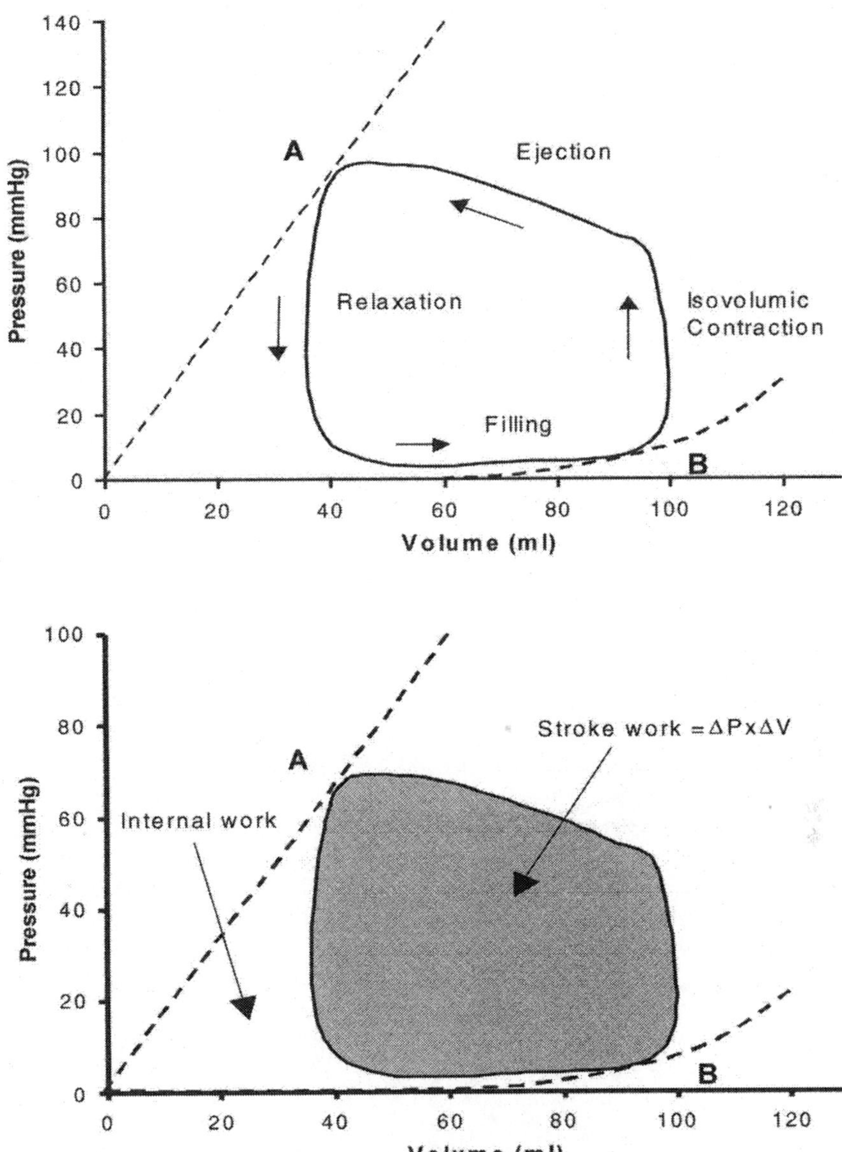

Figure 2. A Pressure-volume loop of the left ventricle describing one cardiac cycle. The four phases are depicted: isovolumic contraction, ejection, isovolumic relaxation and filling. B. Types of cardiac work. The pressure-volume is a measure of stroke work while potential energy represents internal work.

Stroke work is reflected by the size of the pressure-volume loop and is proportional to both the pressure generated by the ventricular contraction and the stroke volume. The same stroke work can be performed by various combinations of stroke volume and intraventricular pressure.

Ejection fraction is often regarded as a measure of the inotropic state of the ventricle. This is largely a misconception since it is influenced by loading conditions. Changes in left ventricular contractility can be appreciated by the counterclockwise rotation of the slope of the end-systolic pressure-volume relation (Figure 3) (Suga and Sagawa, 1974, Sagawa et al., 1988, Kass and Maughan, 1988).

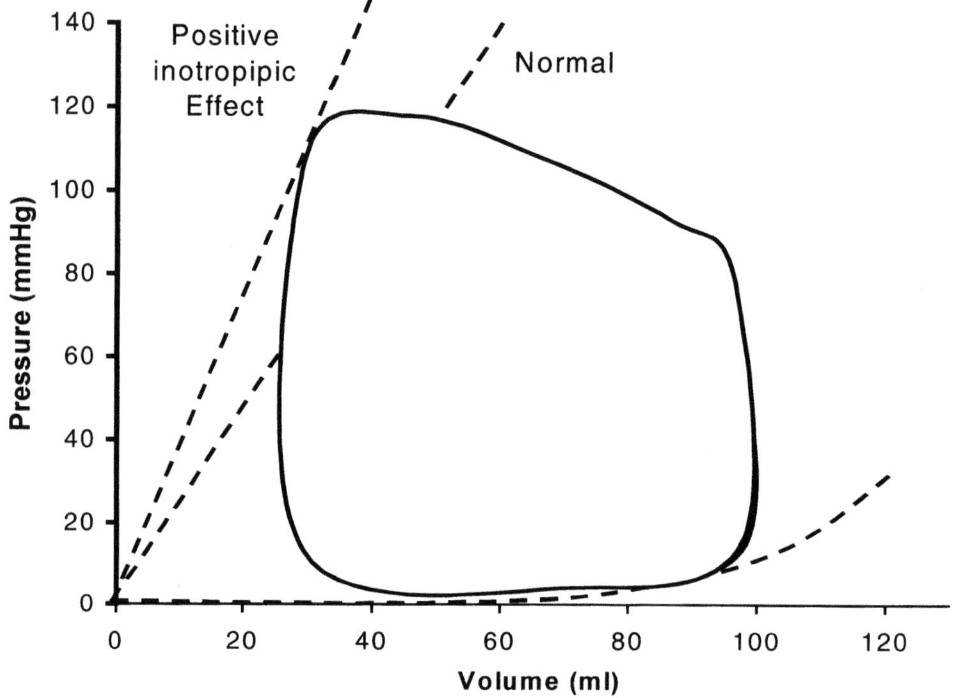

Figure 3. Effect of increased inotropism on the slope of the end-systolic pressure-volume relation of the left ventricle.

The amount of external work per minute (minute work) is given by the product of the stroke work and heart rate:

$$\textit{Minute work} = \textit{stroke work} \times \textit{heart rate}$$

Internal work is needed to stretch elasticities and rearrange myofibrillar structure during tension development. This work is degraded in heat during diastole. A greater portion of total cardiac expenditure is due to internal work: 75-85% of total work. Internal work is directly proportional to wall stress, so that the progressive dilation generally seen in failing hearts increases the quantity of internal work. Tachycardia also increases internal work simply by increasing the time spent by the heart in systole.

The gross mechanical efficiency of any machine is the external work achieved per unit of energy expended. In hemodynamic terms, the efficiency of the heart is defined as the ratio of stroke work (SW) to myocardial oxygen consumption (MVO_2):

$$Efficiency = SW/MVO_2$$

The efficiency of the heart ranges between 5% and 20%; it is about the same for all mammals. This indicates that a large mammal would provide a greater amount of mechanical, external work, in accordance with the larger metabolism it generates (Gibbs, 1974). The efficiency is greater when the stroke volume is high and the heart rate is low. Conversely, for a given stroke volume, a larger the ventricle is characterized by a lower efficiency because of the elevated energy cost due to increased internal work and oxygen requirements of the myocardium (Katz, 2000).

4 Basics of Heart Failure

Heart failure may be defined as the pathophysiologic state in which an abnormality of cardiac function is responsible for the failure of the heart to pump blood at a rate commensurate with the metabolic needs of the body or can do so only from an abnormally elevated filling pressure (Braunwald, 1997). It is above all a clinical syndrome – that is characterized by circulatory congestion or inadequate tissue perfusion - that can complicate virtually any form of heart disease. The most common causes of heart failure in industrialized countries are the following:

1. Coronary artery disease.
2. Idiopathic dilated cardiomyopathy.
3. Systemic hypertension.
4. Valvular heart disease.

Loss of functioning myocytes is one of the hallmark of heart failure. This may be segmental, secondary to coronary artery disease, or diffuse. Many types of cardiac diseases are due to excessive mechanical loading upon the myocyte cells. The two most common types are those resulting from an increased resistance to ventricular emptying of increased afterload (e.g., aortic stenosis, systemic hypertension) and those resulting from an increase preload (e.g., mitral regurgitation, aortic regurgitation).

In the so-called systolic heart failure, i.e., classical heart failure, an impaired inotropic state conducts to a weakened systolic contraction, which leads, ultimately, to a reduction in stroke volume and ventricular enlargement. While the normal ventricle is able to maintain a relatively normal stroke volume in response to an increased afterload, the failing ventricle is markedly affected by an increased afterload.

In most patients the clinically overt heart failure is preceded by periods of myocardial dysfunction, during which stroke volume and cardiac output may be maintained by compensatory mechanisms, such as the Frank-Starling's law, sympathetic adrenergic stimulation, fluid retention by the kidney and myocardial hypertrophy. The impaired left ventricular performance may become apparent during an exercise test. Failure indicates the limits and imperfections of the adaptational processes.

Cardiac remodeling denotes alterations of size, shape and performance of the left ventricle that develops after myocardial injury or overload. Ventricular hypertrophy, that is an integral constituent of the remodeling process, is characterized by increased muscular mass and wall thickness and tends to restore systolic stress toward normal values (Cohn, 2000). Ventricular hypertrophy must be regarded as a fundamental useful adaptation, which enables ventricular chambers to sustain chronic hyperfunction. The character of the initial stimuli (increased preload, increased afterload, primary loss of myocytes as in myocardial infarction, or primary depression in contractility as in cardiomyopathy) responsible for inciting the hypertrophy appears to play a critical role in determining the nature of the response. Grossman and colleagues (Grossman, 1975) proposed that increased wall stress signals myocyte growth to continue until wall stress returns to normal. This hypothesis requires that the myocardium can detect increased wall stress occurring with pressure and volume overload, respectively.

Laplace's law can have major implications for an injured ventricle that dilates to permit an increase in an adequate stroke volume. A dilated left ventricle is capable of maintaining systolic pressure at the cost of higher systolic wall stress. Initially, changes of ventricular shape, wall thickness and volume may help to normalize wall stress and uphold cardiac performance. Augmenting cavity size may permit the ventricle to propel a much greater fraction of diastolic volume with the same amount of fiber shortening or to eject the same volume with a reduced degree of shortening. This mechanism is especially apparent in a large spherical ventricle with poor ejection fraction, whose capacity of preserving stroke volume as well as cardiac output depends upon the size and shape of the ventricle. Since the volume of blood ejected at each beat is the product of the ejection fraction and the end-diastolic volume, it is clear that the same stroke volume may result from different values of the variables. This may be visualized by an hyperbolic function - that is identified by the equation: ejection fraction × end-diastolic volume = constant - describing all possible combinations of the two variables resulting in the same value of stroke volume (line of iso-stroke). As systolic performance deteriorates, an increase in ventricular dimension associated with left ventricular remodeling readily moves the ventricle along the curve allowing it to propel the same volume of blood but with reduced fiber shortening and ejection fraction (Figure 4) (Barsotti and Dini, 2002).

Despite that changes in left ventricular geometry and wall thickness may be temporarily useful in maintaining muscular pump function, this occurs at significant high cost and is commonly followed by the unfavorable consequences of dilatation. The more important of these is the need for midwall fibers of a dilated ventricle to develop greater force to exceed end-diastolic wall stress before it can generate pressure and shorten against vascular resistance. As a result, ventricular enlargement leads to a higher systolic wall stress for any given systolic pressure.

A greater left ventricular asynchrony index during cardiac cycle was observed in patients with idiopathic dilated cardiomyopathy because of a greater systolic wall stress due to marked chamber enlargement. Polyfasic time-length curves of one of more of ventricular hemiaxes – that possibly reflect abnormalities in ventricular dynamics and in the temporal sequence of contraction – have been observed in more than 50% of the patients with dilated cardiomyopathy, while they were virtually absent in the other groups (Barsotti et al., 1980, Zucchi et al., 1987). A greater regional systolic wall stress may account for the occurrence of these polyfasic curves; which may also reflect incoordinate wall motion with the worsening of global ventricular function to such a degree that exceeds what it could be expected for the specific cardiac disease state.

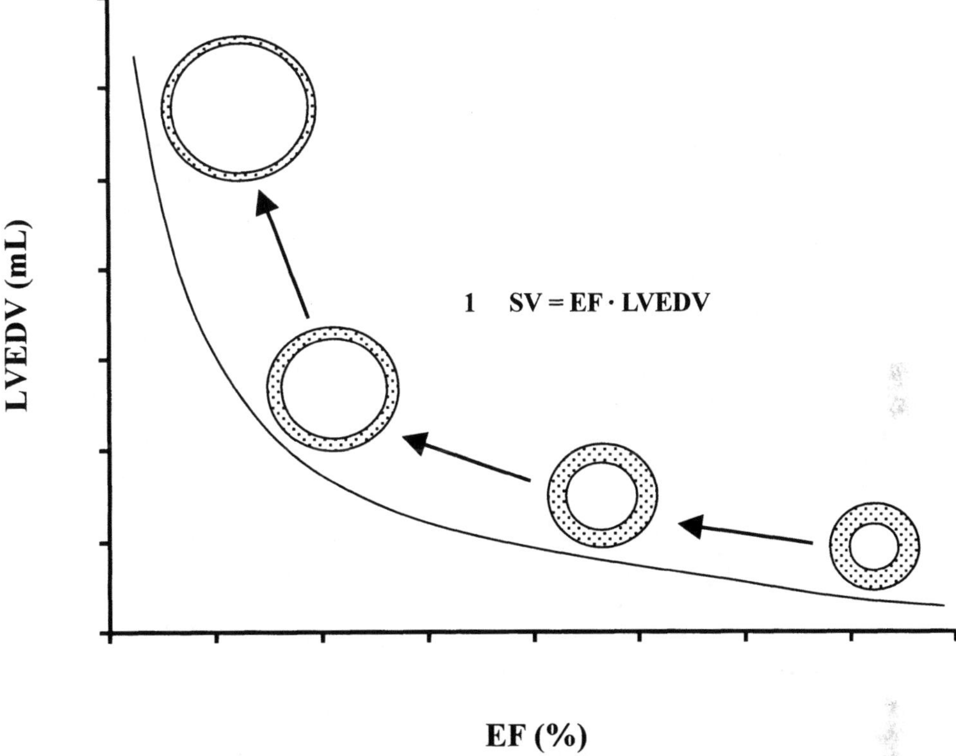

Figure 4. To maintain a constant stroke volume the following relation must be fulfilled: ejection fraction × end-diastolic volume = constant. In a graph plotting end-diastolic volume against ejection fraction, this relationship identifyes an hyperbolic function (line of iso-stroke) that - on the assumption that the ventricle is a sphere – demonstrated that the same stroke volume may be obtained by different combinations of the two variables. At high end-diastolic volumes, the volume ejected by the ventricle at each systole is produced by a very low ejection fraction at the expense of a remarkable increase in myocardial wall stress and oxygen consumption.

The basic mechanisms of myocardial failure remain a very active area of investigation. Morphologic studies have provided detailed information on the changes of myocardial tissue occurring in response to increased wall stress usually following myocardial injury or overload. In post-infarct ventricular remodeling, distortion and thinning of the area of infarction is generally associated with dilatation of the entire left ventricle with eccentric hypertrophy of noninfarcted segments. Inadequate eccentric hypertrophy of viable myocardium appears to be a critical factor capable **of** affecting the evolution toward heart failure. It has been clearly demonstrated that the inability of hypertrophy to keep pace with an abnormally high ventricular wall stress may be responsible for depressed left ventricular ejection fraction in many patients with cardiac valve diseases or cardiomyopathies (Strauer, 1987).

When inadequate hypertrophy occurs, because of an elevation in ventricular wall stress, myocardial oxygen consumption rises and this may lead to the exhaustion of coronary blood flow reserve in the subendocardial layers (Hittinger et al., 1995). Increased biochemical stress on cardiac myocytes, either trough genetic abnormalities or through excessive stress on the chamber wall due to myocytes loss or severe hemodynamic loading, generates a persistent signal for ventricular growth. Biochemical signals that promote myocyte cell growth may lead to cell death or apoptosis (Olivetti et al., 1997). A direct relationship between the extent of myocyte apoptosis and systolic wall stress has been recently documented especially in the subendocardial regions of the failing human heart (Di Napoli et al., 2000). The progressive mechanical overload of the spared hypertrophied myocytes could explain the initiation of a vicious circle that leads to progressive ventricular decompensation (Barsotti et al., 1998).

As remodeling further progresses, the ventricle loses its characteristic elliptical shape, becoming more spherical. The development of a spherical left ventricle in patients with dilated cardiomyopathy is associated with a further increase in wall stress and may cause functional mitral regurgitation. Elevation in ventricular preload, wall stress and stroke volume are all documented ventricular adaptations to severe mitral regurgitation. The importance of the law of Laplace during the progression of ventricular remodeling to heart failure lies in the fact that wall stress is a major determinant of both cardiac expenditure and efficiency; a higher wall stress increases the first and decreases the second. With chamber enlargement, the spherical shape needs more energy because of the greater diameter that increases the force that is needed to produce a given pressure; this marked rise in energy expenditure is associated with an important decrease in cardiac efficiency.

References

Weber KT, Brilla C (1991) Pathological hypertrophy and cardiac interstitium: fibrosis and renin-aldosterone system. *Circulation 83:1849-1865.*

Opie LH: *The Heart. Physiology, from Cell to Circulation.* Philadelphia: Lippincott, Williams & Wilkins, 1998.

Levick JR: *An Introduction to Cardiovascular Physiology*. Butterworth-Heinemann Ltd, Oxford 1995.

Ross J, Jr, Covell JW, Sonnenblick EH, Braunwald E (1966) Contractile state of heart characterized by force-velocity relations in variably afterloaded and isovolumic beats. *Circ Res 18:149.*

Ross J, Jr, Braunwald E (1974) The study of left ventricular function in man by increasing resistance to ventricular ejection with angiotensine. *Circulation 29:739-749.*

Shepherd JT, Vanhoutte PM: *The Human Cardiovascular System. Facts and Concepts.* New York: Raven Press, 1979.

Katz AM: *Physiology of the Heart.* Philadelphia: Lippincott, Williams & Wilkins, 2001.

Mirsky I (1969) Left ventricular stresses in the intact human heart. *Biophys 9:189-208.*

Hutchins GM, Bulkley BH, Moore GW et al (1978) Shape of the human cardiac ventricles. *Am J Cardiol 41:646-654.*

Streeter DD, Sponitz HM, Patel DP et al (1969): Fiber orientation in the canine left ventricle during diastole and systole. *Circ Res 24:339-347.*

Dodge HT, Baxeley WA (1969) Left ventricular volume and mass and their significance in heart disease. *Am J Cardiol 23:528-537.*

Chapman CB, Baker O, Reynolds I, Bonte F (1958) Use of biplane cinefluorography for measurement of ventricular volume. *Circulation 18:1105-1117.*

Fifer MA, Grossman W: Measurement of ventricular volumes, ejection fraction, mass, wall stress, and regional wall motion. In: Baim DS, Grossman W, eds. *Grossman's Cardiac Catheterizaton, Angiography, and Intervention.* Philadelphia: Lippincott, Williams & Wilkins, 2000:353-356.

Gunther S, Grossman W (1979) Determinants of ventricular function in pressure-overload hypertrophy in man. *Circulation 4:679-688.*

Suga H, Hisano R, Hirata S, et al (1982) Mechanism of higher oxygen consumption rate: pressure-loaded versus volume-loaded heart. *Am J Physiol 242:H942-H948.*

Carabello BA: Abnormalities in cardiac contraction: Systolic dysfunction. In: Hosepund JD, Greenberg BH, eds. *Congestive Heart Failure. Pathphysiology, Diagnosis, and Comprehensive Approach to the Management.* New York: Springer-Verlag: 1993:54-67.

Suga H, Sagawa K (1974) Instantaneous pressure-volume relationship and their ratio in the excised supported canine left ventricle. *Circ Res 35:117-126.*

Sagawa K, Maughan L, Suga H, Sunagawa K: *Cardiac Contraction and the Pressure-Volume Relationship.* Oxford University Press, New York 1988.

Kass DA, Maughan WL (1988) From Emax to pressure-volume relations: a broader view. Circulation 77:1203-1212.

Gibbs CL: Cardiac energetics. In: Langer GA, Brady AJ, eds. *The Mammalian Myocardium.* New York: Wiley:, 1974:105-133.

Katz AM: *Heart Failure: Pathophysiology, Molecular Biology, and Clinical Management.* Philadelphia: Lippincott, Williams & Wilkins, 2000.

Braunwald E: *Heart Disease. A Textbook of Cardiovascular Medicine.* Philadelphia: WB Saunders Company, 1997.

Cohn JN, Ferrari R, Sharpe N (2000) Cardiac remodeling: Concepts and clinical implications: a consensus paper from an international forum on cardiac remodeling. Behalf on an International Forum on Cardiac Remodeling *J Am Coll Cardiol 35:569-582.*

Grossman W, Jones D, McLaurm LP (1975) Wall stress and patterns of hypertrophy in the human left ventricle. *J Clin Invest 56:56-64.*

Dini FL (2002) Le basi cardiodinamiche del rimodellamento chirurgico del ventricolo sinistro. *Il Cardiologo 2:35-41.*

Barsotti A, Mariotti R, Balbarini A, Mariani M (1980) Quantitative evaluation of the regional left ventricular function in normal subjects by means of cineangiocardiography. *Cardiovasc Res 14:30-40.*

Zucchi R, Barsotti A, Mariotti R, Biadi O, Balbarini A, Mariani M (1987) Asynergy and left ventricular performance in dilated cariomyopathy. *Clin Cardiol 10:153-158.*

Strauer BE (1987) Structural and functional adaptation of the chronically overloaded heart in arterial hypertension. *Am Heart J 114:948-957.*

Hittinger L, Mirsky I, Shen YT, Patrick TA, Bishop SP, Vatner SF (1995) Hemodynamic mechanisms responsible of reduced subendocardial coronary reserve in dogs with severe left ventricular hypertrophy. *Circulation 92:978-986.*

Olivetti G, Abbi R. Quaini F, Kajstura J, Cheng W, Nitahara JA, Quaini E, Di Loreto C, Beltrami CA (1997) Apoptosis in the failing human heart. *N Engl J Med 336:1131-114.*

Di Napoli P, Taccardi AA, Vianale G, Soccio M, Gallina S, Calafiore AM, Barsotti A, De Caterina R (2000) Systolic left ventricular wall stress modulates cariomyocytes apoptosis in patients with severe dilated cardiomyopathy (Abstr). *Eur Heart J (Suppl):366.*

Barsotti A, Dini FL, Di Napoli P, Gallina S (1998) Interstitial remodeling and chronic dysfunction of viable noninfarcted myocardium in postischemic heart failure. *J Cardiovasc Diagn Proc15:17-20.*

Element of Physiology and Mechanics of Human Arteries

Robert S. Reneman, Arnold P.G. Hoeks and Lilian Kornet[1]

Departments of Physiology and Biophysics, Cardiovascular Research Institute, Maastricht University. Maastricht, the Netherlands

Abstract. Some elements concerning the physiology and the corresponding mechanical properties of human arteries are introduced. The non-invasive ultrasound techniques developed to assess strain, distensibility and compliance and IMT, velocity and wall shear rate/stress distribution in arteries are introduced. Artery wall properties, and intima-media thickness (IMT) change with age and disease are discussed with the relation with artery wall function and structure and role in the genesis of atherosclerosis. In hypertension and with aging elastic arteries become stiffer (loss of distensibility and compliance), the degree of stiffening also varies along artery bifurcations and differs in elastic and muscular arteries. These variations give differences in intima-media thickness and sites of preference for atherosclerosis. Alterations in structure and composition of the arterial wall responsible for the loss of distensibility and compliance and the increase in intima-media thickness (IMT) in disease and aging. Reasons and effects are discussed.

1. Non-Invasive Ultrasound Techniques to Assess strain, Distensibility and Intima-Media Thickness (IMT) in Human Arteries

1.1 Introduction

Under normal circumstances arteries, especially the larger elastic ones, distend during systole, the pump phase of the heart, and, hence, can store volume energy during this phase of the cardiac cycle. This elastic behavior of the artery wall limits the increase in blood pressure during systole for a given amount of volume flow. The amount of volume energy that can be stored during systole does not only depend on the degree of distension of the artery, but also on its initial diameter. With increasing age and in diseases, like borderline and established hypertension and atherosclerosis, the elastic arteries become stiffer, processes not necessarily affecting all parts of the arterial tree to the same extent. Moreover, arteries become thicker at older age and when exposed to high blood pressure or low wall shear stress. To get insight into these vascular processes and to be able to follow them adequately, in the past decades several ultrasound techniques have been developed to non-invasively assess the elastic properties of artery walls and IMT in humans. These techniques and the parameters employed to characterize properties of arteries will be discussed in this lecture. More detailed information

[1] Chapters 4 and 5, only.

can be found in the following selected references (Bots et al., 1997, Glagov et al., 1992, Hoeks, 1993, Hoeks et al., 1985, 1990, 1997, Hokanson et al., 1972, Lehmann et al., 1993, Milnor, 1989, Patel and Fry, 1966, Pignoli et al., 1986, Reneman et al., 1985, Riley et al., 1992, 1996, Salonen et al., 1991, Van Merode et al., 1989, 1996, Wells, 1969, Willekes et al., 1999).

1.2 Parameters to Characterize the Elastic Behavior of Arteries

1.2.1 Distensibility and Compliance

Parameters commonly used are distensibility and compliance, defined as the relative (ΔV/V) and absolute (ΔV) change in arterial volume (V) for a change in pressure (Δp), assuming a linear relationship between the time-dependent change in lumen cross-sectional area and pressure. The error arising from this assumption is small for the changes in pressure and lumen cross-sectional area during the cardiac cycle, especially in elastic arteries. Therefore, in medicine the increase in volume and pressure in the systolic phase of the cardiac cycle is commonly used to describe arterial distensibility and compliance. Since it is practically impossible to accurately determine volume and volume changes non-invasively in vivo, distensibility and compliance are generally expressed in terms of changes in lumen cross-sectional area, A, (ΔA/A/Δp) and ΔA/Δp, respectively). The terms area compliance and area distensibility are often used for these relations. This simplification is allowed because in unimpeded arteries length is constant so that the change in volume during the cardiac cycle is caused by a change in lumenal cross-sectional area alone. The constant length can be explained by the fact that arteries are longitudinally tethered at their in vivo length and by the experimental finding that at in vivo length, isolated vessels neither lengthen nor shorten with pressure changes. The expression for distensibility coefficient (DC) in terms of luminal cross-sectional area is

$$DC = \frac{\Delta A/A}{\Delta p} \quad \text{in Pa}^{-1} \tag{1.1}$$

For practical purposes this relation can be rewritten as a function of diameter (d) rather than area. This is allowed if it is assumed that the artery lumen is circular in cross-section and that Δd is small relative to d, reasonable assumptions in humans. Then, this relation becomes

$$DC = \frac{2\Delta d/d}{\Delta p} \quad \text{in Pa}^{-1} \tag{1.2}$$

Similarly, compliance can be expressed in terms of luminal cross-sectional area, providing

$$CC = \frac{\Delta A}{\Delta p} \quad \text{in m}^2\text{Pa}^{-1} \tag{1.3}$$

This relation can also be rewritten in terms of diameter

$$\text{CC} = \frac{\pi d \Delta d}{2\Delta p} \quad in\, \text{m}^2.Pa^{-1} \tag{1.4}$$

We usually measure Δd as the increase in arterial diameter during the systolic phase of the cardiac cycle. The "reference" diameter should be the average diameter, but in practice end-diastolic diameter is commonly used. The pulse pressure (systolic minus diastolic pressure) is generally taken as Δp.

1.2.2 Strain

Strain is the change in length relative to the mean length. For arteries (radial) strain gives a measure of the relative deformations to which arteries are exposed. It can be calculated as Δd/d. However, one should keep in mind that this strain is not a material property and that it depends on pulse pressure and wall thickness.

1.2.3 Pulse Wave Velocity

Another parameter often used as an index of arterial stiffness is pulse wave velocity (PWV); the stiffer the artery is, the higher this velocity will be. In relation to distensibility, PWV can be written as

$$\text{PWV} = \frac{1}{\sqrt{\rho \text{DC}}} \quad in\, \text{m.s.}^{-1} \tag{1.5}$$

where ρ = the density of blood.

1.2.4 Elastic Modulus

The Young's modulus (E), which is a characterization of wall material, is the ratio of stress and strain in the vessel wall. Assuming pressure independence, we can derive E through the Moens-Korteweg equation of pulse wave velocity

$$PWV = \sqrt{Eh / \rho d} \qquad \text{(where h is wall thickness)}$$

The Young's modulus can then be written as

$$E = \frac{d}{hDC} \quad \text{in Pa} \tag{1.6}$$

1.3 Discussion and Interpretation of the Parameters

The use of DC and CC has the advantage that d and Δd can be measured accurately and reliably non-invasively by means of ultrasound (see below), but requires the determination of Δp at the site of measurement of d and Δd or at a representative site elsewhere. We showed that for the assessment of DC and CC in the carotid arteries Δp in the brachial artery is a good substitute. However, studies in our institute also showed that brachial artery Δp and Δp as assessed in a finger are not representative of Δp in the femoral artery. An additional advantage of the use of DC and CC is the possibility to obtain information about artery wall distensibility and compliance locally (a few millimeters apart) along the arterial tree. In subjects without occlusive atherosclerotic disease variations in distensibility along arterial bifurcations, if any, can be assessed by determining Δd/d alone, because in these subjects Δp may be assumed to be the same at each site in a bifurcation. In this way an impression can be obtained about the degree of strain to which arteries are exposed locally. The possibility to obtain information about changes in distensibility locally is required to follow adequately artery wall properties in ageing, hypertension and atherosclerosis, because under these circumstances different parts of arterial bifurcations are affected differently.

PWV can be determined by assessing the foot-to-foot delay between two simultaneously recorded pressure or distension waves in the artery under investigation. The use of PWV has the advantage that no arterial pressure recordings are necessary, but the simultaneous recording of the two waves requires some skill. Moreover, depending on the technique used to estimate the foot-to-foot delay, the time resolution is on the order of a few milliseconds requiring averaging over a larger number of cardiac cycles for precise measurement, thereby discarding the short-term variations in physiological conditions. Even more important, PWV provides an average value of the artery segment studied and does not allow the assessment of local changes in arterial distensibility and compliance, which can be considered to be a limitation (see above).

The Young's modulus describes artery wall properties perse, because this parameter also considers wall thickness, which is increased, for example, in hypertension and at older age. The Young's modulus cannot be measured directly, but is derived from arterial distensibility, artery wall thickness and arterial diameter. When a decrease in distensibility is associated with an increase in wall thickness and a slight or no change in arterial diameter, the Young's modulus does not necessarily have to change. This may, for example, occur when wall mass increases without a change in wall composition. In such a situation one might conclude that the elastic material properties of the artery are not changed, but that the reduced arterial distensibility results from geometric changes. Therefore, systolic arterial blood pressure, which is determined by arterial compliance, may increase despite the absence of a change in the Young's modulus. It is essential to derive the Young's modulus before drawing conclusions regarding the elastic properties of the artery wall.

In summary the Young's modulus has to be determined if information about the elastic properties of the artery wall is required. If one is interested in the functional aspects of the artery wall, such as its effect on systolic and diastolic arterial blood pressure, compliance and/or distensibility should be studied.

1.4 Ultrasound Techniques Used to Assess Non-Invasively d and Δd in Humans

In the past decades several ultrasound techniques have been developed to determine non-invasively the elastic properties of the artery wall. One of the first techniques made use of amplitude tracking, i.e., tracking of the crossing of the leading edge of the received wall signals in A-mode. In this approach both d and Δd are overestimated, mainly because the echogenicity depends on the distance between transducer and artery wall, which varies with the phase of the cardiac cycle. An alternative approach makes use of the (change in the) instantaneous phase of the received radio frequency (RF) signal; the instantaneous position of the first zero-crossing with respect to a point of reference within the sample windows is determined, the position of the sample windows being adjusted accordingly. Later developments along this line aimed at improvements in the detection and processing methods. To facilitate the identification of the arterial segment and the positioning of the sample windows at the reflections from the arterial walls the displacement estimation technique was combined with 2D-echo systems. The use of dedicated signal processing techniques allows automatic positioning of the sample windows, simply by locating the lumen of the artery. In a separate development the detection of the change in position of the artery wall was based on assessment of the average phase of the received signal over the sample window (Doppler processing) rather than the instantaneous phase. In an early application, where echo tracking was combined with on-line assessment of the velocity distribution over the cross-section of an artery as a function of time, both sample volumes were fixed in size and depth. A fixed position of a sample window limits adequate tracking of the walls, especially when their displacement is large. Therefore, in later systems, designed only for diameter and wall motion detection, the sample windows were allowed to track the moving position of the artery wall. In more recent systems the displacement detection algorithm is based on RF correlation tracking rather than Doppler processing. This type of tracking is less sensitive to phase interference by RF signals returned from closely spaced reflectors. Moreover, the output of the RF correlator is presented in a fraction of the sample distance instead of a fraction of the assumed wavelength of the received signal, as in Doppler processing. This makes the result independent of the actual carrier frequency of the received signal which will be modified by the depth dependent attenuation, especially for the emission of wide bandwidth signals (short bursts) used to obtain a high resolution.

The equipment developed by us and currently in use in our institute consists of a two-dimensional B-mode imager attached to a vessel wall moving detector system (Pie Medical, Maastricht, The Netherlands). Based on the B-mode image, an M-line perpendicular to the artery of interest is selected and the induced RF signals within a sample volume coinciding with the artery walls are processed. To this end the RF signals are stored in a large memory and line after line the data acquired are transferred to a personal computer (PC, Laser 386) with a mathematical processor. The first line acquired is graphically presented on a display, allowing manual identification of the anterior and posterior wall boundaries by placing two markers representing the sample windows for data processing. Once the walls are identified the remaining data are transferred and processed on the fly. To extract the change in position of either the anterior or posterior wall, averaged over a few RF lines, the approach based on the cross-correlation model for corresponding segments of subsequent RF lines is applied. The width of the windows used is equivalent to 2 cycles of the RF-signal (equivalent to 240μm at a carrier frequency of 7 MHz). The position of these windows is updated for each subsequent line using the detected cumulative change in phase of the RF-signal with respect to the sampling frequency averaged over the window. Computation takes less than 5 seconds depending on the length of the sample window and the number of input lines. After processing of all lines available the displacement of the anterior and posterior vessel walls, and the difference between both signals reflecting the change in arterial diameter as function of time, are displayed. Post-processing yields Δd/d * 100% for each subsequent cardiac cycle, the RR interval and end-diastolic diameter. With this system arterial wall displacements of a few micrometers can be resolved. This resolution is better than that obtained with systems using zero-crossing detectors in which the resolution is limited by the sample frequency.

For good accuracy the M-line has to be positioned perpendicular to the artery axis under visual control on the B-mode, because deviations from normal or the centerline position will introduce errors. One should realize, however, that the influence of deviations from normal on the d and Δd readings is relatively small. For example, a deviation of 20° results in an error of only about 5% in the values for d and Δd. The real-time B-mode image is essential to establish the plane of movement of the axis of the artery under investigation. To avoid erroneous estimation of the change in diameter the plane of movement and the line of observation should match. The intra- and inter-observer variability for the assessment of Δd/d is about 8% (coefficient of variation) for the carotid arteries and about 12% for the femoral arteries. The arterial diameter can be determined with an intra- and inter-observer variability of 2.8 - 4.5% for the carotid arteries and of 2.0 - 3.0% for the femoral arteries.

1.5 Non-Invasive Assessment of Intima-Media Thickness (IMT) in Humans

Two-dimensional echo systems are also used to determine artery wall thickness, a parameter necessary to calculate the Young's modulus. Besides, wall thickness has gained increasing attention because of its involvement in adaptive responses to physiological and pathophysiological processes as, for example, in ageing and hypertension. Moreover, increase in wall thickness in the carotid arteries has been proposed as a marker of atherosclerosis elsewhere in the arterial system. The non-invasive assessment of wall thickness by means of ultrasound is limited to IMT, because the advetitia cannot be distinguised reliable from the surrounding structures. On a B-mode image only the intima is visible as a distinct layer and the media appears as a relatively dark narrow band, because of its low echogenicity. Since the echogenicity of the adventitia is relatively high, as compared with the other layers in the wall, in the anterior wall the large trailing echos from the adventitia tend to obscure the weaker signals from the closely spaced intima. Therefore, measurements of IMT are usually made on the posterior wall of arteries.

Since the first IMT measurements made by Pignoli and colleagues, using callipers, a variety of techniques has been developed, making use of automated digitization of the B-mode image, automated interpolation of the intima-media complex, automated spatial averaging procedures or improved display and zoom techniques. All these techniques provide a mean IMT over a length of 10-20 mm. When interested in the use of IMT as an indicator of a specific disorder, averaging over a certain segment may be appropriate. When studying the relation between IMT on the one hand and artery wall properties or wall shear rate/stress on the other, however, local assessment of IMT is indicated, because in an artery segment significant differences in IMT can be observed at short distances from each other. Recently we developed an automated method to assess IMT locally, which is based on processing of the received RF signals, using the system for the assessment of d and Δd (see above). After RF data acquisition the first M-line activated is displayed on a personal computer screen allowing identification of a window of 3 mm covering the posterior lumen-wall and inter-wall transitions. Data from this window over time are then stored on hard disk for further off-line processing. In this processing the amplitude envelope of the RF-signals is taken and after phase alignment of the signals the time average of the envelope is determined to reduce speckle interference. An edge detection algorithm with preselected threshold is applied to the time averaged envelope to assess IMT. With this technique IMT can be assessed in vivo with a precision of on the order of 45 μm. The results obtained with this method compare favorably with those obtained with other methods currently in use. The method is independent of the B-mode imager used and allows the assessment of differences in IMT in short artery segments.

2. Non-Invasive Assessment of Velocity and Shear Rate Distribution in Human Arteries by Means of Ultrasound

2.1 Introduction

Blood flowing through a vascular segment exerts a tangentially directed shear stress on the luminal surface of endothelial cells, the inner layer of the vascular wall. Wall shear stress is the product of wall shear rate, i.e. the radial derivative of the velocity near the wall, and local blood viscosity. Wall shear stress has been shown to be an important determinant of the release of vasoactive substances from the endothelial cells and, hence, of vessel wall function. Several of the vasoactive molecules stimulate the expression of adhesion molecules and chemokines involved in intima-media thickening.

Since in large arteries the velocity profile, i.e. the velocity distribution over the cross-sectional area, is a flattened parabola, shear is low in the center of a vessel and high near the vessel wall. Reliable non-invasive assessment of wall shear rate requires accurate measurement of low blood flow velocities close to the vessel wall.

More detailed information can be found in the following selected references (Brands et al., 1995, Busse and Fleming, 1966, 1998, Frangos et al., 1985, Hoeks et al., 1993, Kornet et al., 1999, Oyre et al., 1997, 1998, Perktold et al., 1994, Reneman et al., 1986, Rubanyi et al., 1990, Sakamoto et al., 1979, Samijo et al., 1997).

2.2 Theoretical Considerations

To estimate local wall shear rate, the velocity distribution v(r) as function of the radial position r generally is modeled as follows

$$v(r) = v_m\left[1-\left(\frac{r}{R}\right)^n\right] \qquad 0 \leq r \leq R \tag{2.1}$$

where v_m is maximum velocity at the center position r = 0 and R is the radius of the artery. The bluntness factor n controls the shape of the velocity distribution. For n = 2, the result will be a parabolic velocity profile, as may be observed for fully developed flow under stationary conditions (constant flow velocity). For increasing n the velocity profile will become more and more blunt, as may happen in early systole and midsystole. The shear rate SR at the wall is given by the derivative of v(r) with respect to r at r = R

$$\mathrm{SR} = \left.\frac{\mathrm{dv(r)}}{\mathrm{dr}}\right|_{r=R} = nv_m \left.\frac{r^{n-1}}{R^n}\right|_{r=R} = \frac{nv_m}{R} \tag{2.2}$$

Equation 2.2 can be given as function of volume flow Q = vA, deriving from Equation 2.1 the velocity averaged over the cross-sectional area A = πR^2:

$$\bar{v} = \frac{1}{\pi R^2}\int_0^R 2\pi r v(r)dr = \frac{nv_m}{n+2} \tag{2.3a}$$

$$v_m = \frac{(n+2)\bar{v}}{n} \tag{2.3b}$$

Substituting (2.3b) into (2.2) yields

$$\mathrm{SR} = \frac{(n+2)\bar{v}}{R} = \frac{(n+2)\bar{v}A}{\pi R^3} = \frac{(n+2)Q}{\pi R^3} \tag{2.4}$$

For n = 2 (parabolic velocity distribution), Equation 2.4 converts to the well-known Hagen-Poiseuille relation, which is applicable for steady flow through a rigid tube. The inverse relation of shear rate to the cubed power of the radius implies a high sensitivity: any change in mean volume flow rate can be compensated for by a relatively small change in vessel radius.

In the past, the Hagen-Poiseuille relation was frequently used to estimate wall shear rate in the circulation. This requires the assessment of volume flow Q and the local vessel radius R. Alternatively, one may use the law of conversion of blood volume at branches to estimate wall shear rate. In both cases, assumptions are made regarding the shape of the velocity distribution. In these approaches the velocity profile is assumed to be a fully developed parabola, which is less appropriate because of the bluntness of the profile in human arteries. Moreover, neither approach takes into account the effect of elasticity. It has been shown that the distensibility of the wall markedly reduces local wall shear rate. A direct method based on assessment of the radial derivative of the velocity distribution with a high-resolution multigate pulsed Doppler ultrasound system has the advantage that no assumptions are involved. This requires the development of an ultrasound system capable of measuring low blood flow velocities close to the wall with high precision and high spatial resolution.

Multigate pulsed Doppler systems can assess instantaneous velocity distributions along a single line of observation known as M-mode. High resolution along the ultrasound beam can be achieved if the duration of the emitted ultrasound bursts is short and the processing is adjusted to these short emissions (wide signal bandwidth with respect to the carrier frequency). Because of the wide signal bandwidth, combined with the frequency-dependent attenuation, the carrier frequency of the signal returned by a scatterer or reflector at some depth will deviate from the emitted center frequency in an unpredictable way, rendering conventional ultrasound Doppler systems inadequate. To reduce the inherent effect of the width of the ultrasound beam on the axial resolution the angle of observation should be obtuse (greater than 60°). This has the adverse effect of causing more reverberations, resulting in spurious echo signals (clutter) that apparently originate from within the lumen. Moreover, the vessel walls will return signals with a higher amplitude, which will add to the clutter and severely mask the signals returned by the red blood cells close to the (anterior) wall. By use of a high-pass filter with a cutoff frequency related to the anticipated maximal Doppler frequency induced by the clutter, the blood Doppler signals at a relatively high Doppler frequency can be regained at the expense of a reduction of the available Doppler frequency (velocity) range. Appropriate processing of the received radio frequency signal reduces the shortcomings of conventional pulsed Doppler systems, enabling the high-precision assessment of low blood flow velocities close to the wall. The mean of the maximum of the radial derivatives of the velocity distributions at the anterior and posterior walls is used as an estimate of local axial wall shear rate.

2.3 Ultrasound Technique Used to Non-Invasively Assess Wall Shear Rate and to Estimate Wall Shear Stress

As mentioned before, the basic requirements for reliable estimation of the velocity distribution in an artery, which enable shear-rate assessment, are (1) a large velocity range extending to low velocities, (2) low sensitivity for vessel-wall artifacts, (3) high spatial resolution and (4) high velocity resolution. Direct processing of the radio frequency signal based on the cross-correlation of corresponding segments (depth window) of the radio frequency signal over subsequent observations makes the velocity estimation insensitive to the carrier frequency. Moreover, the sensitivity to noise (velocity resolution) can be made low arbitrarily by extending the length of the depth window or of the window in time (number of subsequent observations or radio frequency lines). However, a long window in depth reduces depth resolution and, therefore, the assessment of velocities close to the wall, whereas a long window in time reduces the time resolution. The location of the peak of the cross-correlation function, representing the mean velocities of the scatterers within the sample window, can be estimated by exploring the cross-correlation function over some depth shift interval. This procedure can be accelerated, while at the same time the precision of the result is made independent of the sample frequency of the radio frequency signal by modeling the spectral power density distribution of the radio frequency signal. If one assumes that the spectral distribution has a gaussian shape, the shape of the cross-correlation function is known. A few points of the cross-correlation function in the vicinity of a depth and a time lag of zero sample points suffice to solve for the mean velocity and for the signal-to-noise ratio. To give the estimated velocity distribution a smooth and detailed shape, the length of the sample windows should match the resolution of the ultrasound system in echo mode (0.2 to 0.3 mm) along the ultrasound beam, while subsequent windows may partially (50%) overlap.

The cross-correlation model method is presently executed off-line in software on a depth and time matrix of radio frequency data points. In M-mode, conventional echo-Doppler systems do not provide a high pulse repetition frequency in combination with a large signal bandwidth to achieve a wide range to measure velocity with a good axial resolution. Therefore, during data acquisition the emitter/receiver of the echo system is replaced by a separate combined emitter/receiver, which also supplies the reference signal for sampling. This implies that there is no (visual) control over the beam direction and signal level during acquisition. Sampling in depth is synchronized with the time of emission at a sample frequency of approximately four times the emission frequency. The dynamic range of the sample points is 10 bits (60dB). To allow for the assessment of low amplitude values, a signal overload of the analog-to-digital converter of 12 to 18 dB (2 to 3 bits) is tolerated. Conversion starts synchronously with an electrocardiographic trigger. The size of the internal memory of the data acquisition system (presently four M samples) limits data acquisition to a depth range of 10 mm and a time range of 1 second at a 6-kHz pulse repetition frequency of the ultrasound system (emission frequency, 5 MHz). To extend the velocity range to low velocities, the following processing scheme is used. First, the signal-to-noise ratio and the velocity of the raw signal are estimated. If the signal-to-noise ratio is high and the observed velocity is low, then a significant clutter component is present within the signal window. In that case, signals in time direction are corrected for the observed velocity, resulting in a phase alignment of all signal segments. A high-pass filter with a low cutoff frequency is then used to eliminate the clutter component. Because of the phase alignment, the cutoff frequency is now dictated by the bandwidth of the clutter component rather than by the anticipated maximal velocity of the slowly moving structures, allowing for a relatively low cutoff frequency independent of the phase of the cardiac cycle. The apparent result is an extension of the velocity range to low velocities of the order of a few centimeters per second. After clutter removal the signal-to-noise ratio and mean velocity are estimated again by use of the cross-correlation model method. If the signal-to-noise ratio is sufficiently high (>-6 dB), the remaining signal originates from moving blood cells and the estimated velocity (after correction for the imposed frequency correction related to clutter suppression) is accepted; otherwise it is set to zero. This procedure is repeated for all partially overlapping sample windows, each with a length of 0.3 mm, spaced along the ultrasound beam at 0.15 mm range intervals, resulting in a reliable and detailed estimate of the instantaneous velocity distribution. Also, subsequent windows in time (with a length of 5 or 10 milliseconds) overlap by 50%. The resulting two-dimensional velocity distribution is smoothed by a 3x3 median filter to remove the occasional extreme velocity value. To further reduce fluctuations of the velocity estimates the median filter is followed by a 3x3 sliding-window averaging filter.

The time-dependent shear rate distribution directly follows from the observed velocity distribution by taking the absolute value of the radial first derivative. This provides high spatial peaks at sites with a high spatial velocity gradient. The shear rate distribution obtained is corrected for the angle of observation (60° to 70°, estimated from the B-mode), affecting both velocity and spacing along the beam. For each line in depth the mean of the peak values at the anterior and posterior luminal sites is considered to be the estimate of wall shear rate. Repeating the procedure for all available lines results in the shear-rate waveform.

Because of the limited resolution due to the finite size of the sample volume, velocities can not be determined at the wall. Therefore, the maximum value of the radial derivative of the velocity is considered as the estimate of instantaneous wall shear rate. In general the maximum shear rate is assessed about 300 μm from the blood-intima boundery. This implies that the measured shear rate has to be considered as a least estimate, because shear rate may be higher near the wall than at the site of assessment. From the shear distribution mean wall shear rate, the time-averaged shear rate over one cardiac cycle (mean wall shear rate) and peak wall shear rate, the value at peak systole, can be determined.

In the common carotid artery the intra-subject intra-session variability on different days varies between 13 and 15 % for peak wall shear rate and between 10 and 12 % for mean wall shear rate (coefficients of variation) and the inter-subject intra-session variability on different days varies between 16 and 19 % for peak wall shear rate and between 11 and 17% for mean wall shear rate. In the femoral artery these values are somewhat higher. Therefore, about 16 measurements have to be performed to obtain reliable values of mean and peak wall shear rate.

Because in large arteries the plasma layer is only 3-7 μm, which is substantially smaller than the spatial resolution of our wall shear rate system, the effect of this layer on blood viscosity can be neglected and local whole blood viscosity can be used for the determination of wall shear stress.

Beside ultrasound techniques, Magnetic Resonance Imaging (MRI) is used to assess wall shear rate in vivo. The results obtained with MRI are promising, and for the carotid artery compare favorably with those obtained with ultrasound. MRI has the advantage that three-dimensional information can be obtained in a rather direct way.

3. Changes in Artery Wall Properties and Arterial Intima-Media Thickness (IMT) with Ageing and in Hypertension

3.1 Changes in Artery Wall Properties and IMT with Increasing Age

Arteries, mainly the larger elastic ones, generally become stiffer with increasing age, even when compliance is assessed at similar pressure. From the third age decade onwards distensibility and compliance of the elastic common carotid artery decrease linearly with age, the reduction in compliance being less steep than the reduction in distensibility. The less pronounced decrease in compliance can be explained by the increase in arterial diameter observed with increasing age. In this way the ability of the arteries to store volume energy is reduced less than expected on the basis of the loss of distensibility, limiting the increase in systolic arterial blood pressure at older age. The Young's modulus of the carotid artery and the pressure-strain modulus of the femoral artery increase with age, indicating loss of elastic properties of the artery wall. The distensibility of the common femoral artery is reduced at older age, but the compliance of the brachial artery and the distensibility of the deep and superficial femoral arteries are not.

Even when studied at similar distending pressure with increasing age the changes in Young's modulus are not in direct relation with the changes in compliance and distensibility. The changes in the latter two variables also depend on the changes in diameter and wall thickness with age. In both normotensive and hypertensive rats, at similar diastolic arterial blood pressure aortic distensibility was found to be reduced with increasing age, and aortic diameter to be increased, resulting in no significant change in aortic compliance. Aortic wall thickness was not influenced by age in these experiments. The Young's modulus was found to be increased, suggesting loss of elasticity of artery wall material.

The increase in artery wall thickness with increasing age results mainly from adaptive intimal thickening, a physiological adaptation to mechanical stresses, secondary to variations in flow, wall tension or both. Beside intimal thickening, which is caused by enhanced contents of reticulated non-fibrous connective tissue and smooth muscle cells, there is progressive fibrosis of the media.

It should be realized that the loss of distensibility with increasing age is not necessarily homogeneous along arterial bifurcations. In the carotid artery bifurcation, the carotid artery bulb, where predominantly the baroreceptors , which play an important role in the acute control of blood pressure, are located, is more severely affected by age than the remainder of the bifurcation. The pronounced reduction in carotid artery bulb distensibility likely contributes to the decreased baroreceptor sensitivity observed at older age. The femoral artery bifurcation is differently affected by age as far as the reduction of distensibility is concerned. In the femoral artery bifurcation the feeding common femoral artery is affected by age, while the deep and superficial femoral arteries are not. With increasing age the distensibility becomes more homogeneously distributed along the femoral artery bifurcation, while it becomes more heterogeneously distributed along the carotid artery bifurcation.

The strains $\Delta d/d$ to which arteries are exposed are different in elastic and muscular arteries and are also differently affected by age. For example, at younger age strain is only 2.5-6% in the muscular common femoral artery and 9.5-11% in the elastic common carotid artery. At older age the values of strain are reduced to 1.7-4% in the common femoral artery and to 5-5.5% in the common carotid artery. These data indicate that the wall is exposed to lower strain in muscular arteries than in elastic arteries at both younger and older age and that the reduction in strain to which the wall is exposed with increasing age is slightly more pronounced in elastic than in muscular arteries (– 50 vs – 33% reduction). It is of interest to note that with increasing age the wall strain along the carotid artery bifurcation becomes inhomogeneous, while it becomes homogeneous along the femoral artery bifurcation. It should be noted that differences in strain along bifurcations imply inhomogeneous deformation during the cardiac cycle.

The above observations show that it is necessary to assess artery wall properties locally along arterial bifurcations to be able to follow adequately changes in arterial distensibility and compliance with increasing age. These local changes can now be studied accurately with the ultrasound technique described before, since it allows the determination of d and)d/d along bifurcations at distances of a few millimeters apart.

Loss of compliance of the elastic arteries will lead to an increase in systolic arterial pressure, an independent risk factor for cardiovascular disease, and will expose increased load on the heart.

3.2 Changes in Artery Wall Properties and IMT in Hypertension

3.2.1 Established Hypertension (Arterial Pressure > 160/100 mmHg)

It has been known for quite some years that arteries show a stiffer behavior in patients with established hypertension than in normotensive control subjects. More recent studies, however, have shown that the loss of arterial distensibility and compliance is not a generalized phenomenon along the arterial tree. For example, in the elastic carotid artery at ambient mean arterial pressure both distensibility and compliance are significantly lower in untreated mild and moderate essential hypertensive patients than in age-matched control subjects. In the radial artery, however, no significant differences in distensibility and compliance could be detected between untreated hypertensive patients and age-matched controls. As compared to normotensive subjects, in patients with established hypertension the diameter was found to be significantly increased in the brachial artery, but not in the carotid artery and the radial artery. It has been proposed that the loss of distensibility and compliance is caused by degenerative wall processes induced by stress fatigue.

Wall thickness, artery mass and the wall thickness-lumen diameter ratio are significantly larger in patients with established hypertension than in age-matched control subjects, at least in the radial artery. This observation is consistent with the hypothesis that arteries try to keep wall stress constant. Whether this adaptational process takes place also in other human arteries is as yet unknown. In the radial artery the Young's modulus was found to be similar in established hypertensive patients and in age-matched control subjects, despite the increase in wall mass.

Although structural changes have been observed in arteries of hypertensive patients, it is still a matter of debate whether in hypertension the reduction of artery wall distensibility is mainly caused by the increase in arterial blood pressure or whether structural changes in the wall contribute to this process. The results of Laurent and colleagues are in favor of a dominant role of increased arterial blood pressure, but observations in patients with borderline hypertension (see below) and recent findings in spontaneously hypertensive rats (SHR) argue against a dominant role of elevated blood pressure. In 6 week-old SHR compliance and distensibility are significantly reduced and media mass is significantly increased, as compared with normotensive Wistar Kyoto rats, while arterial blood pressure is not significantly different between these strains at this age. These findings indicate that in some forms of genetic hypertension alterations in artery wall properties are not dependent on elevated blood pressure.

3.2.2 Borderline Hypertension (140/90 mmHg > Arterial Blood Pressure < 160/100 mmHg)

Not only in established hypertension, but also in borderline hypertension elastic arteries are less distensible and less compliant than in age-matched control subjects. Because in the carotid arteries these differences in distensibility and compliance are already apparent at relatively young age, it has been proposed that these arteries age more quickly in patients with borderline hypertension than in normotensive subjects, especially since the decrease in arterial distensibility and compliance is substantial and the difference in arterial blood pressure between patients and normotensives is limited. If this hypothesis is correct, in borderline hypertensives the decrease in distensibility of the carotid artery should be most pronounced in the carotid artery bulb, because in this part of the carotid artery bifurcation artery wall distensibility is most affected by age. This is indeed the case, the proximal part of the bulb being significantly less distensible than the rest of the bulb. Whether this local wall stiffening leads to disturbed baroreceptor sensitivity in these patients is as yet unknown. If reduced wall distensibility hampers proper functioning of the baroreceptors, which have to regulate pressure on information derived from a smaller range of distension than in the normal more distensible carotid artery bulb, it may quite well be that this process plays a role in the development of borderline hypertension. The observations that distensibility and compliance are substantially reduced in borderline hypertensives, as compared with normotensives, despite relatively small differences in arterial blood pressure and that the bulb is more affected than the remainder of the carotid artery bifurcation, although these parts are exposed to the same mean blood pressure, also indicate that the changes in artery wall properties in these patients cannot be explained by increased blood pressure alone.

More detailed information about the points introduced in this chapter can be found in the following selected references (Boutouyrie et al., 1995, Girerd et al., 1994, Green et al., 1966, Gribbin et al., 1971, 1979, Hayoz et al., 1992, Laurent et al., 1994, 1994b, Learoyd and Taylor, 1966, Menotti et al., 1998, Mozersky et al., 1972, O'Rourke, 1990, Reneman et al., 1985, 1986, Riley et al., 1992, Safar et al., 1981, Sagie et al., 1993, Stary et al., 1992, Van Gorp et al., 1995, Van Gorp et al., 2000, Van Merode et al., 1988, 1993, 1996, Ventura et al., 1984).

4. Wall Shear Rate/Stress in Human Arteries

4.1 Introduction

Wall shear stress, the product of wall shear rate and blood viscosity, is the drag exerted by flowing blood on the endothelial cells. It has been shown in a variety of studies that wall shear stress in an important determinant of endothelial cell function: for example, it induces the release of vasoactive compounds affecting blood vessel diameter.

Based upon the theory of minimal energy expenditure, mean wall shear stress should be regulated via diameter adaptation and be more or less the same along the vascular tree. Several authors indeed found that if flow rate, and thus mean wall shear stress, is forced to change from its physiological state by a shunt or by an arterial ligation positioned downstream, arterial diameter adapts and mean wall shear stress is restored towards its baseline value. If viscosity, and thus mean wall shear stress, is increased, mean wall shear stress is reduced towards its baseline level, also by an increase in arterial diameter. According to theory the optimal value of mean wall shear stress in arteries should be 1.5 Pa ($\pm$ 25%).

The theory of minimal energy expenditure assumes that blood flow is continuous and non-pulsatile and that minimizing energy expenditure to circulate blood is the only design criterion for the circulatory system. The vascular system, however, must also be able to respond quickly to enhanced flow demands and to distribute blood over the different compartments, depending on the local requirements, and to store potential energy associated with vessel distension, because of the pulsatile character of blood flow. This potential energy is stored predominantly in the more central elastic arteries, while the more peripheral muscular arteries have mainly a conductive function. The conductive muscular arteries in the legs transport highly varying amounts of blood due to the large variations in the requirements of the peripheral vascular bed (compare, for example, exercise and rest). Therefore, the theoretical predictions, as supported by some experimental data, are not necessarily representative of the situation in the arterial system.

More detailed information can be found in the following selected references (Ando et al., 1994, 1996, Chapell et al., 1998, De Keulenaar et al., 1998, Frangos et al., 1985, Kamiya et al., 1984, Kornet et al., 1998, 1999, 2000, Ku et al., 1985, La Barbera, 1990, Levesque and Nerem, 1985, Murray, 1926, Reneman et al., 1985, Rossitti and Lofgren, 1993, Rubanyi et al., 1990, Samijo et al., 1998, Tsao et al., 1996, Zarins et al., 1983).

4.2 Wall Shear Rate/Stress in Human Arteries

In recent studies it was shown that mean wall shear stress in the human common carotid artery is close to the theoretical value of 1.5 Pa and that this value decreases significantly with increasing age, in both males and females, reaching values around 1.1-1.2 Pa at the age of 60 years. These values are still within 25% of the optimal value predicted by the model of minimal energy expenditure. The decrease in mean wall shear stress with increasing age can be explained by the increase in arterial diameter to reduce the loss of compliance at older age; a good example of compromise between two parameters to be regulated. At rest, in the femoral artery bifurcation, however, mean wall shear stress is substantially lower in both the common (0.35 Pa) and the superficial (0.49 Pa) femoral artery. The lower mean wall shear stress values in the femoral artery bifurcation can likely be explained by reflections from the periphery. Unlike in the common carotid artery, in the femoral artery bifurcation mean wall shear stress does not change with increasing age.

4.3 Wall Shear Rate/Stress in Artery Bifurcations and Their Relation to Intima-Media Thickness (IMT)

In studies relating flow measurements in scale models of human arteries and intima-media thickening in corresponding series of autopsy specimens, it has been shown that in such arteries as the aorta ,coronary arteries and carotid arteries, the intima is thicker in areas with low than with high wall shear stress. Direct assessment of the relation between wall shear stress and IMT in arteries has been hampered by the lack of techniques to determine this relation non-invasively in humans. With the recent developments in ultrasonography, as described in the previous lectures, it has become possible to assess non-invasively wall shear rate, and, hence, wall shear stress ,and IMT in humans at the same site in arteries. This allows us to study the relation between these parameters along arteries and near artery bifurcations where flow patterns are complicated. In artery bifurcations flow velocities are high near the flow divider and flow separation with reversed flow occurs opposite to this divider where mean wall shear stress is relatively low. With the use of these ultrasound techniques we were able to show that peak systolic and mean wall shear rate are significantly lower near the carotid artery bifurcation than 20-30 cm more upstream in the common carotid artery; the relative differences between both locations being independent of age. In this bifurcation IMT is larger at the site of lower wall shear rate. Near the carotid artery bifurcations IMT is negatively correlated with peak systolic and mean wall shear rate. The lower wall shear rate near the bifurcation than more upstream can likely be explained by reflections from the external carotid artery. The influence of these reflections greatly disappears upstream, where the velocity profile is more blunted than near the bifurcation.

In the femoral artery bifurcation mean wall shear stress is significantly lower in the feeding common femoral artery than in the superficial femoral artery. These differences in wall shear stress are also associated with differences in IMT; the thickness being larger at the site of lower wall shear stress. The relative differences in mean wall shear stress and IMT between the common and superficial femoral arteries are independent of age. The difference in mean wall shear stress between both arteries can be explained by reflections from the periphery, being more pronounced in the common than in the superficial femoral artery. In the former artery the velocity profile is affected by reflections from both the deep and the superficial femoral artery. The difference in wall shear stress between both sites disappears when the vascular bed is dilated and the influence of reflections is reduced.

4.4 The Relation Between Wall Shear Stress and Intima-Media Thickness

It has been well established that intima-media thickening occurs in areas of low mean wall shear stress. These areas are also the site of preference of atherogenesis. In general atherogenesis starts in these areas to further progress along the arterial tree. Although the mechanism underlying the relation between low mean wall shear stress on the one hand and intima-medial thickening and atherosclerosis on the other is still incompletely understood, the evidence for a relation between wall shear stress and biological processes is increasing. For example, regions of enhanced shear stress are characterized by more elongated endothelial cells, whereas regions of relatively low shear stress are associated with more rounded endothelial cells. During cell turnover, intercellular junctions become leaky, The leakiness will be enhanced by the downregulation of intercellular adhesion molecule –1 (ICAM-1) at low wall shear stress, increasing the width of the clefts in between endothelial cells. These mechanisms may be held responsible for the enhanced influx of macromolecules as lipoproteins and albumin, phenomena known to occur in atherosclerosis. Moreover, at low wall shear stress vascular adhesion molecule –1 (VCAM-1) is upregulated, causing (mononuclear) leukocytes to adhere to the vessel wall, one of the early steps in atherogenesis. In addition, low wall shear stress induces the upregulation of various growth factors.

5. The Relation Between Artery Wall Structure and Composition and its Function.

The arterial wall can be subdivided into three layers, from interior to exterior: the intima, the media and the adventitia. The intima consists of a single layer of endothelial cells supported by a limited number of smooth muscle cells. The intima is separated from the media by the lamina elastica interna, consisting solely of elastic lamellae. The media mainly consists of elastin and collagen fibers and smooth muscle cells, while proteoglycans are also present in this layer of the vascular wall. The number of elastin fibers decreases relative to the number of collagen fibers from the heart towards the peripheral arteries, whereas the number of smooth muscle cells increases downstream. This implies that the proximal arteries, closest to the heart, are more elastic and have a buffering function and that the distal arteries, in the periphery of the arterial network, are more rigid and mainly have a conductive function. The adventitia is composed of fibroblasts, proteoglycans and collagen arranged in bundles.

It is generally accepted that the resistance to stretching of the artery wall at low pressure is almost entirely due to elastin fibers, at physiological pressure due to both elastin and collagen fibers, but mainly the latter ones, and at high pressure almost entirely due to collagen fibers. This implies that the Young's modulus and compliance are not constant, but depend on arterial pressure or stretch. Recently, it has been shown that under control conditions the contribution of collagen fibers to the incremental elastic modulus increases with pressure at the cost of the contribution of elastin fibers. During administration of phenylephrine, a vasoconstrictor, however, the contribution of collagen fibers to the incremental modulus with increasing pressure is reduced, in an absolute sense and relative to the contribution of elastin fibers, probably because the activation of vascular smooth muscle cells induces contraction and reduction of arterial diameter. Whether the collagen fibers and the smooth muscle cells are connected in parallel or in series is still a matter of debate. The observation that arterial distensibility is reduced when dilation is caused by passive stretch due to increased arterial blood pressure, but increased when dilation is caused by reduced arterial smooth muscle tone, argues in favor of collagen fibers in series with smooth muscle cells, both being in parallel with elastin fibers.

The non-linear behavior of especially the muscular arteries has an important practical implication. Comparisons of Young's modulus and distensibility are only permitted when the mean distending pressure (i.e. arterial pressure) is the same. Thus comparison of patients with hypertension and healthy subjects, as far as the elastic properties of the artery wall are concerned, is only allowed at equal arterial pressures, a situation difficult to achieve in vivo.

An alternative is to construct and compare full pressure-diameter relations, but at the present state of the art in vivo these relations can reliably be obtained only in the radial artery.

The changes in artery wall composition that can be held responsible for the functional changes of the arteries, in terms of loss of distensibility and compliance, are still incompletely understood. The increase in artery wall thickness with increasing age results mainly from adaptive intimal thickening, a physiological adaptation to mechanical stresses, secondary to variations in flow, wall tension or both. Besides intimal thickening, which is caused by enhanced contents of reticulated non-fibrous connective tissue and smooth muscle cells, there is progressive fibrosis of the media. Total cellularity of both intima and media decreases as extracellular material accumulates. In a recent study on spontaneously hypertensive and control rats we were able to show that the loss of elasticity and distensibility of the arterial wall with increasing age and in hypertension can be ascribed neither to changes in the amount of elastin and collagen in the arterial wall nor to the degree of collagen cross-linking. The increase in wall thickness in hypertensive patients may be explained by the elastin fragmentation observed in these patients, leading to smooth muscle cell proliferation and migration towards the intimal layer.

It should be realized that artery wall thickening, which may lead to a change in Young's modulus, indicating a change in elastic properties of the artery, does not necessarily result in a change in arterial compliance.

Also the possible role of proteoglycans in artery wall function has drawn attention. Under normal circumstances chondroitin-sulfate and dermatan-sulfate proteoglycans and the glycoaminoglycan hyaluronic acid are present in arteries. Loss of proteoglycans from the arterial wall has been proposed to be responsible for the stiffening of arteries with increasing age. This is supported by the observation that removal of chondroitin-dermatan sulfate from the vessel wall increases artery wall stiffness. A role for proteoglycans in artery wall function is further supported by the observation that in patients with pseudo-xanthoma elasticum accumulation of proteoglycans in the media is associated with pronounced elasticity of the arteries – a significant smaller Young's modulus as compared to control subjects – which is maintained throughout life. (Kornet et al, manuscript by our group under review).

One should realize that relating artery wall composition and structure to its function is difficult, because it is practically impossible to obtain artery wall specimen from the site of function assessment in humans. Therefore, one has to rely on indirect relations, i.e. function assessment in vivo and composition and structure determination in autopsy specimen, or on representative animal models. Transgenic animals are likely to be an asset in this type of studies.

Detailed information in this field can be found in the following selected references (Barra et al., 1993, Gandley et al., 1997, Greenwald and Berry, 1978, Harkness et al., 1957, Kohn,

1977, Laurent, et al., 1994, Li et al., 1998, Nichols and O'Rourke, 1990, Stary et al., 1992, Roach and Burton, 1957, Robert et al., 1970, Van Gorp et al., 2000).

References

Ando, J., Ohtsuka, A., Katayama, Y. et al. (1994). Intracellular calcium response to directly applied mechanical shearing force in cultured vascular endothelial cells. *Biorheology.* 31:57-68.

Ando, J. Tsuboi, H., Korenaga, R., et al. (1996). Down-regulation of vascular adhesion molecule-1 by fluid shear stress in cultured mouse endothelial cells. *Annuals of the New York Academy of Sciences* 748:148-157.

Barra J G, Armentano R L, Levenson J., et al. (1993). Assessment of smooth muscle contribution to descending thoracic aortic elastic mechanics in conscious dogs. *Circulation Research* 73: 1040-1050.

Bots, M.,L., Hofman, A., Grobbee, D.E. (1997). Increased common carotid intima-media thickness. Adaptive response or a reflection of atherosclerosis? Findings from the Rotterdam study. *Stroke* 28:2442-2447.

Boutouyrie, P., Laurent S., Girerd, X. et al. (1995). Common carotid artery stiffness and patterns of left ventricular hypertrophy in hypertensive patients. *Hypertension* 25:651-659.

Brands, P.,J., Hoeks, A.,P.,G., Hofstra, L., Reneman, R.,S. (1995). A non-invasive method to estimate wall shear rate using . *Ultrasound in Medicine and Biology* 21:171-185.

Busse, R., Fleming, I. (1966). Endothelial dysfunction in atherosclerosis. *Journal of Vascular Research* 33:181-194.

Busse, R., Fleming, I. (1998). Pulsatile stretch and shear stress: physical stimuli determining the production of endothelium-derived relaxing factors. *Journal of Vascular Research* 35:73-84.

Chapell, D.,C., Varner, S.,E., Nerem, R., M., et al. (1998). Oscillatory shear stress stimulates adhesion molecule expression in cultured human endothelium. *Circulation Research* 82:532-539.

De Keulenaar, G.W., Chappell, D.C., Ishizaka N. et al. (1998). Oscillatory and steady laminar shear stress differentially affect human endothelial redox state. *Circulation Research* 82:1094-1101.

Frangos, J.A., Eskin, S.G., McIntyre L.V., Ives, C.L. (1985). Flow effects on prostacyclin production by cultured human endothelial cells. *Science* 227:1477-1479.

Gandley, R.E., McLaughin, M.K., Koole, T.J. et al. (1997). Contribution of chondroitin-dermatan sulfate-containing proteoglycans to the function of rat mesenteric arteries. *American Journal of Physiology* 42:H952-H960.

Girerd, X., Mourad, J-J., Acar, C., et al. (1994). Noninvasive measurement of medium-sized artery intima-media thickness in humans: in vitro validation. *Journal of Vascular Research* 31:114-120.

Glagov, S., Vito, R., Giddens, D.P, Zarins, C.,K. (1992) Microarchitecture and compostition of artery walls: relationships to location, diameter and the distribution of mechanical stress. *Journal of Hypertension* 10:S101-S104

Green, M.,A., Friedlander, R., Boltax, A.,J. et al. (1966). Distensibility of arteries in human hypertension. *Proceedings Society of Experimental Biology in Medicine* 121:580-585.

Greenwald, S.E., Berry, C.L. (1978). Static mechanical properties and chemical composition of the aorta of spontaneously hypertensive rats: a comparison with the effects of induced hypertension. *Cardiovascular Research* 12: 364-372.

Gribbin, B., Pickering, T.,G., Sleight, P., (1979). Arterial distensibility in normal and hypertensive man. *Clinical Science* 56: 413-417.

Gribbin, B., Pickering, T.,G., Sleight, P., Peto, R. (1971). Effect of age and high blood pressure on baroreflex sensitivity in men. *Circulation Research.* 29:424-431.

Harkness M.L.R., Harkness, R.D., McDonald, D.A. (1957). The collagend and elastin content of the arterial wall in the dog. *Proceedings of the Royal Society* 146B:541-551.

Hayoz, D., Rutschmann, B., Perret, F. et al. (1992). Conduit artery compliance and distensibility are not necessarily reduced in hypertension. *Hypertension* 20:1-6

Hoeks, A.,P.,G. (1993). Non-invasive study of the local mechanical arterial characteristics in humans. In: Safar M E, O'Rourke M F (eds). *The arterial system in hypertension*. The Netherlands: Kluwer Academic Publishers 119-134.

Hoeks, A.,P.,S., Arts, T.,H.,J., Brands, P.,J., Reneman, R.,S. (1993). Comparison of the performance of the cross-correlation and Doppler autocorrelation technique to estimate the mean velocity of stimulated ultrasound signals. *Ultrasound in Medicine and Biology* 19:727-740.

Hoeks, A.,P.,G., Brands, P., J., Smeets, F.,A.,M., Reneman, R.,S. (1990). Assessment of the distensibility of superficial arteries. *Ultrasound in Medicine and Biology* 16:121-128.

Hoeks, A.,P.,G., Ruissen, C.,J., Hick, P., Reneman, R.,S. (1985). Transcutaneous detection of relative changes in artery diameter. *Ultrasound in Medicine and Biology* 11:51-59.

Hoeks, A.,P.,G., Willekes, C., Boutouyrie. P. et al. (1997). Automated detection of local artery wall thickness based on M-line signal processing. *Ultrasound in Medicine and Biology* 23:1017-1023.

Hokanson, D.,E., Mozersky, D.,J., Summer, D.,S., Strandness, D.,E., (1972). A phase locked echo-tracking system for recording arterial diameter changes in vivo. *Journal of Applied Physiology* 32:728-733.

Kamiya, A., Bukhari, R., Togawa, T. (1984). Adaptive regulation of wall shear stress optimizing vascular tree function. *Bulletin of Mathematical Biology* 46:127-137.

Kohn R R. (1977). Heart and cardiovascular system. In: Finch, C.E., Hayflick, L. (eds). *Handbook of the biology of aging*. New York: Van Nostrand Reinhold Company 281-317.

Kornet, L., Hoeks, A.P.G., Lambregts, J., Reneman, R.S. (1999). In the femoral artery bifurcation, differences in mean wall shear stress within subjects are associated with different intima-media thicknesses. *Artheriosclerosis, Thrombosis and Vascular Biology* 19:2933-2939.

Kornet, L., Hoeks, A.P.G., Lambregts, J., Reneman, R.,S. (2000). Mean wall shear stress in the femoral arterial bifurcation is low and independent of age at rest. *Journal of Vascular Research* 37:112-122.

Kornet, L. Lambregts, J., Hoeks, A.P.G., Reneman, R.S. (1998). Differences in near-wall shear rate in the carotid artery within subjects are associated with different intima-media thicknesses. *Artherioclerosis, Thrombosis and Vascular Biology* 18:1877-1884.

Ku, D.N., Giddens, D.P., Zarins, C.K., Glagov, S. (1985). Pulsatile flow and atherosclerosis in the human carotid bifurcation. *Arteriosclerosis* 5:293-302.

LaBarbera M. (1990). Principles of design of fluid transport systems in zoology. *Science* 249:992-1000.

Laurent, S., Girerd, X., Mourad, J-J. et al. (1994). Elastic modulus of the radial artery wall material is not increased in patients with essential hypertension. *Arteriosclerosis and Thrombosis* 14: 1223-1231.

Laurent, S., Caviezel, B., Beck, L. et al. (1994b). Carotid artery distensibility and distending pressure in hypertensive humans. *Hypertension* 23:878-883.

Learoyd, B.,M., Taylor, M.,G. (1966.) Alterations with age in the viscoelastic properties of human arterial walls. *Circulation Research* 18:278-292.

Lehmann, E.,D., Hopkins, K.,D., Gosling, R.,G. (1993). Aortic compliance measurements using Doppler ultrasound: in vivo biochemical correlates. *Ultrasound in Medicine and Biology* 19:683-710.

Levesque, M.J., Nerem, R.M. (1985). The elongation and orientation of cultured endothelial cells in response to shear stress. *Journal of Biomechanical Engineering* 107:341-347.

Li, D.Y., Brooke, B., Davis, E.C., et al. (1998). Elastin is an essential determinant of arterial morphogenesis. *Nature* 393:276-280.

Menotti, A., Seccareccia, F., Giampaoij, S., Giuli, B. (1998). The predictive role of systolic, diastolic and mean blood pressure on cardiovascular and all causes of death. *Hypertension* 7:595-599.

Milnor, W.,R. (1989). *Hemodynamics.* Baltimore: Williams and Wilkins

Mozersky, D.,J., Sumner, D.,S., Hokanson, D.,E., Strandness Jr., D.,E. (1972). Transcutaneous measurement of the elastic properties of the human femoral artery. *Circulation* 46:948-955.

Murray, C.D. (1926).The pysiological principle of minimum work. I: The vascular system and the cost of blood volume. *Proceedings of the National Academy of Sciencse USA* 12:207-214.Nichols, W.W., O'Rourke, M. F. (1990). *Mc Donald's blood flow in arteries.* London, Melbourne, Auckland: Edward Arnold 424-425.

O'Rourke, M. (1990). Arterial stiffness, systolic blood pressure, and logical treatment of arterial hypertension. *Hypertension* 15:339-347.

Oyre, S., Pedersen, E.,M., Ringgaard, S. et al. (1997) In vivo wall shear stress measured by magnetic resonance velocity mapping in the normal human abdominal aorta. *European Journal of Vascular and Endovascular Surgery* 13:263-271.

Oyre, S., Ringgaard, S., Kozerke, S. et al. (1998). Accurate noninvasive quantitation of blood flow, cross-sectional lumen vessel area and wall shear stress by three-dimensional paraboloid modeling of magnetic resonance imaging velocity data. *Journal of the American College of Cardiology* 32:128-134.

Patel, D.,J., Fry, D.,L. (1966). Longitudinal tethering of arteries in dogs. *Circulation Research* 19:1011-1021.

Perktold, K., Thurner, E., Kenner, T. (1994) Flow and stress characteristics in rigid walled and compliant carotid artery bifurcation models. *Medicine and Biology in Engineering and Computing.* 32:19-26.

Pignoli, P., Tremoli, E., Poli, A. et al. (1986). Intimal plus medial thickness of the arterial wall: a direct measurement with ultrasound imaging. *Circulation* 74:1399-1406.

Reneman, R.S., Van Merode, T., Hick, P., Hoeks, A.,P.,G. (1985). Flow velocity patterns in and distensibility of the carotid artery bulb in subjects of various ages. *Circulation* 71:500-509.

Reneman, R.,S., Van Merode, T., Hick, P., Hoeks, A.,P.,G. (1986). Cardiovascular applications of multi-gate pulsed Doppler systems. *Ultrasound in Medicine and Biology.* 12:357-370.

Reneman, R.,S., Van Merode, T., Hick, et al. (1986). Age-related changes in carotid artery wall properties in men. *Ultrasound in Medicine and Biology* 12:465-471.

Riley, W.A., Barnes, R.W., Evans, G.W., Burke, G.L. (1992). Ultrasonic measurement of the elastic modulus of the common carotid artery. *Stroke* 23:952-956.

Riley, W.,A., Craven, T., Romont, A., Furberg, C. (1996). Assessment of temporal bias in longitudinal measurements of carotid intimal-media thickness in the asymptomatic carotid artery progression study (ACAPS). *Ultrasound in Medicine and Biology* 22:405-411.

Roach, M.R., Burton, A.C. (1957). The reason for the shape of the distensibility curves of arteries. *Canadian Journal of Biochemical Physiology* 35: 681-690.

Robert, L., Robert ,B., Robert, A.M. (1970). Molecular biology of elastin as related to aging and atherosclerosis. *Experimental Gerontology* 5:339-356.

Rossitti, S., Lofgren, J. (1993).Vascular dimensions of the cerebral arteries follow the principle of minimum work. *Stroke* 24:371-377.

Rubanyi, G.M., Freay, A.D., Kauser, K., et al. (1990). Mechanoreception by the endothelium: mediators and mechanisms of pressure- and flow-induced vascular responses. *Blood Vessels* 27:246-257.

Salonen, J.,T., Salonen, R. (1991). Ultrasonographically assessed carotid morphology and the risk of coronary heart disease. *Arteriosclerosis Thrombosis* 11:1245-1249.

Samijo, S.K., Willigers, J.M., Barkhuysen R., et al. (1998). Wall shear stress in the common carotid artery as function of age and gender. *Cardiovascular Research* 29:515-522.

Safar, M.,E., Peronneau, P.,A., Levenson, J.,A. et al. (1981). Pulsed Doppler: diameter, blood flow velocity and volumic flow of the brachial artery in sustained essential hypertension. *Circulation* 2:393-400.

Sagie, A., Larson, M.G., Levy, D. (1993). The natural history of borderline isolated systolic hypertension. *New England Journal of Medicine* 329:1912-1917.

Sakamoto, K., Kanai, H. (1979). Electrical characteristics of flowing blood. *IEE Transactions Biomedical Engineering* 26:686-695.

Samijo, S.,K., Willigers, J.,L., Brands, P.,J. et al. (1997). Reproducibility of shear rate and shear stress assessment by means of ultrasound in the common carotid artery of young human males and females. *Ultrasound in Medicine and Biology.* 23:583-59.

Stary, H.C., Blankenhorn, D.H., Chandler, A.B. (1992). A definition of the intima of human arteries and of its atherosclerosis-prone regions. *Arteriosclerosis and Thrombosis* 12:120-134.

Tsao, P.S., Buitrago, R., Chan, J.R., Cooke, J.P. (1996). Fluid flow inhibits endothelial adhesiveness, nitric oxide and transcriptional regulation of VCAM-1. *Circulation* 94:1682-1689.

Van Gorp, A.W., Van Ingen Schenau, D.S., Hoeks, A.P.G., et al. (2000). In spontaneously hypertensive rats alterations in aortic wall properties precede development of hypertension. A*merican Journal of Physiology* 287: H1241-H1247.

Van Gorp, A.,W., Van Ingen Schenau, D.,S., Hoeks, A.,P.,G. et al. (1995). Aortic wall properties in normotensive and hypertensive rats of various ages in vivo. *Hypertension* 26:363-368.

Van Merode, T., Brands PJ, Hoeks APG, Reneman RS (1993) Faster ageing of the carotid artery bifurcation in borderline hypertensive subjects. *Journal of Hypertension* 11:171-176.

Van Merode, T, Brands, P.,J., Hoeks, A.,P.,G., Reneman, R.,S. (1996) Different effects of ageing on elastic and muscular bifurcations in men. *Journal of Vascular Research* 33:47-52.

Van Merode, T., Hick, P.,J.,J., Hoeks, A.,P.,G. et al. (1988). Carotid artery wall properties in normotensive and borderline hypertensive subjects of various ages. *Ultrasound in Medicine and Biology* 14:563-569.

Van Merode, T., Lodder, J., Smeets, F.,A.,M. et al. (1989). Accurate noninvasive method to diagnose minor atherosclerotic lesions in carotid artery bulb. *Stroke* 20:1336-1340.

Ventura, H., Messerli, F,H., Oigman, W. et al. (1984). Impaired systemic arterial compliance in borderline hypertension. *American Heart Journal* 108:132-136.

Wells, P., N., T., (1969). *Physical principles of ultrasonic diagnosis.* London, New York: Academic Press.

Willekes, C., Brandts, P.,J., Willigers, J.,M. et al. (1999). Assessment of local differences in intima-media thickness in the human common carotid artery. *Journal of Vascular Research* 36:222-228.

Zarins, C.K., Giddens, D.P., Bharadvaj, B.K., et al. (1983). Carotid bifurcation atherosclerosis: quantitative correlation of plaque localization with flow velocity profiles and wall shear stress. *Circulation Research* 53:502-514.

Zeitfracht Medien GmbH
Ferdinand-Jühlke-Straße 7
99095 Erfurt, Deutschland
produktsicherheit@kolibri360.de